ADVANCES IN UNDERWATER TECHNOLOGY AND OFFSHORE ENGINEERING

Volume 3

Offshore Site Investigation

***ADVANCES IN
UNDERWATER TECHNOLOGY
AND OFFSHORE ENGINEERING***

Vol. 1   Developments in Diving Technology
Vol. 2   Design and Installations of Subsea Systems
Vol. 3   Offshore Site Investigation
Vol. 4   Evaluation, Comparison and Calibration of
        Oceanographic Instruments.

ORGANISING COMMITTEE

Mr Dennis Ardus – chairman
Dr C. D. Green
Mr F. E. Toolan
Mr Tim Freeman

ADVANCES IN UNDERWATER TECHNOLOGY AND OFFSHORE ENGINEERING

Volume 3

Offshore Site Investigation

Proceedings of an international conference, (Offshore Site Investigation), organized by the Society for Underwater Technology, and held in London, UK, 13 and 14 March 1985

Published by
Graham & Trotman Ltd.

First published in 1985 by
Graham & Trotman Limited
Sterling House
66 Wilton Road
London SW1V 1DE

British Library Cataloguing Publication Data

Offshore Site Investigation (Conference: London)
Advances in underwater technology and offshore engineering:
proceedings of an international conference (Offshore Site Investigation).
Vol. 3: Offshore Site Investigation
1. Ocean engineering
I. Title II. Society for Underwater Technology 627'.7 TC0000
ISBN-13: 978-94-011-7360-5   e-ISBN-13: 978-94-011-7358-2
DOI: 10.1007/978-94-011-7358-2

© Society for Underwater Technology

Softcover reprint of the hardcover 1st edition 1985

Typeset in Great Britain by Spire Print Services Ltd, Salisbury

ROBERT HARTNOLL (1985) LTD., BODMIN, CORNWALL

Contents

PART II
IN-SITU TESTING, SAMPLING AND
THE MEASUREMENT OF GEOTECHNICAL PROPERTIES

PART III
EVALUATION OF DESIGN PARAMETERS

Offshore Positioning Requirements, Systems and Quality Control: A Review

J. G. Riemersma, Shell Internationale Petroleum Maatschappij, The Hague, The Netherlands

SUMMARY

Offshore site surveys require the positioning of a survey platform (survey vessel, underwater vehicle or towed body) with a high relative accuracy. This accuracy depends on the purpose of the survey but lies normally between 1 and 5 m. It is also important to realize that a site survey does not stand on its own but is always related to previous work carried out in the same area (e.g. geophysical surveys, drilling operations) and serves as a preparation for future work (e.g. platform installation, pipeline laying). The positionings of all these surveys must be properly related to each other. With these requirements in mind, the positioning aspects of offshore site surveys are reviewed along the lines of setting up shore stations, the calibration of positioning systems, quality control during the survey, the post-processing of data and final reporting.

The paper is divided into the following sections:

- Shore control for offshore positioning.
- Offshore radio locationing systems.
- Underwater acoustic positioning systems.
- Calibration of offshore radio locationing systems.
- Calibration of underwater acoustic positioning systems.
- Quality control during the survey.
- Post-processing and reporting.

A few of the positioning systems used in site surveys are reviewed and their different calibration procedures discussed.

This paper is intended to serve as a guide for the preparation of offshore site surveys, which may form the basis for future, more detailed, offshore engineering investigations. Continuity, recoverability and repeatability of survey control are essential ingredients in this process.

SHORE CONTROL FOR OFFSHORE POSITIONING

Offshore radio locationing systems require

transponders which can be placed onshore or on permanent offshore structures. In most cases the co-ordinated positions of these transponders are already known from previously executed surveys. The quality control is then limited to the confirmation that the transponders are placed on undisturbed survey markers and that their co-ordinates are based upon the same geodetic network and datum. When re-establishing survey markers, care should be taken that proper station descriptions are used to avoid ambiguity in their identification.

When there is doubt concerning the validity of the co-ordinates, or when new stations have to be established, then a survey has to be carried out. Co-ordinates are mostly established by satellite-Doppler observations or, less frequently, by terrestrial methods. If the survey area is well known, and the local triangulation of verified good quality (local errors less than 1 m), then terrestrial methods can be used. However, such prior confidence and knowledge does not normally exist and it is consequently necessary to either establish new stations, or verify old ones, by Doppler observations. The risks involved in neglecting this, with regard to possible survey mispositioning and vessel downtime, are too high to take.

When satellite-Doppler is used, it is clearly necessary to establish the datum transformation between satellite and local datum. Existing information may be available and there may be a local legal requirement to use it. However, in areas where it is felt necessary to make Doppler observations, it is advisable either to verify the existing transformation parameters or to establish new ones by carrying out observations on one or more local higher-order triangulation points.

The Doppler observations must be made in translocation mode. To achieve the required accuracy, a minimum of 20 common and balanced (EW- and NS-going) passes should be used in the final solution. Computations can generally be performed by the receiver's microprocessor in the field, as a full network post-processed solution is not necessary. It provides accuracy of better than 1 m, which is satisfactory for offshore positioning. The usual individual Doppler count and pass acceptance criteria apply, e.g. passes above 15° elevation, at least ten minutes of common Doppler data and a standard deviation of the 3D single pass solution of less than 40 m in latitude, longitude and height. Specific attention should be paid to achieving the required number of common passes. This requires a pre-survey selection of the satellite passes to be observed, and good communication procedures during the Doppler survey to ensure the satisfactory observation of the programmed passes at master and remote sites.

The computed and raw satellite-Doppler data are always of potential long-term value and should therefore be regarded in this context, rather than just as a one-off survey application. The raw Doppler data cassettes should be safely stored to enable later reprocessing if required. Survey documentation should be supplied in a suitable format and should contain all information which will ensure its long-term usefulness. It must include details of the computation parameters and results (for example, using a Doppler positioning summary sheet as given in the Ordnance Survey of Great Britain (1981)), and particular attention should be given to avoiding ambiguity in the recovery of survey markers. The description of a survey marker's position — its access, the method used in determining its co-ordinates and the datum parameters — is best entered on a 'Station Description' form. This form should at least contain the following information:

- survey marker identification name or number;
- photograph and sketch of station and surroundings;
- date of survey;
- geographical co-ordinates of marker;
- geodetic datum pertaining to these coordinates, including geoidal model;
- elevation of marker;

- datum pertaining to the elevation;
- method by which station was surveyed and its relation to any network;
- measurements to reference markers, e.g. magnetic bearing and distance;
- information on the access to the station site;
- information on the property of the station site;
- in case the marker is determined by satellite-Doppler—
  cartesian coordinates X, Y, Z of translocated Doppler antenna in satellite datum,
  translocation master station identification,
  translocation master station adopted coordinates,
  transformation parameters, satellite to local datum,
  height of antenna above marker,
  relation between marker and Doppler antenna, e.g. magnetic bearing and distance;
- in case rectangular coordinates are used—
  rectangular coordinates,
  projection system parameters pertaining to these.

An example of a Station Description form is included as an Appendix.

OFFSHORE RADIO LOCATIONING SYSTEMS

The surface positioning systems required for high-accuracy offshore site surveys (better than 5 m) are mainly radio locationing systems. They can be divided into several categories, e.g. into short-, medium- and long-range systems, into their method of signal comparison (pulse measurement or phase comparison), or into their different modes of pattern transmission (range/bearing or circular/linear, range/range or circular, and hyperbolic patterns). In this paper the last-mentioned classification is used. A few of the most common positioning systems

are reviewed; more detailed information on these and other systems can be found in Munson (1977).

When selecting a positioning system for offshore site surveys, the following criteria should be observed:

(1) it should have sufficient range to cover the survey area;
(2) it should provide at least three independent position lines simultaneously;
(3) in case an offshore platform has to be used as station site, the systems station (transponder) should not be affected by moving cranes on the platform or by other electromagnetic disturbances (e.g. reflections);
(4) it should be capable of using an omnidirectional antenna with the mobile;
(5) it should, optionally, have space diversity antennae to cancel multipath effects.

Irrespective of the system chosen, the selection criteria for shore station sites are as follows:

(1) free of any obstruction which can cause signal deflections;
(2) land–seawater boundary changes should be limited to a minimum, especially for 2 MHz systems;
(3) position lines should intersect between 30 and 150° in the survey area.

Circular and Linear Systems

A range/bearing positioning system has normally one shore station with one mobile. It creates a circular and a linear positioning pattern. The mobile transmits a pulse to the shore transponder to measure the distance. Together with the distance measurement pulse, the shore transponder returns the angular value between an orientation reference and the mobile. The positioning accuracy of the system is dependent on the distance (angular accuracy equals two minutes of arc). There are only a few systems of this type, most of which are very restricted in their maximum range (up to 5 km); only the Artemis system has a range of about 24 km. This, along with the limitation of two lines of position, i.e. no redundancy, are the disadvantages in using the system. Furthermore, the system suffers from loss of signal at certain distances due to multipath effects which cause signal cancellation. The area where these occur can be calculated as they are a function of the antennae heights. The system is often used as a back-up or in dynamic positioning of vessels near platforms.

Circular Systems

This group of positioning systems is the most common one used in site surveys. The mobile transmits a signal to the shore stations (three or more), and on receipt of the return signals the distances to each of the shore stations are displayed. Some systems allow only the use of one mobile, while others have time-sharing facilities for several mobiles. Distances are determined either by measuring travel-times of pulses or by phase comparison.

Phase comparison systems have the disadvange of ambiguity, equal to half the wavelength of the lowest modulation frequency (lane). This requires a continuous check on the lane-settings and lane-counts; it also requires avoiding situations leading to loss of signal, causing loss of lanes. Phase comparison systems, especially in the 2 MHz band, have the advantage of long ranges, up to 500 km, and have a high accuracy, in the order of 5 m. They suffer from night effect (skywave) causing very unstable readings and have therefore an increased chance of losing lanes. Argo, Hifix/6 and Hyperfix are typical examples of phase comparison systems in the 2 MHz band. Their lane-width is approximately 80 m, allowing for a high accuracy — 0.02 of a lane — but making it more difficult to check lost lanes (80 m ambiguity). Present techniques to avoid lane losses have improved considerably, but occasionally lane-slips still occur.

Pulse measurement systems, including the long coded pulse measurement systems,

are mostly non-ambiguous (Syledis has an ambiguity of 10 km), and display their reading directly in 'metres'. They are not affected by skywave effects and have a high accuracy (better than 3 m). The disadvantage is that their maximum range is normally limited to one to two times the line of sight.

Examples of pulse measurement systems are Syledis, Maxiran, Trisponder and Motorola. The last two systems are limited to a maximum range normally of the order of 40 km due to their high frequency (9 GHz). Both systems are often used for calibrating other radio locationing systems.

Maxiran, operating in the 450 MHz band, is presently not much in use for site surveys. One of the main problems is obtaining an operating licence for this system, as the bandwidth is 5.7 MHz and the maximum transmitting power is 20 kW. The 450 MHz lies in a very congested frequency spectrum. Another problem is the heavy onshore aerial tower and the need for a second aerial on board the mobile when working with three ranges, as the mobile aerial is directional as opposed to omnidirectional. On the other hand, ranges of over 500 km have been reached.

Syledis, operating in the same frequency band of 420–450 MHz, with a bandwidth of 2 MHz and a transmitting power of 20 W, is a more appropriate system for site surveys. The maximum range of Syledis is 80 km, but occasionally ranges of up to 150 km have been recorded with good accuracies (5 m). Syledis uses an omnidirectional antenna on board the mobile and has the option to use space-diversity antennae onshore to limit signal cancellation. Time-sharing with other mobiles using the same shore transponders is possible. The system has also an option to use a passive mobile receiver, the SR3, which uses the principle of 'pseudo' ranges. Up to now, there has been a reluctance to use this equipment for site surveys as some problems with this receiver have not yet been solved (van Kuijk, 1984). When these problems have been solved, the use of SR3 will become a possibility, especially in areas where a Syledis chain is intensively used (e.g. the North Sea).

Hyperbolic Systems

A hyperbolic system requires, per positioning pattern, two shore stations, one master and one slave transponder. All shore stations are synchronized with the signal transmitted from the so-called master station. The mobile consists of a passive receiver measuring differences in the arrival time of signals from two stations, either by phase comparison or by timing pulse arrivals. A disadvantage of the hyperbolic system is the wide lane expansion when moving away from the master–slave baseline. The lane-width on the baseline is approximately 80 m for the 2 MHz systems, while at the edges of the coverage the lane-width can expand to 500 m. All the 2 MHz equipment mentioned above under 'Circular Systems' have a hyperbolic option, as does Syledis.

UNDERWATER ACOUSTIC POSITIONING SYSTEMS

Underwater acoustic systems can be divided into long, medium, short and ultra-short baseline categories. The purpose of the long and medium baseline systems is the direct positioning of a survey vessel or an underwater vehicle. The short and ultra-short baseline systems are used from a survey vessel for vehicle and towed-body tracking, positioning a guide-base while drilling soil-borings, and for dynamic positioning.

Long and Medium Baseline Systems

The difference between medium and long baseline systems lies in their frequency band, which affects the maximum range of the system. The low-frequency transponders (8–16 kHz) have a maximum range of 5 to 10 km, the medium-frequency transponders (24–36 kHz), a maximum range of 1 to 3 km, and the high-frequency transponders (50–100 kHz), a maximum range of 700 m. There is no difference, however, in the method of deploying and operating them. The majority can be switched on or off and can be automatically released from the sea-bed. There are three different types of transponders: common or dual-mode, relay and intelligent.

The common transponder listens on a common frequency and, upon receipt of a signal from the acoustic control unit, re-transmits the signal, on a selected frequency, back to the transducer of the control unit. Knowing the velocity of sound in seawater, the slant range between transducer and transponder can be calculated from the travel-time of the signal. Systems using this type of transponder are mainly for the direct positioning of survey vessels or manned submersibles.

The procedure, in which the relay transponder is used, is often called the 'sing-around' method. The control unit transmits a signal on the relay transponder's frequency, and upon receipt the relay transponder re-transmits this signal to the transducer of the control unit and to the transponders in the array. The control unit then receives the direct signal from the relay transponder and the array transponder return signals, which have travelled via the relay transponder. From the observed travel-times the distances from the array transponders to the vessel's transducer and the relay transponder can be calculated. The relay transponder is used for remote positioning of underwater vehicles and other objects.

The latest type is the intelligent transponder, which, in addition to its normal functions, can also receive, gather, process, store and transmit data via telemetry. These transponders also often carry other sensors such as temperature and salinity probes, depth sensors and inclinometers. With transponders of this kind it is possible to measure directly the baseline distances between transponders in an array as well as measuring the distance to the survey vessel's transducer (Kelland et al., 1982).

One of the major problems in underwater acoustic positioning is the propagation of the signals through seawater. Due to the difference in temperature and salinity, the water column is layered, causing deflections of the signal path (Snell's law). Another major problem is noise, caused for instance by propellers and engines, which results in distortion of the acoustic signal. All other sources, such as pingers operating in the same frequency band, can interfere and even cause complete disruption of the survey. This is especially the case with the intelligent transponders, as interference on a command frequency can block the transmission of data. It is therefore essential that during the array calibration the frequency band is monitored for unwanted signals. Examples of long and medium baseline equipment are the Sonardyne PAN II and Micronav, Racal's Aqua-fix 4 and Wimpol's Marax systems.

Short Baseline Systems

Two transducers are mounted on the vessel's hull, usually on a baseline of 10 to 20 m perpendicular to the ship's axis. They transmit acoustic pulses which are received and returned by a transponder on the target. The distances from the two transducers to the transponder are determined by measuring the round-trip travel time of the acoustic pulses. Using the known length of the baseline and its bearing, determined by the vessel's gyro, the position of the target can be computed.

Low-frequency systems (8–16 kHz) have a maximum range of 2 to 3 km, a range accuracy of 3 to 5 m and a bearing accuracy of 2 to 3 degrees. High-frequency systems (50–100 kHz) have a maximum range of 1 km, a range accuracy of 30 to 50 cm and a relative bearing accuracy of 0.5 of a degree. The systems are able to position several transponders simultaneously, e.g. up to 16 with the Oceano system. The short baseline system has mainly been used with submersibles but, with the development of new systems, it is now hardly used.

Examples are the Ferranti ORE and Oceano systems.

Ultra-short Baseline Systems

Ultra-short baseline systems work with an acoustic transmit/receive array element housed in one single transducer unit of 20 to 30 cm width. The position calculation is based on range and direction measurements, giving the relative position between the transducer and the transponder. The distance is determined by measuring the round trip of the acoustic pulse. The horizontal and vertical direction is determined from differences in pulse arrival times at several acoustic sensor elements distributed over the transducer head. The transducer is usually mounted in the vessel's hull and can be lowered, through a gate-valve, to such a depth that it is free of propeller and thruster noise. The system also includes pitch and roll sensors (Vertical Reference Unit) for the vertical control, and a gyro for the horizontal orientation. Standard transducers remain in a fixed position whereas tracking transducers, which are automatically directed to the transponder by turning the transducer head horizontally and vertically, work with a wide beam for search and a narrow beam for tracking. It is possible to track several transponders simultaneously, e.g. up to 9 with the Simrad HPR 209. The system's accuracy is of the order of 0.5% of the observed slant ranges and the maximum range is 1000 to 1500 m. Two frequently used systems are the Simrad HPR and the Honeywell RS-7.

CALIBRATION OF OFFSHORE RADIO LOCATIONING SYSTEMS

Calibration of a radio locationing system is needed to determine the systematic error sources inherent in the system and to check that the equipment is functioning properly before the start of the survey. The main error sources are the propagation velocity of the signal over seawater, and the zero or index error, caused by delays in equipment and cabling. The two methods used to assess these errors are a long and a short baseline calibration respectively. There is a slight difference in calibrating phase comparison and pulse measurement systems. Differences also occur between systems producing hyperbolic, circular and linear patterns (Riemersma, 1979). In this paper the calibration methods are divided according to the latter distinction.

Calibration of Linear Systems

The problem of propagation speed is irrelevant to this system. The position of the orientation reference marker should follow the same criteria as for shore stations. The calculation of the reference orientation (RO) value should be checked and it should be very clear whether this value refers to a True azimuth or to a Grid azimuth. This value should be entered in the on-board computer to avoid mistakes and to ensure that it is logged in the computer. The RO value in the transponder should be set to zero when aiming at the orientation reference marker. During the calibration, the orientation value should be checked against theodolite observations. During the survey, the aiming at the orientation reference marker should be checked regularly.

Calibration of Circular Systems

As mentioned before, circular patterns can be generated by both pulse and phase comparison systems. In the latter case the systems have to be installed at the station site and on board the survey vessel before calibration because of the effect of the ground conductivity on the signal transmission. The short baseline calibration takes place on board the survey vessel situated a few kilometres offshore from the station site. It is essential with 2 MHz systems that the minimum distance between mobile and transponder is not below that recommended by the manufacturer. A distance of 10 km is often used. A highly accurate Electronic Distance Measuring (EDM) instrument, such as a tellurometer, is used for comparison. The delay value is derived from the difference in distance measured with the radio locationing system and the EDM equipment, taking into account the offset distance between the two antennae. At least 20 observations should be taken at regular time intervals.

Pulse measuring systems can be calibrated prior to their installation on a shore baseline, the distance of which is known to an accuracy better than 0.1 of a metre. In this case it has to be ensured that all equipment is installed at the shore station and on the vessel in the same manner as it was calibrated.

The shore baseline should have a length of not more than 5 km to avoid the effect of propagation speed errors and should be longer than the minimum distance recommended by the manufacturer in order to avoid distortion of the signal (although with some equipment it is possible to use an attenuator). At least 20 observations, taken at regular time intervals of about 10 seconds, should be made for each set of equipment. Experience has taught that it is best not to touch the setting of the mobile equipment at all during the survey but to enter the delay values in the computer. It will provide the user with a record — computer printout and tape-log — of delays used, and it facilitates easy switching from one pattern to another.

The long baseline calibration is identical for both the pulse measurement and the phase comparison systems. This calibration can be done either on a known baseline over seawater or by crossing the line between two shore stations. Where feasible, more than one crossing should be done to ascertain redundancy in data.

Calibration of Hyperbolic Systems

The calibration of hyperbolic systems, using either pulse measurement or phase comparison techniques, is carried out after the system has been installed on the shore station sites and on board the survey vessel. Observations of all position lines are taken near the shore station sites, taking into consideration the minimum distance as specified by the manufacturer. The co-ordinates of these calibration positions have to be known with an accuracy better than that supplied by the system. Alternatively, the calibration can be done by crossing baseline extensions and by checking against known offshore platforms (transit fixes). The delay values for each pattern and the propagation speed over seawater can be calculated from the least squares solution of the observations. The delay value includes the error caused by a baseline crossing totally or partly over land. At least 20 observations should be taken at each position and should be logged on a computer.

It is essential that during the calibration the stability of the patterns is closely monitored, for example, pattern values for Pulse-8 should not vary more than 0.05 milliseconds during any 1-second interval.

CALIBRATION OF UNDERWATER ACOUSTIC POSITIONING SYSTEMS

One of the parameters required in an underwater acoustic calibration is the speed of sound in seawater. This can be measured either directly or indirectly, and should have an accuracy of better than 2 m/s. Sufficient redundancy should be provided so as to ensure that there is no ambiguity in the resulting velocity profile determined by the temperature/salinity bridge or seawater sound velocity meter. The equipment used must be properly calibrated against an industrial standard high-precision thermometer and with a known saline solution. Calibration of underwater acoustic transponders should always be carried out using a high-precision surface positioning system, which itself should be properly calibrated.

Calibration of Long and Medium Baseline (LBL) System

The purpose of the calibration is to obtain, without ambiguity, the co-ordinates of the transponders in an array and to determine possible systematic error sources. For deploying transponders in an array, the following criteria are to be applied:

(1) The distance between low-frequency transponders should be no more than 5000 m, for medium-frequency transponders no more than 1500 m, and for high-frequency transponders no more than 700 m. However, the maximum ranges depend very much on the water depth.
(2) The maximum number of transponders in one array should be no more than 6.
(3) Transponders should be deployed so as to avoid impeding their acoustic line of sight.

The method of calibration depends on the type of transponder used. For intelligent transponders the 'boxing-in' method, in a dynamic calibration mode, is used. The survey vessel circumnavigates at least two of the array's transponders in a circle, the radius of which should be no more than twice the water depth. At least 100 data sets of acceptable quality, evenly distributed around the sailed figure, should be collected. In addition, a maximum of 20 direct transponder baseline measurements should be made between all transponders, especially between the calibrated transponders.

If non-intelligent transponders are used, the survey vessel should sail around all the transponders in a cloverleaf type of figure. This method is discussed in detail in Riemersma (1977). To verify the transponder calibration, the 'boxing-in' method should be carried out over at least two of the transponders. At least a 100 data sets should be observed and recorded.

All equipment must be interfaced to a minicomputer to allow real-time calculation, display and logging of data. Sufficient redundancy in the positioning data should be acquired for the computation of a mathematical and statistical solution in order to derive the size and sign of the velocity error, the range index error and the offset error. The vertical separation between the vessel's transducer and the transponder on the seabed should be determined from depth sensors in the transponder or from echosounder readings, together with the fixed offsets for the transducer depth and the height of the transponder above the seabed.

For each fix the following data should be printed and logged on cartridge or floppy disc:

- time of fix;
- surface positioning raw ranges (unfiltered);
- antenna northing and easting;
- standard deviation of the single observation;
- gyro compass reading;
- ranges to the transponders;
- transducer northing and easting;
- standard deviation of the single observation;
- baseline ranges between transponders (intelligent mode).

Before the start of the survey, the calibrated array should be verified by simultaneously observing the ranges to the array's transponders and the surface positioning system's pattern data. The vessel's position obtained from these data should be compared, taking into account the offset between aerial and transducer.

Calibration of Short and Ultra-short Baseline (USBL) System

Before the start of a calibration, one should ensure that the vessel's USBL transducer (tracking or fixed) is rigidly mounted through a gatevalve and that its operational position is clear of hull, propellers, thrusters and other possible noise sources.

All equipment must be interfaced to a minicomputer to allow real-time calculation, display and logging of data. Sufficient redundancy in the positioning data should be

acquired for the computation of the size and sign of each of the following error sources:

(1) scaling error:
   (a) velocity error
   (b) range index error
   (c) offset error
(2) horizontal alignment:
   (a) gyro error
   (b) transducer alignment
(3) vertical alignment:
   (a) pitch error
   (b) roll error
   (c) transducer vertical alignment

The gyro error is determined in the gyro calibration. To determine the other errors, two calibration procedures are executed, a dynamic and a static calibration.

Dynamic Calibration

Deploy a transponder ensuring that it is clear of all structures and pipelines. The survey vessel should sail around this transponder maintaining a constant heading, steering either a triangle or a rectangular figure. The range of the transponder should vary from one to three times the water depth. At least 100 data sets of acceptable quality, evenly distributed around the sailed figure, should be collected. All position fixes should be computed and displayed in real time to verify the systematic distribution of fixes around the transponder. For each fix, the following data should be printed and logged on cartridge or floppy disc:

- time of fix;
- surface positioning raw ranges (unfiltered);
- antenna northing and easting;
- standard deviation of the single observation;
- gyro compass reading;
- X, Y, Z offset of the USBL;
- pitch and roll of the vertical reference unit.

Static Calibration

The survey vessel takes up position and

remains stationary over the transponder to collect data for the precise assessment of the depth. At least 100 data sets should be collected, displayed, printed and logged. In addition to the data collected in the dynamic calibration, the echo-sounder depths should be recorded with each fix.

Computation of Calibration

The computation of the data is done by three-dimensional least squares adjustment and should provide the following data:

- transponder northing and easting;
- transponder average depth (from the static calibration);
- standard deviation of the transponder position;
- standard deviation of the transponder depth;
- pitch error;
- roll error;
- transducer vertical alignment error;
- transducer horizontal alignment error;
- scale error;
- a new velocity value to adjust for the scale error.

All data must be recalculated with the new velocity value, but careful consideration must be given to the velocity calculated from the temperature and salinity profile. If the velocity value is unrealistic, then all data must be closely inspected and, when necessary, re-observed.

To carry out a proper quality control, a printout or plot of the following data must be available:

- setting-up parameters;
- observations with their residuals;
- final results;
- fix distribution around the transponder;
- residual frequency distribution;
- residual/range scatter plot;
- angular error analysis.

QUALITY CONTROL DURING THE SURVEY

The quality control of the positioning during a survey is divided into two categories: that done prior to and at intervals during the survey, and that done in real time for each position fix.

The quality control carried out at the start of the survey, and that carried out at intervals during the survey when equipment or positioning parameters are changed, should consist of the following checks:

(1) Are the correct geodetic datum and projection parameters used in the navigation computation?
(2) Are the calibration values, as determined in the calibration, properly used in the navigation calculation?
(3) Are the offset parameters for the different types of equipment, antennae, transducers, echo-sounder and geophysical sensors correctly applied?
(4) Are all positioning data and other survey data properly recorded and logged on a computer medium, and is the format in which the data are recorded according to specifications?
(5) Are all position calculations done with the proper formulae?

The real-time quality control is needed to control the performance of the positioning systems, to check the positioning accuracy and to detect errors in the data and computation. The calculation of the fix position from redundant positioning data is done using the least squares variations of coordinates method (Cross, 1981).

Detecting errors in time is essential when using systems with ambiguity in their position lines (lane-slips). Methods have been developed to correct in real time for these lane-slips (Nicolai, 1983).

The main items scrutinized in real-time quality control are:

(1) The standard deviation in the single observation. This value is independent of the position line configuration; it is

solely a measure of the stability of the positioning system and will give an indication on the presence of systematic errors. It should be noted, however, that these errors will not be noticeable if the vessel stays in one area (Riemersma, 1979).

(2) The standard deviation in the position, with parameters defining the standard error ellipse. These values are correlated with the configuration of the position lines and give, in addition to the performance of the positioning system and possible systematic errors, a measure of the accuracy of the fix (Houtenbos, 1981).

(3) The residuals in the raw data after a least squares position calculation.

Upon completion of each survey line and after specific parts of the survey a statistical summary in the form of histograms is required. The histograms should cover the standard deviations in the single observations, the standard deviations in the position, the offsets from the planned survey lines and, where required, the interval between the fixes.

POST-PROCESSING AND REPORTING

Irrespective of the quality of the positioning data collected and computed in real time during the survey, all data should be post-processed after the completion of the survey. In the case of poor real-time data quality, post-processing takes advantage of two-sided smoothing techniques to improve the reliability of the co-ordinates. If the original data are good, they act as a quality control check on the real-time acquisition. Furthermore, this post-processing quality control should not be treated as an automated batch process. Unlike real-time data, it provides information on potential systematic positioning errors affecting the whole survey; it allows for more detailed investigation into effects which may have degraded real-time data quality and that

may not have been originally possible due to operational pressures; and it provides a necessary independent check on the work performed during the survey.

The procedures for this quality control and some potential pitfalls are set out in detail in McLintock (1984). Further checks especially suited to the densely surveyed grids used for site surveys are comparisons of water depth at line intersections and the positioning of either co-ordinated or non-coordinated seabed features, e.g. wellheads, pipelines, rocks.

As was discussed for the reporting of shore control data, reporting the positioning aspects of a survey is fundamental to its successful completion. Such reports will always differ according to the type and area of the survey but they should:

(1) allow easy access to the final results of the survey;

(2) give recommendations that could improve the execution of similar future jobs;

(3) be clearly marked with project title, contract number and start and completion date of survey.

Also they should contain the following information:

(4) summarized statement of the contracted survey requirements, together with the contractor's assessment of results against objectives;

(5) chronological report of major events;

(6) time summary, with total time allocated to port or transit time, weather downtime, equipment downtime or surveying time;

(7) list of survey personnel involved in the survey;

(8) report of equipment performance stating usage, basic parameters such as station coordinates, survey accuracy, breakdown, and including any remarks or recommendations;

(9) dated tabulated summary listings of results, for example, northing, easting, depth, etc.;

(10) cross-reference list of recording channels and patterns;
(11) graphical results of major features — such as side-scan sonar results — reduced to a manageable volume of A4 folded sheets, annotated with coordinates and other parameters to allow proper interpretation;
(12) maps, properly labelled with all geodetic datum and projection parameters;
(13) survey summary map showing the survey location, coverage of all maps, data tapes and other records;
(14) co-ordinate and raw data tapes in standard formats, for example, HP discs or cartridges, 9-track tapes in UKOOA or SEG formats;
(15) lists of all tapes, reports and records, cross-referenced to line numbers, fix numbers, etc. and fully annotated with project title, acquisition date and contract number.

ACKNOWLEDGEMENTS

The author would like to thank his colleagues in the topographical departments of Shell Expro, London, and Shell Internationale Petroleum Maatschappij, The Hague, for many useful discussions and suggestions on this paper. Without their help there would have been no paper.

REFERENCES

1. Cross, P. A. 1981. The computation of position at sea. *Hydrogr. J.* (20), April 1981.

2. Houtenbos, A. P. E. M. 1982. Quality control in offshore positioning. *Hydrogr. J.* (21), July 1981.

3. Kelland, N. C., Partridge, C. J. and Lawes, D. C. 1982. Applications and advantages of microprocessors in underwater acoustic positioning equipment — hydro 82. *Hydrographic Society Symposium*, Southampton, UK.

4. McLintock, D. N. 1984. A coordinated approach to the quality control of offshore positioning. *Second International Hydrographic Technical Conference*, Plymouth, UK.

5. Munson, R. C. Rear Admiral NOS, 1977. Positioning systems — Reporting on the work of working group 414b. *XVth International Congress of Surveyors FIG*, Stockholm, Sweden.

6. Nicolai, R. 1983. Lane control by least squares adjustment. *Hydrogr. J.* (28), April 1983.

7. Ordnance Survey of Great Britain, 1981. Report of an investigation into the use of Doppler-satellite positioning to provide coordinates on the European datum 1950 in the area of the North Sea. Professional Papers, New Series No. 30.

8. Riemersma, J. G. 1977. The use of underwater acoustics for the positioning of offshore structures. *XVth International Congress of Surveyors FIG*, Stockholm, Sweden.

9. Riemersma, J. G. 1979. Quality control of offshore positioning surveys. *First International Hydrographic Technical Conference*, Ottawa, Canada.

10. van Kuijk, E. 1984. The establishment, calibration and monitoring of a permanent combined Syledis/ARGO navigation system offshore NW Borneo. *Second International Hydrographic Technical Conference*, Plymouth, UK.

Appendix

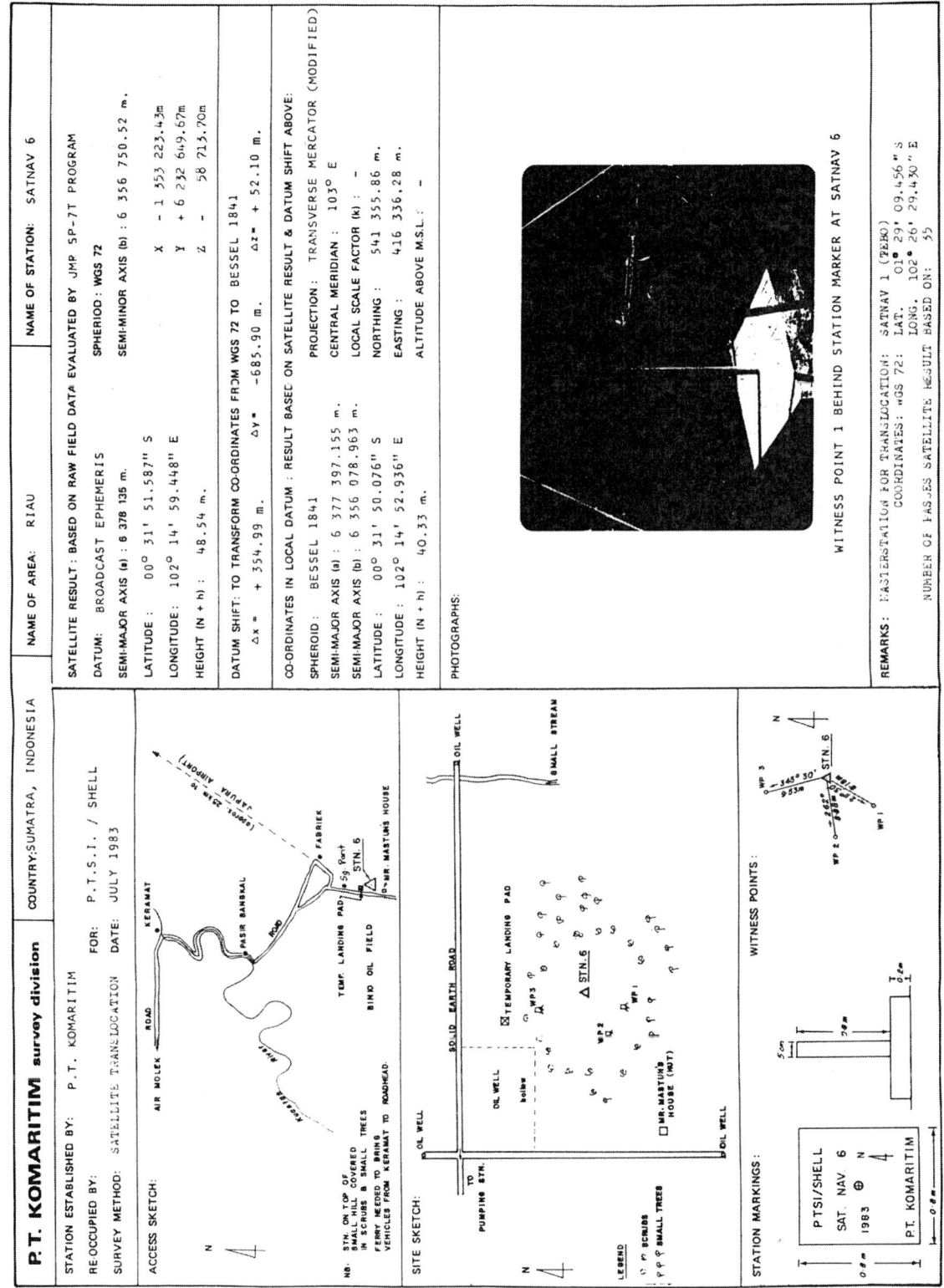

P.T. KOMARITIM survey division

| COUNTRY: SUMATRA, INDONESIA | NAME OF AREA: RIAU | NAME OF STATION: SATNAV 6 |

STATION ESTABLISHED BY: P.T. KOMARITIM

RE-OCCUPIED BY:

SURVEY METHOD: SATELLITE TRANSLOCATION   FOR: P.T.S.I. / SHELL   DATE: JULY 1983

ACCESS SKETCH:

NB. STN. ON TOP OF
SMALL HILL COVERED
IN SCRUBS & SMALL TREES

FERRY NEEDED TO BRING
VEHICLES FROM KERAMAT TO ROADHEAD.

SITE SKETCH:

LEGEND
o o SCRUBS
φ φ SMALL TREES

STATION MARKINGS:

PTSI/SHELL
SAT. NAV. 6
1983
P.T. KOMARITIM

WITNESS POINTS:

SATELLITE RESULT: BASED ON RAW FIELD DATA EVALUATED BY JMR SP-7T PROGRAM

DATUM: BROADCAST EPHEMERIS   SPHERIOD: WGS 72

SEMI-MAJOR AXIS (a): 6 378 135 m.   SEMI-MINOR AXIS (b): 6 356 750.52 m.

LATITUDE : 00° 31' 51.587" S   X = - 1 353 223.43m
LONGITUDE : 102° 14' 59.448" E   Y = + 6 232 649.67m
HEIGHT (N + h) : 48.54 m.   Z = - 58 713.70m

DATUM SHIFT: TO TRANSFORM CO-ORDINATES FROM WGS 72 TO BESSEL 1841

Δx = + 354.99 m.   Δy = -685.90 m.   Δz = + 52.10 m.

CO-ORDINATES IN LOCAL DATUM : RESULT BASED ON SATELLITE RESULT & DATUM SHIFT ABOVE:

SPHEROID : BESSEL 1841   PROJECTION: TRANSVERSE MERCATOR (MODIFIED)

SEMI-MAJOR AXIS (a): 6 377 397.155 m.   CENTRAL MERIDIAN : 103°
SEMI-MAJOR AXIS (b): 6 356 078.963 m.   LOCAL SCALE FACTOR (k): -

LATITUDE : 00° 31' 50.076" S   NORTHING : 541 355.86 m.
LONGITUDE : 102° 14' 52.936" E   EASTING : 416 336.28 m.
HEIGHT (N + h) : 40.33 m.   ALTITUDE ABOVE M.S.L.: -

PHOTOGRAPHS:

WITNESS POINT 1 BEHIND STATION MARKER AT SATNAV 6

REMARKS: MASTERSTATION FOR TRANSLOCATION: SATNAV 1 (PEKO)
COORDINATES: WGS 72: LAT. 01° 29' 09.456" S
LONG. 102° 26' 29.430" E
NUMBER OF PASSES SATELLITE RESULT BASED ON: 55

2

Oahu OTEC Preliminary Design: Sea-Floor Surveys

A. W. Niedoroda, A. C. Palmer, J. R. Pittman, R. J. Brown and Associates, J. P. Vandermeulen, NOAA, US Department of Commerce, and J. F. Campbell, Institute of Geophysics, University of Hawaii, USA

INTRODUCTION

The Ocean Thermal Corporation and its subcontractors are engaged in a multiphase programme to build and operate an ocean thermal energy conversion (OTEC) facility off Kahe Point, Oahu. Ocean Thermal Corporation's OTEC power plant functions as a closed-cycle heat engine. Large quantities of sun-warmed water, drawn from the surface of the ocean, are used to vaporize ammonia. As the ammonia gas expands, it drives a turbine generator to produce electricity. Cold water pumped up from the depths of the ocean then condenses the gaseous ammonia back to the liquid state. The process is termed 'closed-cycle' because the working fluid ammonia, is pumped continuously around the work cycle.

The design and deployment of the cold-water pipe (CWP), which will have a diameter of 6 m and a length of 3300 m and will extend to a water depth of 670 m, is recognized as the most formidable prerequisite for the development of this island-based

OTEC system. From the outset, it was recognized that the cold-water pipe would be routed across major relief features, and would therefore have to be designed to accommodate a hostile sea-floor environment. The steep bottom slopes and relatively great depths for this ocean structure are representative of the new conditions and sea-floor processes which the offshore industry is facing in several frontier areas.

PROJECT REQUIREMENTS

The conceptual design was completed in early 1983. The bathymetric profile used for the conceptual design phase is shown in Fig. 1. This cross-section was developed through a synthesis of existing bathymetric data. These data sources included bathymetric maps prepared from several US coast and geodetic surveys and NOAA (NOS) surveys of the area and an area-wide geophysical survey with coarse trackline spacing conducted by Normak et al. (1982).

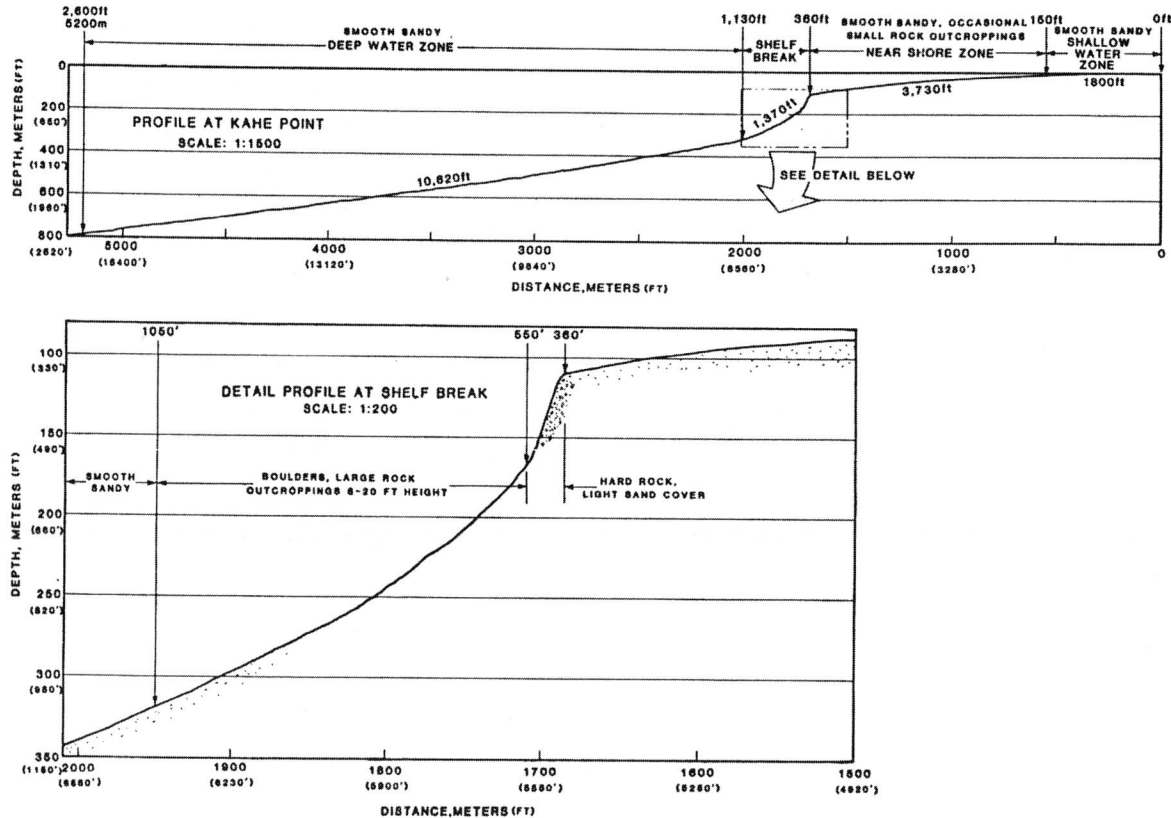

Fig. 1 Conceptual design-phase bathymetric profile

Reports from bottom observations and depth meausrements made by the research submarines *Makali'i* and *Turtle* (Coles, 1982; Offshore Investigations Limited, 1983) were also used.

The cold-water pipe route crosses a narrow (approximately 1.5-km) wide shelf which extends to a water depth of 115 m. Beyond this depth the bottom slope increases rapidly along a near vertical escarpment to a water depth of about 200 m. Beyond 200 m the bottom slope decreases to an average slope of 8° at a depth of 500 m.

The presence of the near vertical submarine escarpment along the conceptual design project depth profile strongly influenced the selection of the cold-water pipe design and the installation methods. The conceptual design was based on a three-part CWP system. A concrete pipe would cross the relatively shallow shelf area. The submarine escarpment section was to be crossed with a double steel wall laminated pipe section which would be connected to a fibre-reinforced plastic (FRP) pipe that was to extend to a depth of 670 m. The 1830 m long FRP section of the cold-water pipe was to be installed using a surface-tow technique. The steel escarpment section was to be installed as a monolithic piece and anchored to the bottom with piles.

Discussions at the beginning of the preliminary design phase centered about improving our knowledge of the shape and composition of the sea floor in the project area so that an adequate preliminary design could be accomplished. It was also recognized that additional survey data could sharply alter the definition of the sea floor and thus affect the overall design concept.

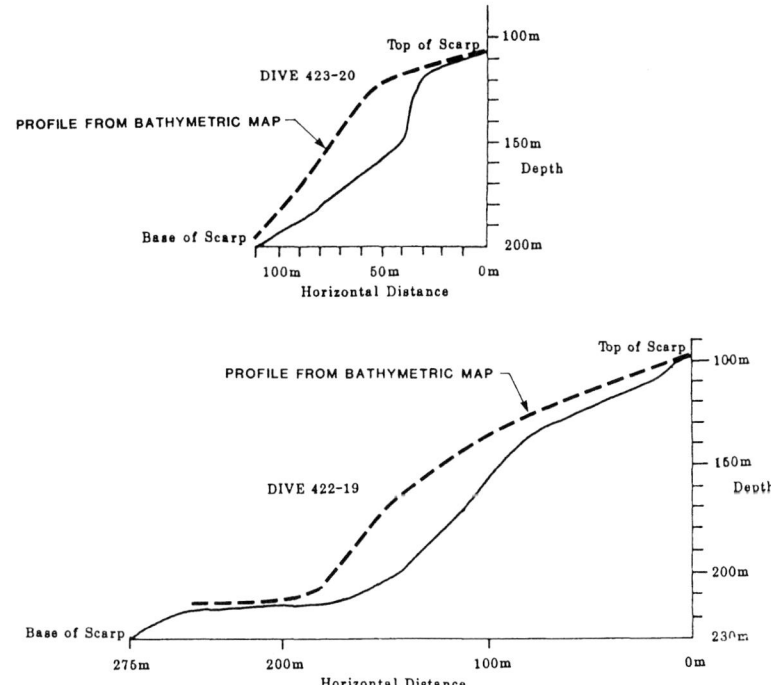

Fig. 2 Comparison of submarine and map-derived bathymetric profiles

SURVEY REQUIREMENTS AND PLANNING

An existing bathymetric map was used to plan the survey, but a comparison with profiles measured from a submarine revealed serious discrepancies (Fig. 2).

The primary aims of the preliminary design survey were to

- map the shelf edge and submarine escarpment to establish whether the near-vertical slopes are continuous features throughout the project area;
- explore areas where conceptual design data indicate possible alternative routes;
- locate intermediate-scale bathymetric features in all water depths across the project area with a resolution adequate for preliminary route selection;
- evaluate the relative dangers posed by sea-floor geo-hazards;
- estimate the composition and thickness of the bottom material underlying the escarpment section;

- evaluate the distribution and thickness of bottom sediments seaward of the escarpment section;
- evaluate relative hazards and foundation requirements posed by sediments shoreward of the shelf edge; and
- obtain sediment and rock samples for the preliminary determination of their mechanical properties.

The preliminary design field survey was planned as a two-phase operation. The first phase would concentrate on collecting high-resolution, precision bathymetric and geophysical sub-bottom data over the project area. It was recognized that a brief period would be needed to make a initial analysis of these data and to adjust the final plans for the second phase of the field survey. The second phase of the field survey was planned to concentrate on mapping the entire project area from several perspectives, using the Sea MARC II bottom imaging system. In addition, the bottom was to be sampled at key locations with conventional oceanographic sampling equipment to evaluate the distribution and mechanical properties of the bottom sediments.

FIELD METHODS

The field survey was conducted in two phases.

The first phase utilized the NOAA Survey Ship, the *Rainier*, to obtain precision bathymetric soundings on a tight grid over the entire project area. This vessel was also used as a platform for sub-bottom profiling using a 10 in³ airgun, an 800 J sparker, and a 3.5 kHz profiler (supplied and operated by NORTEC of Seattle, Washington). The primary tracklines were to be oriented perpendicular to the general bottom contours and spaced at 50 m intervals so that a 1 : 3600 scale map could be prepared. All navigation was based on a Motorola MiniRanger III electronic navigation system. The geophysical and bathymetric data were collected along the same tracklines. Five crosslines were surveyed.

The bathymetric data were collected with a Raytheon DSF 6000 dual-beam precision fathometer. This instrument operates on frequencies of 24 and 100 kHz. The 100 kHz acoustic beam has a beamwidth of 7.5° between its half power points and is intended for precise depth measurements in relatively shallow water. The 24 kHz beam is much wider (27° in the fore-and-aft direction and 47° athwartship). It increases the system's ability to detect shoaling bottom features to the side of the vessel track, and it extends the system's overall depth range capability. Normally, the returns from both beams are displayed on the sounder's graphic record while the return from either the high- or low-frequency beam is selected for digitization and computer recording.

The two-range method of navigation was used with the MiniRanger III line of sight positioning system. Shore transponders were installed over existing geodetic control stations along the coast. The selected sites provided excellent intersection angles between the lines of position (LOP); and the observed ranges did not exceed 7000 meters, so that the system was operated well within its design capability. The system had been calibrated over a fixed baseline ashore, and this calibration was checked before and after survey operations by comparison with sextant fixes to geodetic control. A check angle was measured in addition to the basic two angles to ensure that the correct stations were being observed.

Digital data consisting of one depth input, two navigation LOPs, ship's heading, and time were recorded in real time using the ship's HYDROPLOT system. A sample interval of approximately 6 seconds was selected as suitable for the survey scale and at the ship's speed being used. Depths were recorded in fathoms and tenths, ranges in meters. In addition to the digital data recording, the HYDROPLOT system generated a real-time plot so that progress and preliminary results could be evaluated. This also provided steering information to the helmsman to guide him along the intended survey lines. Seismic data were recorded independently, except that time and event marks were kept synchronized with bathymetric data recordings.

These operations were, for the most part, the same as those which the ship normally conducts as part of its hydrographic survey mission. Since this was the first use of the DSF 6000 echo-sounder aboard the *Rainier*, some problems were encountered in its operation, particularly in controlling the system's gain when the depth changed rapidly. Steering the ship at slow speed (4 knots) accurately down the rather short and closely spaced sounding lines in a tidal cross-current was also a challenge. The slow speed was necessitated by the seismic sensors being towed astern and by the need to resolve the very steeply sloping bottom as sounding lines were run nearly perpendicular to the bottom contours. Except for some caution in turning from one line to another, no maneuvering interference was experienced from the towed seismic gear. Other vessel traffic was, fortunately, very limited in the area, but the very short survey period involved did have to be timed to the last moment to avoid interfering with routine Navy operations in the same area.

Several days after the ship's bathymetric and seismic survey had been completed, one of the *Rainier*'s survey launches was used to tow a Klein 100 kHz short-range side-scan sonar in the area of the steep escarpment. Several lines were run generally parallel to the bottom contours to give full bottom coverage.

During the survey it was discovered that a strong acoustic return, sufficient to trigger the bathymetric digitizing system, could be maintained to a water depth of approximately 330 m. This coverage mapped most of the steeply sloped portion of the project area. However, it did necessitate manual switching to the lower frequency acoustic beam in deeper water. When this was done over a sloped bottom, a distortion in the depth measurements was noted on the records. This was probably due to the different beamwidths associated with the two frequenices. Figure 3 shows the trackline locations for the measurements taken aboard the *Rainier*. The three potential alternative

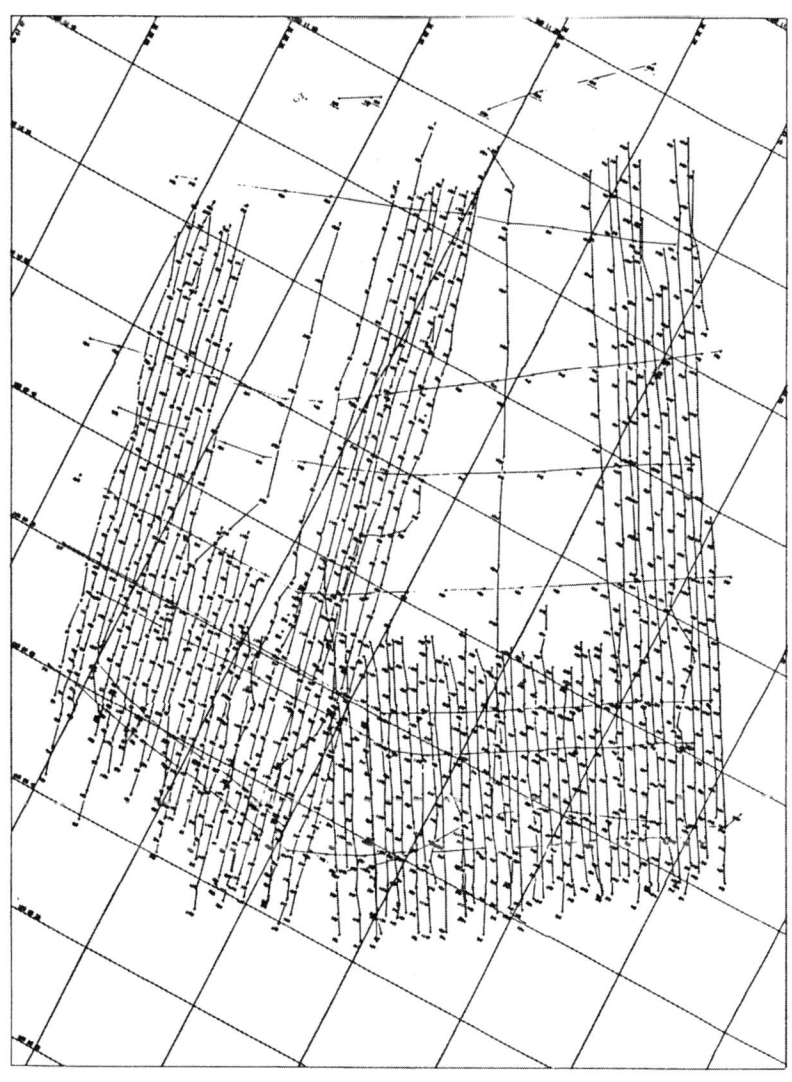

Fig. 3 NOAA ship *Rainier* tracklines

CWP corridors are clearly indicated by the relative density of tracklines in the deep-water areas.

The second phase of the preliminary design field survey used the University of Hawaii's Research Vessel *Kana Keoki*. A brief period (four days) was allocated for the initial reduction of the data collected during the cruise of the *Rainier*. The results of this data analysis were used for the final planning of the *Kana Keoki* cruise. The objectives of this cruise were to image the entire project area sea floor with the SeaMARC II wide-area side-scan system and obtain gravity and box-core samples.

DATA REDUCTION

Much of the data collected during the preliminary design survey required only standard data reduction procedures. However, several unique problems were identified and solved. These problems included adequate removal of artificial features caused by switching between the 100 kHz and 24 kHz precision fathometer beams, establishing the relative relief of features seen on the Sea-MARC II images, and combining data from the precision bathymetric survey data and SeaMARC II images with information from the research submarine dives. Each of these procedures is discussed briefly below.

Bathymetric Data

Bathymetric data were corrected for the sound velocity profile based on a sound velocity cast taken at the conclusion of the survey work. These data were also corrected for the small tide variation based on predicted tidal elevations at Honolulu Harbour, which have been shown to be nearly identical in phase and amplitude to the Kahe Point project area.

Digital depth values were edited manually by scanning the analogue depth-sounding record and picking up missed depths, signif-

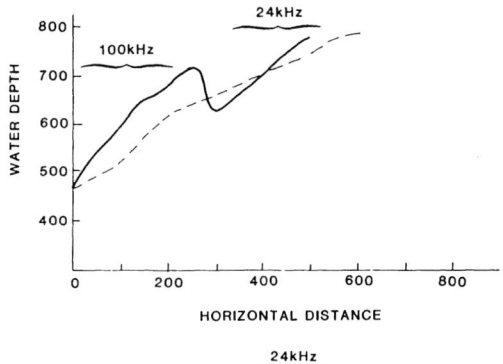

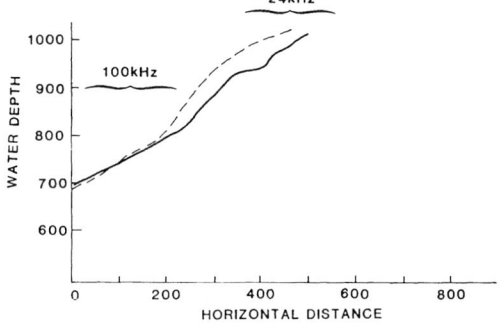

Fig. 4 Horizontal distance effects of changing between the 24 kHz and 100 kHz beams

icant peaks, and deeps not previously selected. The editing tape was then merged with the raw data tape and other corrected information in the HYDROPLOT system aboard the *Rainier* to generate the required position and sounding plots. Position computations were corrected for any range calibration correctors and for the offset between the echo-sounding transducer and MiniRanger antenna locations on the ship. Depth values were corrected for vessel draft, predicted tide, and sound velocity. Plots were made in fathoms and in feet on the modified transverse Mercator projection used by the NOAA Charting and Geodetic Services.

Figure 4 shows a comparison of the relative depth changes associated with switching between the 100 kHz and 24 kHz fathometer beams. These effects were subjectively removed while preparing the final bathymetric contour map.

SeaMARC II Imaging

SeaMARC II is a shallow towed sonar system that uses acoustic energy to produce both a side-scan sonar image and a bathymetric map of the ocean floor. In the deep ocean the system is capable of mapping a swath 10 km wide. For this survey most of the mapping was done with a 5 km swath width.

The electronics for the side-scan sonar subsystem acoustics are located in the tow fish. Signals from the hydrophone arrays are transmitted to the ship and are sampled and corrected for slant-range geometry, assuming a flat bottom at a depth set by the first acoustic arrival received from each ping. Any deviation of the bottom from the assumed flat plane will cause pixels to be mislocated; for example, shallow areas will appear closer to the ship's track than they really are. These data are reduced to 1024 pixels for each side-looking array, displayed in real time on an EPC precision graphic recorder and logged for later playback and analysis. The shipboard recordings are used to guide the survey, to assure total coverage and to allow multiple passes of areas of special interest. The shipboard records were also used to fine-tune decisions on where to do bottom sampling.

Upon return of the data to the laboratory at the Hawaii Institute of Geophysics, the first step in the analysis was to process the navigation data. For this survey the location of the Research Vessel *Kana Keoki* was well documented by a Del Norte trisponder system. The final digital navigation data were then used to produce a ship's speed file, which was merged with the digital side-scan sonar data to pinpoint the along-trace location of each ping. Using the ship's 3.5 kHz bathymetry as a guide, corrections were made to the side-scan data in areas where the bottom was not properly detected by SeaMARC.

After these corrections had been made the data were plotted as a continuous strip, using a 22-inch Versatec plotter. Side-scan sonar data from discrete tracks were then placed over a plot of the final navigation with adjustments made for the distance the fish was towed behind the ship. Minor variations in the ship's tracks were accommodated by cutting the side-scan record and adjusting it to best fit the navigation. Each track was mounted, with bonding wax, on a clear plastic sheet which was keyed to the overall navigation by means of registration tabs. Slight adjustments were made to the location of the tracks to ensure a visual fit of the side-scan sonar images, i.e. to minimize the apparent offset of what is obviously the same feature as that caused by the variations of the real bottom from the flat plane assumption used in placing the pixels.

After final adjustments were made, a composite mosaic of all the side-scan data for the survey was compiled (Fig. 5). Since many of the images overlapped, those that best typified a particular portion of the bottom were chosen to go on the mosaic. Gaps in the data due to effects such as slight bends in the ship's track were filled in with a neutral grey background to enhance the overall visual image.

For the detailed analysis of the bottom needed for this survey both the final mosaic and the individual side-scan sonar tracks were important. Features that appeared in the detailed bathymetry or on the overall mosaic could be viewed from various angles using the set of overlying side-scan images to determine what their real character was. For example, the linear features shown on the SeaMARC II mosaic (Fig. 5) were unexpected as no indication of coherent small-scale relief was evident in the precision bathymetric survey. These acoustic contrasts can result from variations in bottom relief, bottom texture (e.g. ripples versus smooth areas) and sea-floor composition. Two methods were used to sort out these differing effects. The first was based on a detailed mapping of the small-scale features and the second method on comparison of the submarine observations to these maps.

The series of SeaMARC II overlays was used to map the linear bottom features. Because of the overlapping pattern of images it was possible to view each bottom

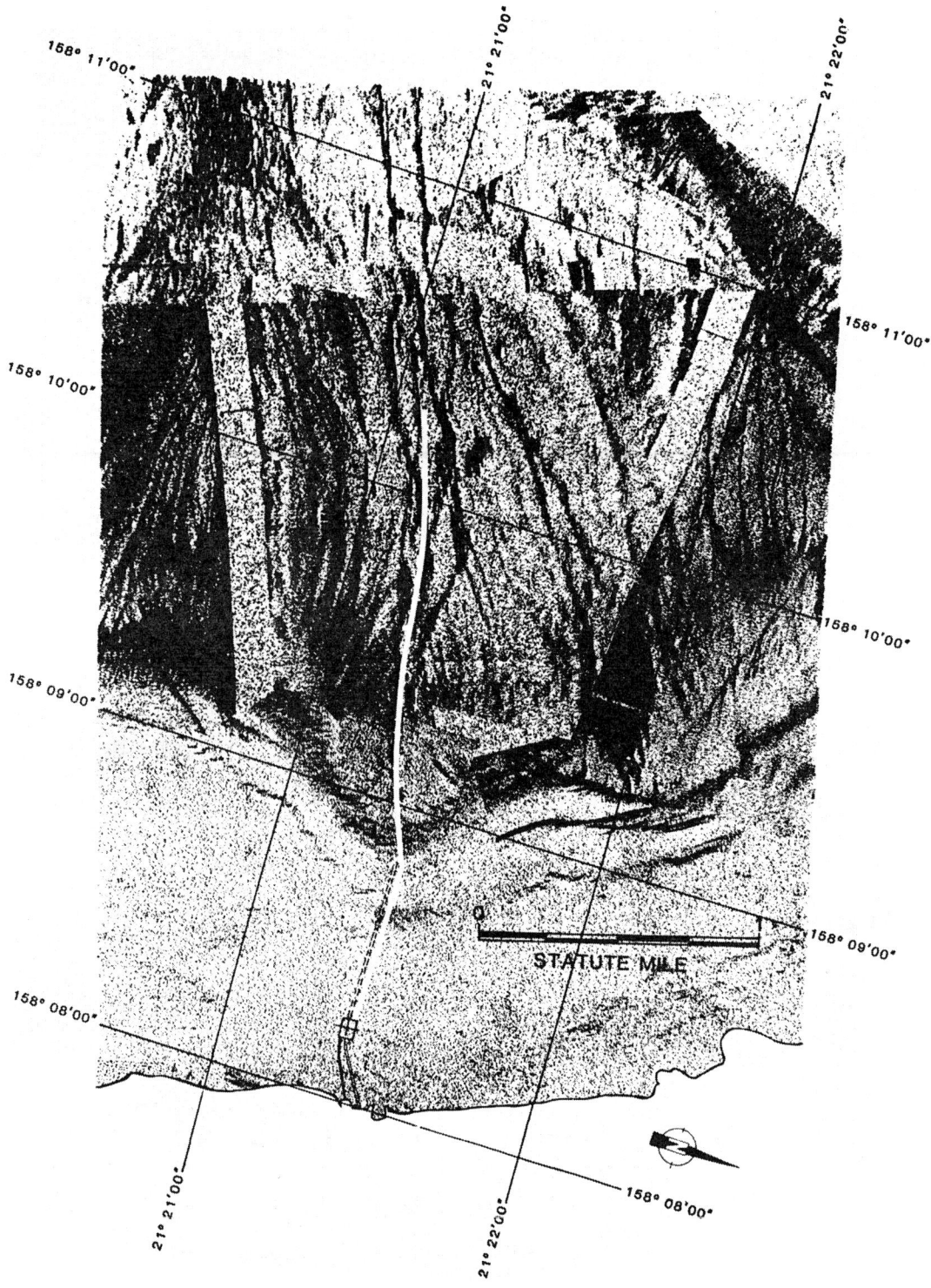

Fig. 5   SeaMARC II mosaic

feature in the project area from the several 'illumination' orientations. This, in turn, made it possible to distinguish changes in the acoustic returns due to small bottom slopes from those due to differences in bottom material or texture. The bottom slopes were marked with a slope hachure notation. The length of the slope was equal to the hachure length. The apparent steepness of the slope was proportional to the hachure density. Tapered hachures were used so that the apparent upslope direction corresponded to the thicker end of the lines. Broad, relatively flat and featureless areas were shaded.

The results of mapping the SeaMARC II data are shown in Fig. 6. Most of the downslope trending linear features which appear

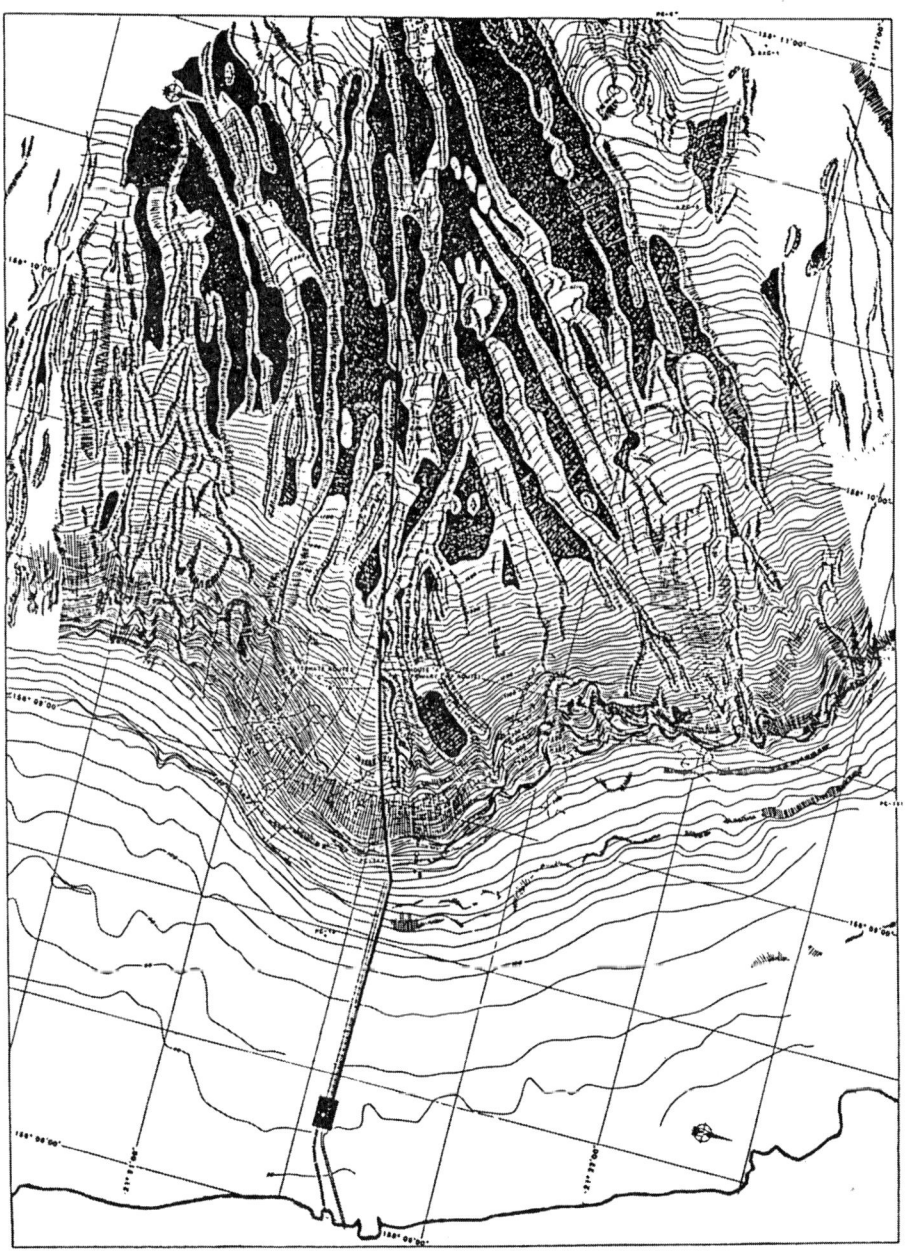

Fig. 6 Composite seaMARC II hachure map and bathymetric map

in Fig. 5 occur as long slope facets oriented across the large-scale bottom slopes. Most of these small linear slope facets occur in pairs and many define long, narrow sea-floor valleys whose axes parallel the downslope direction of the larger-scale bathymetry. These features are particularly noticeable in the northern and southern portions of the deep offshore area (see Fig. 6).

In some places the narrow sea-floor valleys branch so as to resemble a dendritic river pattern. In all cases these long, narrow valleys are discontinuous. That is, unlike most terrestrial stream valleys, these eventually peter out and end.

Even more striking features were detected when mapping the SeaMARC II

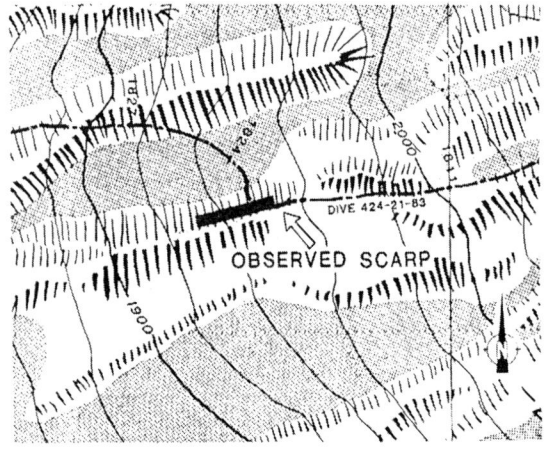

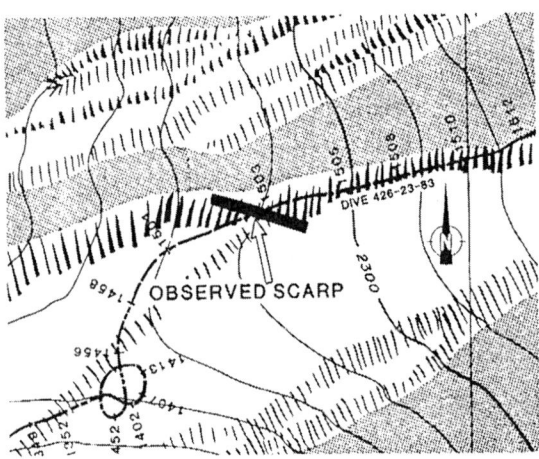

Fig. 7 Examples of submarine observations

images. Careful examination of many of the linear features which appear in Figs 5 and 6 show that they are long, low and positive relief features, not negative ones. These low sea-floor ridges are quite different from anything reported in similar environments. Unlike the narrow sea-floor valleys, these ridges seldom branch. However, they are discontinuous. The best terrestrial analogue to them appears to be the natural levees which occur along alluvial rivers.

The hachure-pattern mapping of the fine-scale relief shown by the SeaMARC II images does not permit evaluation of the vertical scale of the features. Examination of the fathograms showed that these features were too subtle to show up on the 24 kHz or 100 kHz beams. Examination of the records of observations made during reconnaissance dives of the Research Submarine *Turtle* (Noda, 1983) showed that small sedimentary scarps were reported at the exact points where the dive-track crossed some of the long, low and linear sea-floor ridges. Figure 7 shows examples. The relief of most of the linear features which show on the SeaMARC II images, but not on the fathograms, is taken to be in the range of 1 to 4 m.

RESULTS AND ENGINEERING IMPLICATIONS

Bathymetry

The new information about the shape of the sea floor in the project area can be conveniently divided into results pertaining to large- and small-scale bathymetric features. Results concerning both scales of features have influenced the design and the choice of installation method for the CWP.

The bathymetry shown in Fig. 6 shows that the near-vertical slopes reported by the research submarine dives in the 'escarpment' zone are not continuous in the cross-slope direction. This result has led to the location of at least three CWP routes which have relatively moderate slopes with smooth slope gradients. Because the slopes along these routes are not too steep it has been possible to alter substantially the concept for installing the CWP and, in turn, change its design. The steel composite pipe section which was thought necessary for spanning the escarpment zone has been eliminated. It is now envisaged that the portion of the CWP running from the top of the escarpment to its seaward end will consist of fibre-reinforced plastic (FRP). It will be installed using a bottom-tow method.

The shape of the sea floor in the deep offshore zone has been well defined by the preliminary design survey. The most significant features include broad undulations of the bottom with axes parallel to the area-wide slopes. One such undulation provides a gentle, subtle high extending offshore along the centre of the project area. The survey has also located at least two mound-like features with 6 to 20 m of relief (one at a water depth of 630 m and the other at 650 m), whose origins are unknown. The morphology could indicate that they formed as patch reefs in much shallower water. Bottom photographs taken by the *Turtle* of a similar feature located north of the OTEC survey area more clearly suggest that these features are drowned patch reefs. Gregory and Kroenke (1982) found similar features at similar depths on a broad terrace south of Oahu. Drowned reefs have also been discovered off the Island of Hawaii (Campbell, 1984; Moore and Fornari, 1984).

The greatly improved resolution of the bathymetric maps which resulted from combining the high-resolution depth measurements with the detailed image resulting from the SeaMARC II has focused attention on the small-scale relief in the project area. It is recognized that, although some of the elongated features seen in the Sea-MARC II images result from differences in bottom composition and texture, most of these features represent low-relief valleys or ridges. The submarine observations have shown that some of these features have near-vertical relief from 2 to 4 m high. The orientation of these low sediment scarps, parallel

to the low, long ridges, and their association with these features suggests that they have an erosional origin.

The origin of the prevalent small-scale relief features is not immediately apparent. It has been postulated that rapid sediment transport events have occurred within the project area since November 1982 when a series of current meter moorings in the deep offshore area were episodically disrupted and some were moved rapidly downslope by a hurricane (Noda 1983). Dengler (1984a,b) used the pressure records from the current meters which were moved and subsequently recovered to study the temporal pattern of the disrupting events. He concluded that relatively low speed turbidity currents, triggered during the near passage of Hurricane Iwa, were the most likely cause of the episodic mooring movements. Other processes such as submarine debris flows caused by breaking internal waves have been considered. However, the geometry of the small-scale sea-floor relief and the relationship between the probable path of the turbidity currents and current meter movements appear to support the mechanisms proposed by Dengler.

The elongated small-scale relief features are closely aligned with the downslope direction of the next larger scale of bathymetric features. This further implies that they originated through gravity-driven sediment transport processes. Furthermore, it permits the selection of a CWP route which minimizes the chance of impact by turbidity currents and the angle of collision should such an impact occur.

Submarine Observations

The use of submarine observations to estimate the relative relief of the elongated small-scale features has already been discussed. The submarine observations also recorded occasional coralline limestone cobbles in the deep offshore zone and numerous large limestone blocks in a submarine talus area at the base of the escarpment zone (Noda, 1983).

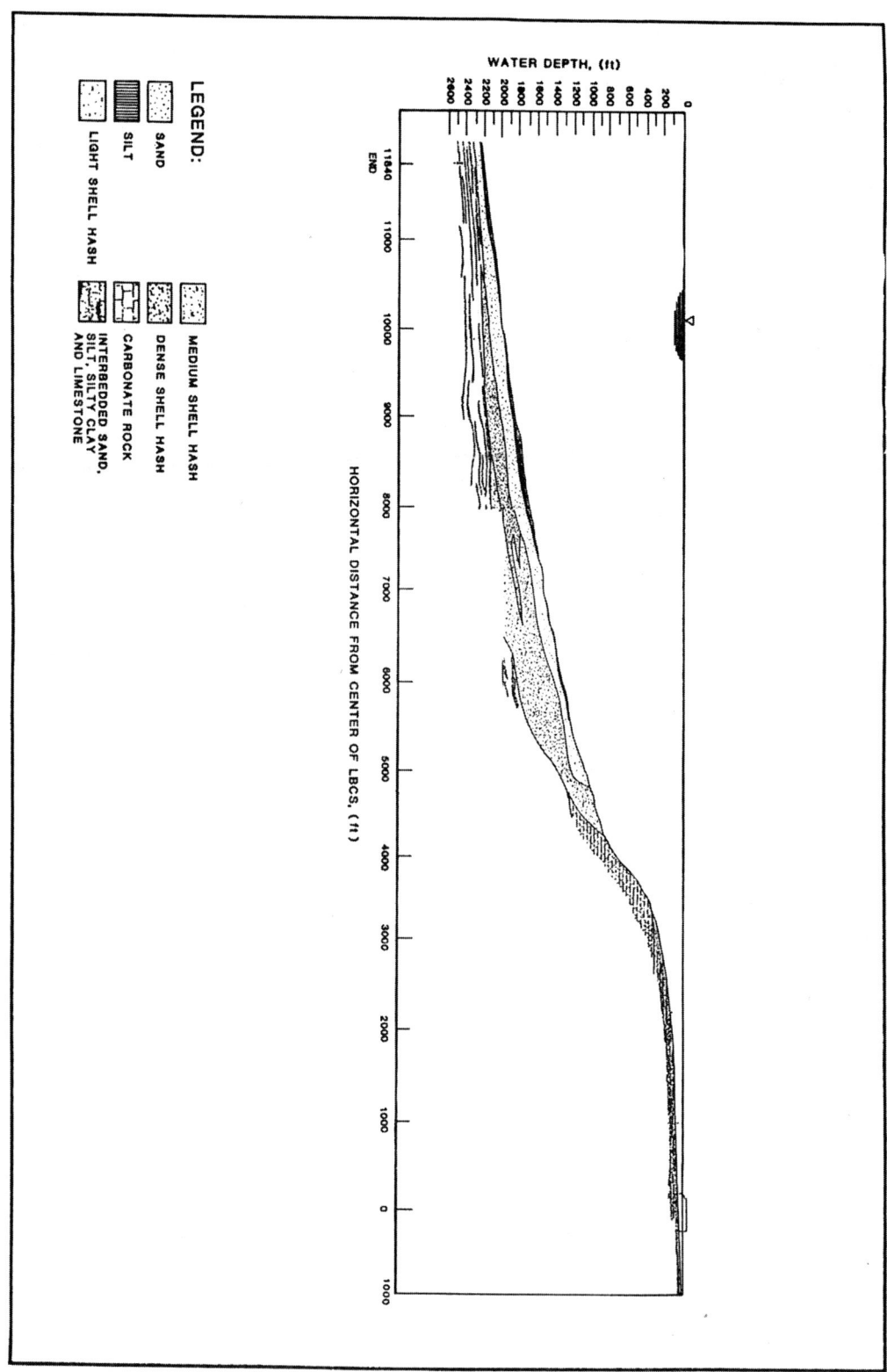

Fig. 8 Shelf composite

Sub-bottom Profiling and Bottom Sampling

The general distribution of sediments is shown on Fig. 8. This figure represents a composite of information from the 3.5 kHz airgun and sparker data, along with data from bottom samples and submarine observations. The composite represents the sediment and rock distribution along the CWP route.

The escarpment is underlain by a thick sequence of what appears to be coralline limestone. The base of the escarpment roughly corresponds to a place where coarse carbonate sand (shell hash) overlaps the limestone. These sediments are overlain by a relatively thin (about 10 m) layer of carbonaceous silt further offshore. This layer is interbedded with graded sands and gravels.

The results of these survey operations showed that adequate foundations can be engineered for the CWP. The portion of the CWP crossing the escarpment will require drilled and grouted piles to resist lateral loads. The section further offshore will utilize lightly loaded gravity foundations.

ACKNOWLEDGEMENTS

The authors wish to thank the following individuals for their assistance in planning and organizing the survey, their special dedication in carrying out the field work, and their help both in the reduction and analysis of the data and in preparing the manuscript: the crew of the *Rainier*, Dick Sylvester and Paul Farley of Nortec, Don Housong and Dan Fornari, Jim Desmond, Toni Smith, and Lydia Elizondo. The authors also would like to thank the Ocean Thermal Corporation, the Hawaiian Institute of Geophysics, the TRW Energy Development Group and R.J. Brown and Associates for their support in publishing this paper. This is Hawaii Institute of Geophysics contribution 1571.

REFERENCES

1. Campbell, J. F. 1984. Rapid subsidence of Kohala Volcano and its effect on coral reef growth. *Geomarine Letters* 4, 31–36.
2. Coles, S. L. 1982. A visual reconnaissance of the bottom between 20 and 365 m depth offshore of Kahe Point, Oahu for the purpose of OTGEC pipelint/cable routing. Internal report. Hawaiian Electric Company, Honolulu.
3. Dengler, A. T., Wilde, P., Noda, E. K. and Normark, W. R. 1984a. Turbidity currents generated by Hurricane Iwa. *Geomarine Letters*.
4. Dengler, A. T., Noda, W. K., Wilde, P. and Normark, W. R. 1984b. Slumping and related turbidity currents along proposed OTEC cold-water-pipe route resulting from Hurricane Iwa. In *Proc. Offshore Tech. Conf.* Paper No. OTC 5702.
5. Gregory, A. E., III and Kroenke, L. W. 1982. Reef development on a mid-oceanic island: reflection profiling studies of the 500-Meter Shelf south of Oahu. *Bull A.A.P.G.* 66(7), 843–859.
6. Moore, J. G. and Fornari, D. J. 1984. Drowned reefs as indicators of the rate of subsidence of the Island of Hawaii. *J. Geol.* 92, 752–759.
7. Noda, E. K. 1982. Site selection bathymetry and sub-bottom seismic surveys in the near-shore zone off Kahe Point, Oahu. Report to RCUH, University of Hawaii.
8. Noda, E. K. 1983. Effects of Hurricane Iwa, November 23, 1982, offshore of Kahe Point, Hawaii. Unpublished report.
9. Normark, W. E., Chase, T. E., Wilde, P., Hampton, M. A., Gutmacher, C. E., Seekins, B. A. and Johnson, K. H. 1982. Geologic report for the OTEC site off Kahe Point, Oahu, Geological Survey Open-file Report 82-468A, US Department of the Interior.
10. Offshore Investigations Ltd., 1983. D.S.V. *Turtle* dives off Kahe Point, Hawaii. Unpublished Report, 27.

3

The Geology of Some Atlantic Abyssal Plains and the Engineering Implications

R. T. E. Schüttenhelm, A. Kuijpers and E. J. Th. Duin, Geological Survey of the Netherlands, Haarlem, The Netherlands

ABSTRACT

Studies of a number of Atlantic abyssal plains in relation to a feasibility study on the sub-seabed disposal of radioactive waste have shown that sedimentation in abyssal plains is caused by various sediment transport mechanisms. The eastern North Atlantic Madeira Abyssal Plain and the western North Atlantic Nares Abyssal Plain may serve as examples of the resulting widely differing sedimentary patterns. An understanding of these patterns is a basis for a successful site investigation in an abyssal plain setting.

INTRODUCTION

Abyssal plains are among the least known parts of our planet. Far more is known about large parts of deserts and polar areas than about these plains, which at first sight seem to be rather monotonous parts of the ocean floor. The AGI *Glossary of Geology* defines an abyssal plain as follows:

'A flat region of the ocean floor, usually at the base of a continental rise, whose slope is less than 1 : 1000. It is formed by the deposition of sediments that obscure the pre-existing topography.' We have since discovered that in places it is difficult to distinguish between continental rises and abyssal plains and, furthermore, that the slope criterion of 1 : 1000 is not always a good one.

Scientists involved in geological ocean research have generally concentrated on plate boundaries, passive margins and ocean history, and commercial firms were not interested in abyssal plains either, apart from Pacific areas with manganese nodules. Therefore there was no demand for surveys of abyssal plain sites. From 1980 onwards this general neglect gave way to some interest from a perhaps rather unexpected quarter.

The feasibility of sub-seabed disposal of radioactive waste is now being investigated in a number of national programmes that are managed by the Seabed Working Group

29

(SWG), a non-exclusive co-operative organization of a number of member states of the Nuclear Energy Agency (NEA), a subsidiary of the Organization of Economic Cooperation and Development (OECD). The objective of the programmes is to provide an international assessment of the technical feasibility and long-term safety of subseabed disposal of high-level radioactive waste as a multimedia concept with marine sediments as the primary isolation barrier. One of the SWG task groups, the Site Assessment Task Group, has to select and study one or more sites which are suitable as a basis for a feasibility assessment.

As described in Auffret *et al.* (1984), two primary criteria — geological stability and barrier capacity — guide the site selection process. Large parts of the oceans have been evaluated by the Site Assessment Task Group on the basis of available archive data, if any, and current opinions. This has resulted in only a limited number of potential areas being of interest (Fig. 1). At present, the on-site studies are concentrated on parts of two North Atlantic abyssal plains. These two abyssal plains are the Madeira Abyssal Plain and the Nares Abyssal Plain, the former in the Eastern North Atlantic Basin, the latter in the Western North Atlantic Basin. They represent, together with the sandy abyssal plains, the different

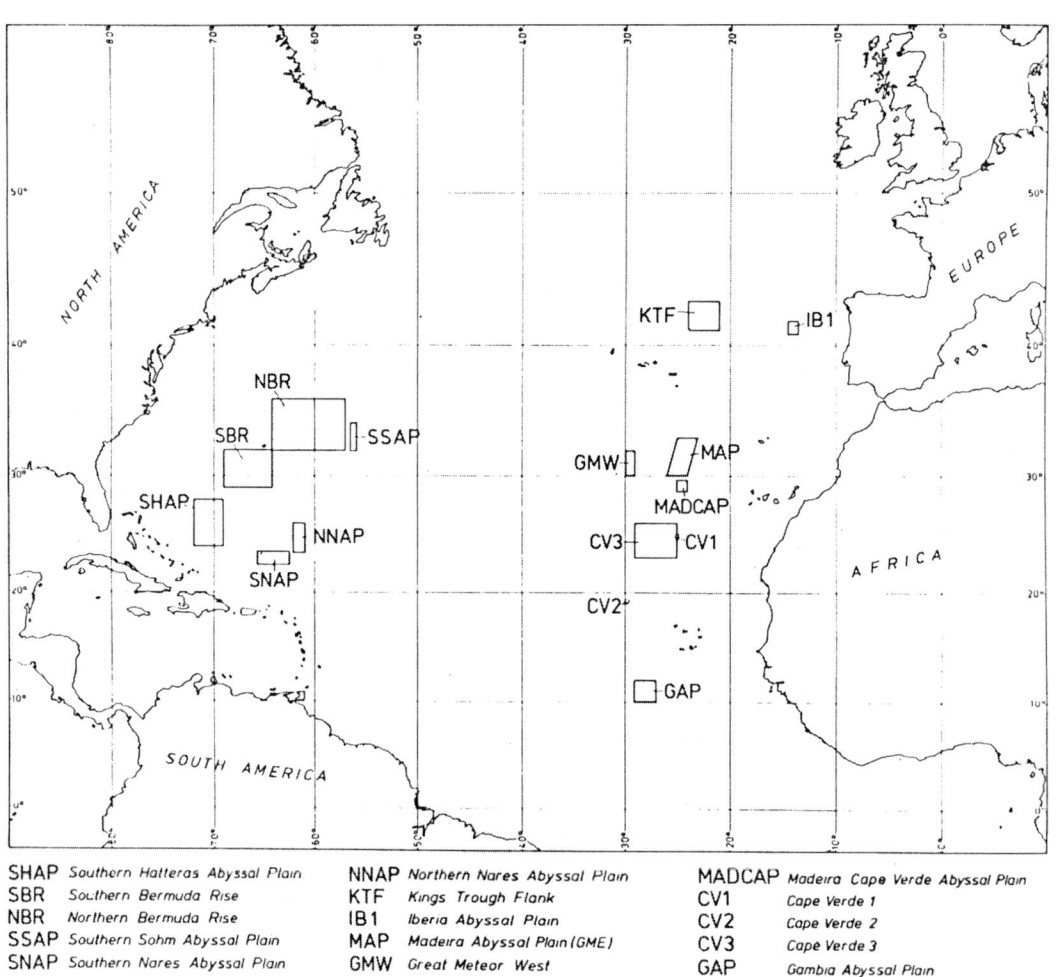

SHAP	Southern Hatteras Abyssal Plain	NNAP	Northern Nares Abyssal Plain	MADCAP	Madeira Cape Verde Abyssal Plain
SBR	Southern Bermuda Rise	KTF	Kings Trough Flank	CV1	Cape Verde 1
NBR	Northern Bermuda Rise	IB1	Iberia Abyssal Plain	CV2	Cape Verde 2
SSAP	Southern Sohm Abyssal Plain	MAP	Madeira Abyssal Plain (GME)	CV3	Cape Verde 3
SNAP	Southern Nares Abyssal Plain	GMW	Great Meteor West	GAP	Gambia Abyssal Plain

Fig. 1 Areas in the Atlantic Ocean studied by the Seabed Working Group. The Cape Verde sites 1 and 3 are part of the Cape Verde Abyssal Plain

types present in the Atlantic Ocean. Abyssal plains in the Atlantic are rather different from many Pacific examples as these are covered by pelagic red clays, a type of sediment that is rare in the Atlantic. So-called typical abyssal plain sediment properties are, therefore, if based on Pacific red clay, of little value for Atlantic settings. The characteristics of the two abyssal plains mentioned earlier will be summarized now with an emphasis on sediments and sediment properties.

MADEIRA ABYSSAL PLAIN

Topography and Bathymetry

The Madeira Abyssal Plain extends in the eastern North Atlantic between 30° and 34°N and 22° and 26°W. The main topographic features around the plain are, in the east, the African Continental Rise and, in the west, the Meteor-Atlantic Seamount Complex and adjacent abyssal hill province. In the far north is the East Azores Fracture Zone with related relief, and in the south a narrow and slightly uneven corridor to the Cape Verde Abyssal Plain.

The part of the Madeira Abyssal Plain studied by SWG is between 30° and 33°N and 23° and 26°W. The abyssal plain is an extremely flat area at a water depth of about 5440 m, situated at the distal extremity of a north-west African sediment transport pathway where it is blocked by abyssal hills (Fig. 2). Depth variations of only a few metres over distances of tens of kilometres occur in the central-western abyssal plain, whereas more to the east the seabed gently rises (Fig. 3). In places, linear arrangements of hills up to several hundred metres high interrupt the monotony of the plain. The abyssal plain extends without major changes to the north as far as the area is surveyed (to 33°N), interrupted only by several hill alignments that are morphological expressions of deeply buried fracture zones. South of 31°N, the sea floor is less flat and the hills are more narrowly spaced.

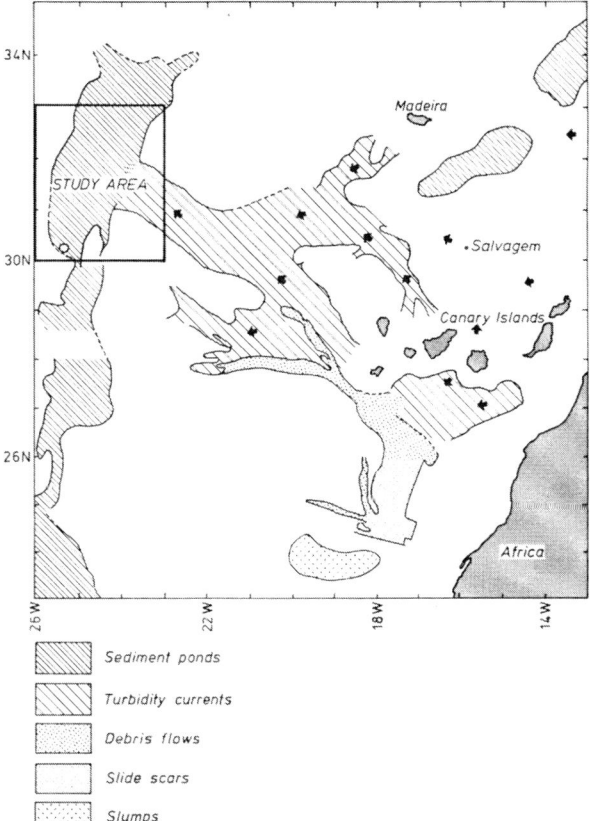

	Sediment ponds
	Turbidity currents
	Debris flows
	Slide scars
	Slumps

Fig. 2 Summary of some sedimentary features and turbidite dispersal paths of the Canary Basin (after Embley, 1982). Recent studies have shown that the sediments in the turbidity current area consist at least in part of debris flows

Sea-floor Imagery

Bottom photographs in turbiditic areas show a smooth sediment surface with numerous well-defined animal tracks (Fig. 4), whereas in a slightly elevated pelagic sediment area the sea floor shows less well preserved tracks of bottom organisms. Rock fragments, including manganese nodules and crusts, are evident on the upper slope of an abyssal hill located in the plain (Auffret *et al.*, 1984).

Sediment Thickness and Structure

The sedimentary cover over a basement of Late Cretaceous age is generally 500–1000 m

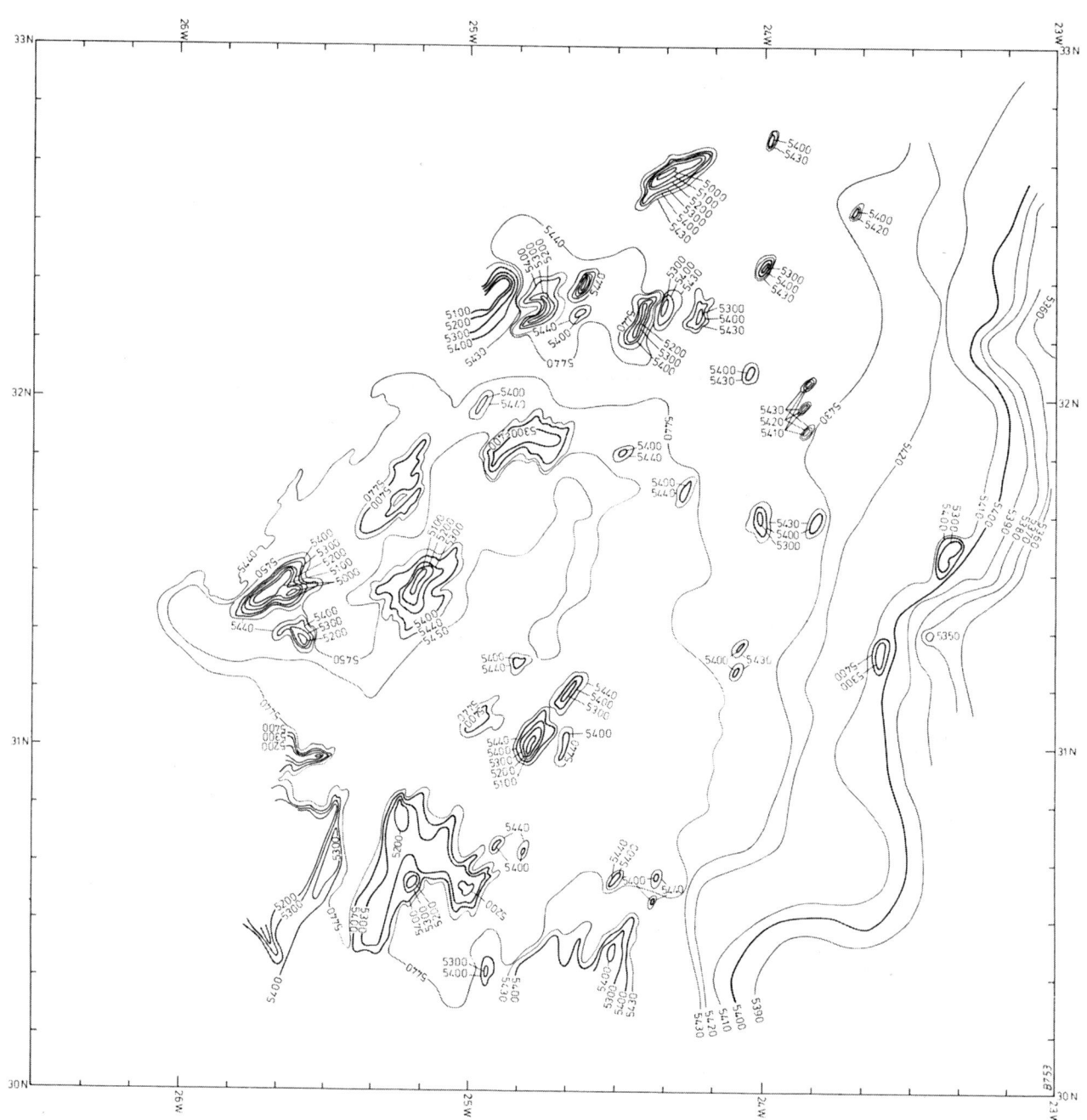

Fig. 3 Bathymetric map of the Madeira Abyssal Plain

Fig. 4   The sediment surface in the central part of the Madeira Abyssal Plain with large animal tracks

with maxima of more than 1750 m over fracture zones. Sub-bottom reflectors are continuous in the western part of the plain (west of 24°W). To the east a lack of continuity and additional reflectors complicates seismostratigraphic interpretations. The sedimentary section can be subdivided into three general intervals: The uppermost 120 m interpreted as alternating turbiditic and pelagic sediments; an intermediate interval of 200 to 400 m of inferred pelagic with some turbiditic sediments; a lower interval comprising very strong but less continuous reflectors (volcanoclastics? limestones?) with varying thickness and a basal pelagic drape. Penetration on the 3.5 kHz records varies across the abyssal plain, depending on sediment texture. In predominantly clayey sediments, penetration generally exceeds 80 m. However, in areas with near-surface sands penetration is much less. Slight undulations in the shallow, parallel-bedded sub-bottom generally reflect the presence of basement culminations deeper in the section. Seismic and 3.5 kHz data have revealed the presence of discontinuous

faults with a small vertical offset, especially in the central and central-northern areas of the plain.

The faults have no surface expression but some remain visible close to the surface. On 3.5 kHz records the offset generally increases with depth. Some continue downward in the lower sedimentary units as observed on watergun records. A fully convincing explanation of the origin of these faults has not yet been presented (Duin *et al.*, 1984).

Stratigraphy

Some 60 piston cores up to 22 m in length, obtained in 1980–1984 during various cruises, consist primarily of correlatable layers of turbiditic marly calcareous ooze (0.2–5 m) separated by thin but continuous beds (c. 0.1 m) of hemipelagic ooze to clay (Fig. 5). Pelagic sediments have accumulated around the carbonate compensation depth (CCD), resulting in highly variable $CaCO_3$ contents. Fine-grained turbidites typically have a 45–55% carbonate content,

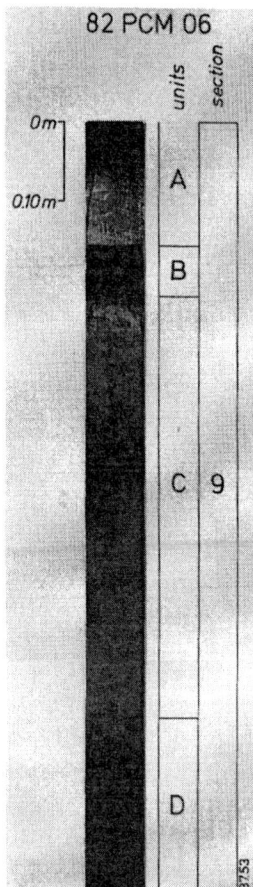

Fig. 5 Photograph of section 9 from between 7 and 8 m depth in core 82PCM06. The section shows, from top to bottom: A, homogeneous basal part of a turbidite (cross striae are artifacts); B, pelagic clay unit; C and D, upper part of turbidite with isolated burrows near the top. (From Kuijpers *et al.*, 1984)

rapid energy dissipation in the moving mass of water and sedimentary material. Deposition appears strongly episodic with an average bulk sedimentation rate in excess of 10 cm/1000 years in the central-western turbidite province. Within the last 200 000 years, six major pulses of rapid sedimentation occurred, separated by periods of slow deposition when biological reworking of the sediment took place (Kuijpers *et al.*, 1984). Based on core and 3.5 kHz data, the area west of 24°W appears free of erosion. The quasi-horizontal and continuous nature of the distal turbidites (Fig. 6) evident in cores and on geophysical records suggests the depositional processes are unaffected by contour currents or diapirism. The westernmost part of the plain contains only minor silt layers and no sand layers. South of 31°N, turbiditic marly oozes only occur locally; alternations of pelagic ooze, marl and brown clay, indicating $CaCO_3$ dissolution cycles and slow deposition rates, are more widespread. Intervals of strong dissolution are thought to correlate with glacial periods. About 9 m long pelagic cores in the southeastern part reach ages of 1.5–1.8 million years, indicating a sedimentation rate of 0.5–0.8 cm/1000 years, which thus is in marked contrast with the age of 150 000 years determined at the base of a 22 m long turbiditic core (Kuijpers *et al.*, 1984).

Physical Properties of Sediment

A number of physical properties were determined in a few turbiditic cores from the central part of the plain (Kuijpers, 1982; Searle *et al.*, 1984). Some were determined in detail throughout core 82PCM29. This core does contain the soupy surficial sediment not sampled in some of the other cores. Measurements in the topmost 1.5 m have been ommitted, as some sediment disturbance due to core-handling in this water-rich interval seems likely. Void ratios in the cores have an average of 2.9 (for fine-grained turbidites) and 2.5 and 3.4 for pelagic oozes and clays respectively. Porosity averages at 73% for fine-grained turbi-

10–12% clay minerals and 0.5–1.5% organic carbon. Sand, composed mainly of angular, volcanogenic minerals and forams, has been deposited at the eastern margin of the plain as sand lobes and on the adjacent part of the plain as sandy turbidites. Maximum grain size in the east is about 450 m with a gradual decrease to the west.

Core study suggests the occurrence of two sediment dispersal paths entering the area from the east. The lateral change from debris-flow deposits to turbidites occurs very rapidly, probably because of a pronounced slope change around 23.5°W and

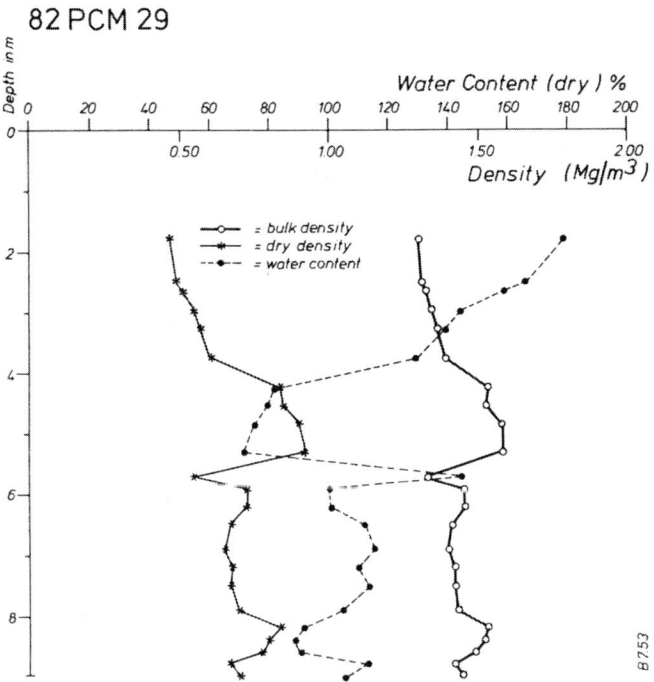

Fig. 6 Isopach pattern of Madeira Abyssal Plain turbidite III deposited between 73 000 and 60 000 years BP. The map illustrates the gradually increasing thickness of an individual fine-grained turbidite to the west, which is a characteristic of all major MAP turbidites. (From Kuijpers *et al.*, 1984)

dites, 68% for sandy and silty turbidites and 71% and 77% for pelagic oozes and clays. Specific gravity varies between 2.46 and 2.73 g/cm³ with an average of 2.64 g/cm³. Water content (dry) is about 110% by weight for fine-grained turbidites at depth, a maximum of 82% for sandy and silty turbidites and around 95% and 129% for pelagic oozes and clays respectively. Wet bulk density averages 1.43 g/cm³ for fine-grained turbidites at depth, 1.58 g/cm³ for sandy and silty turbidites and 1.5 to 1.4 g/cm³ for pelagic ooze and clays (Fig. 7).

Permeability measurements on core samples indicate values averaging 6×10^{-5} cm/s for fine-grained turbidites, values up to 10^{-3} cm/s for sandy and silty turbidites, 10^{-6} cm/s for pelagic ooze and 5×10^{-7} cm/s for pelagic clay. Evidence for open burrows found in a number of cores even at greater depth (12 m) might markedly increase bulk permeabilities.

Shear strength measurements performed with a motor vane and a fall cone device were made in core 82PCM29 perpendicular to bedding planes at intervals of 10 cm. The fall cone measurements included one observation made in the geometric middle of the subsample with nine other measurements that were averaged afterwards. Shear strength plotted against burial depth gives curves in which fluctuations measured with the two different techniques correspond remarkably well (Fig. 8).

The lowest shear strength with values below 2 kPa in core 82PCM29 was measured at the base of the water-rich surficial section. Distinct shear-strength peaks with motor-vane values exceeding 25 kPa can be related to small intervals with pelagic sediment. Turbiditic intervals at depth reach shear strengths of 5–14 kPa. Sensitivities in turbiditic sediments average around 4. Temperature gradients of a few heat flow stations appear linear and suggest normal heat flow fluxes (Noel, 1984).

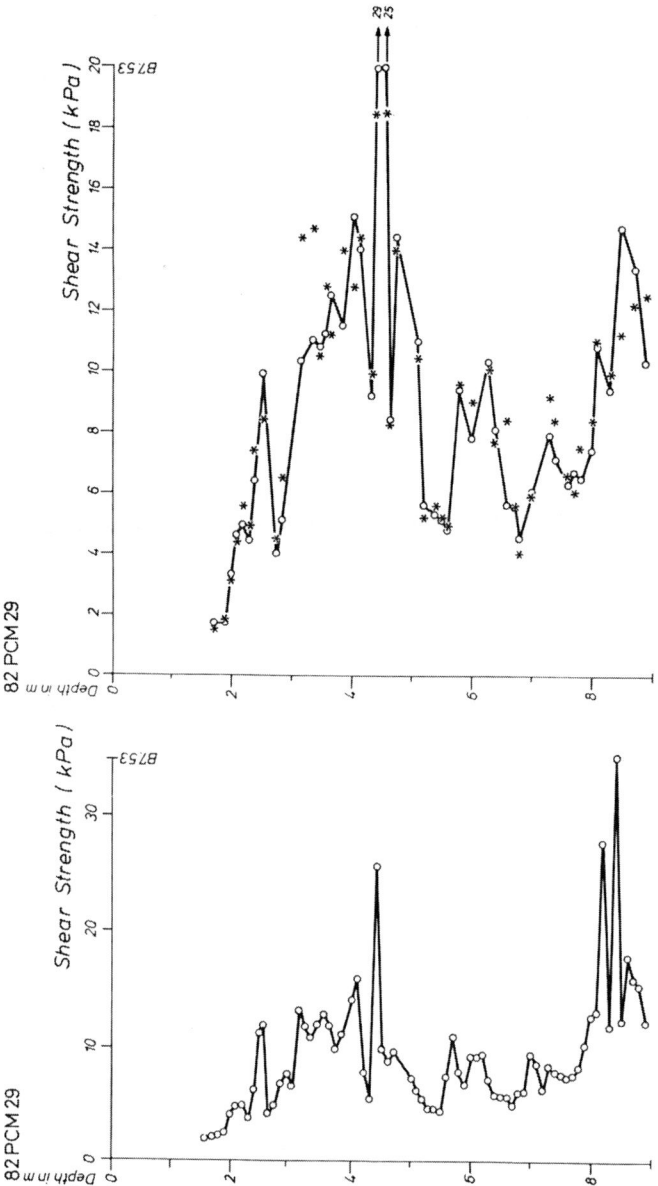

Fig. 7 Madeira Abyssal Plain: bulk density, dry density and water content (dry) of core
82PCM29. (From Kuijpers *et al.*, 1984)

NARES ABYSSAL PLAIN

Topography and Bathymetry

The Nares Abyssal Plain extends in the Western North Atlantic between 22° and 25°N and 60° and 67°W. Main topographic features around the plain are the Bermuda Rise in the north, the Greater Antilles Outer Ridge in the west and south and the lowe flank of the Mid-Atlantic Ridge in the east. In the west, the Nares Abyssal Plain is connected to the Hatteras Abyssal Plain through Vema Gap. The part of the Nares Abyssal Plain studied by SWG is situated between 22° and 24°N and 62° and 65°W.

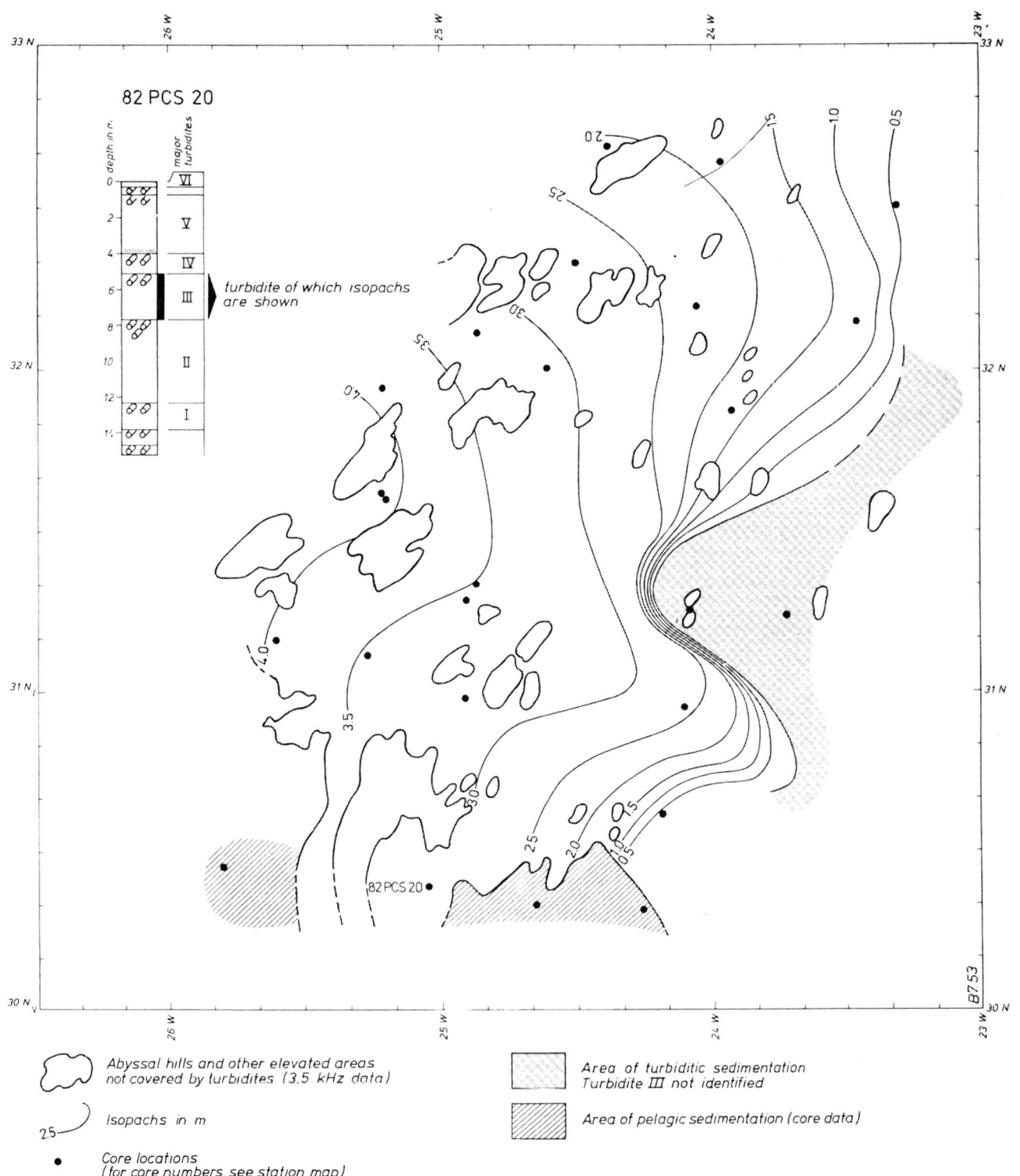

Fig. 8   Madeira Abyssal Plain: shear strength of core 82PVS29 determined with motor vane (left) and fall cone (right). (From Kuijpers *et al.*, 1984)

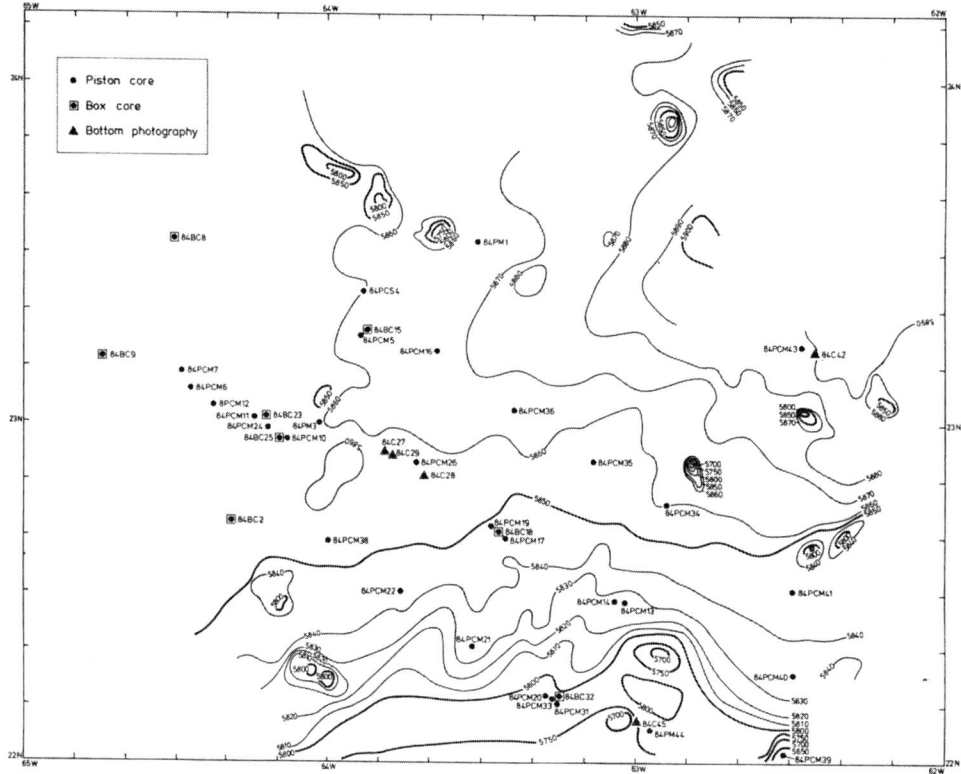

Fig. 9 Nares Abyssal Plain: bathymetry and Dutch sample stations. More than a dozen Canadian
cores (not indicated) are concentrated between cores 84PCM36 and 84PCM19

The water depth in the area is generally
between 5800 and 5900 m (Fig. 9). Apart
from a few scattered abyssal hills, most of
the sea floor is almost flat with an overall
low-angle (0.01°) dip to north-east. South of
22°30'N the seabed rises towards the Greater Antilles Outer Ridge, a broad sedimentary swell north of the Puerto Rico Trench.

Sea-floor Imagery

Camera stations from depths between 5850
and 5900 m show biogenic trails and mounds
without any evidence of recent current
activity. The seabed at a depth between
5700 and 5750 m near the southern margin of
the plain appears to be largely free of
macro-biogenic structures and shows various types of current-induced bedforms (Kuijpers, 1985; Kuijpers and Duin, 1985). These
bedforms are asymmetric ripples with a
wavelength of about 0.15 m and furrows

0.1–0.2 m deep and 0.2–0.4 m wide, with
spacings in excess of 0.8 m. Furrows are
about perpendicular to ripple crests. The
mud clouds stirred up by the camera were
moving in the down stream direction as indicated by the ripple asymmetry. Rather
strong near-bottom currents appear not to
be uncommon on the west side of the Western North Atlantic Ocean.

Sediment Thickness and Structure

The sedimentary cover over an Upper Cretaceous basement generally ranges from 300
to 700 m with values of up to 1000 m in
basement valleys and buried fracture zones.
The lowermost transparent unit consists of a
pelagic clay drape. This unit is at the seabed
on small abyssal hills. On top is a unit with
acoustically laminated sediments that may
consist of cherts, limestones, silts and clays
(Tucholke and Ewing, 1974). An overlying

unit consists of distal turbidites; the upper part of this unit decreases in thickness from west to east and north-east.

On 3.5 kHz records five acoustic sediment types with transitional boundaries are recognized which show the following characteristics (Duin, 1985):

(1) *Acoustic sediment type I* represents predominantly continuous sub-bottom reflectors below a relatively flat seabed. Penetration is moderate (40–70 ms). This acoustic type is typical for the turbiditic area west of approximately 64°W and north of 23°45′N.

(2) *Acoustic sediment type II* is representative of irregular and discontinuous sub-bottom reflectors that can be traced laterally over only a few hundred metres (Fig. 10). These features and the low signal penetration (30–50 ms) probably originate from scattering and attenuation of signal energy due to short-distance changes of sediment properties and bedforms as a result of braided and narrow silt layers in the plain. This echo-type occurs in the generally flat central part of the area and also has extensions in an ENE direction.

(3) *Acoustic sediment type IIIa* is characterized by a 'transparent', low-reflective seabed with parallel and continuous regularly laminated reflectors with good resolution (Fig. 11). Signal penetration is good (70–110 ms). Acoustic sediment type IIIa interpreted as pelagics with turbiditic overbank deposits has a patchy occurrence in the central part of the area, where the sea floor shows minor undulations with vertical amplitudes of up to a few metres.

(4) *Acoustic sediment type IIIb* represents a transparent upper layer with generally continuous sub-bottom reflectors. Thickness

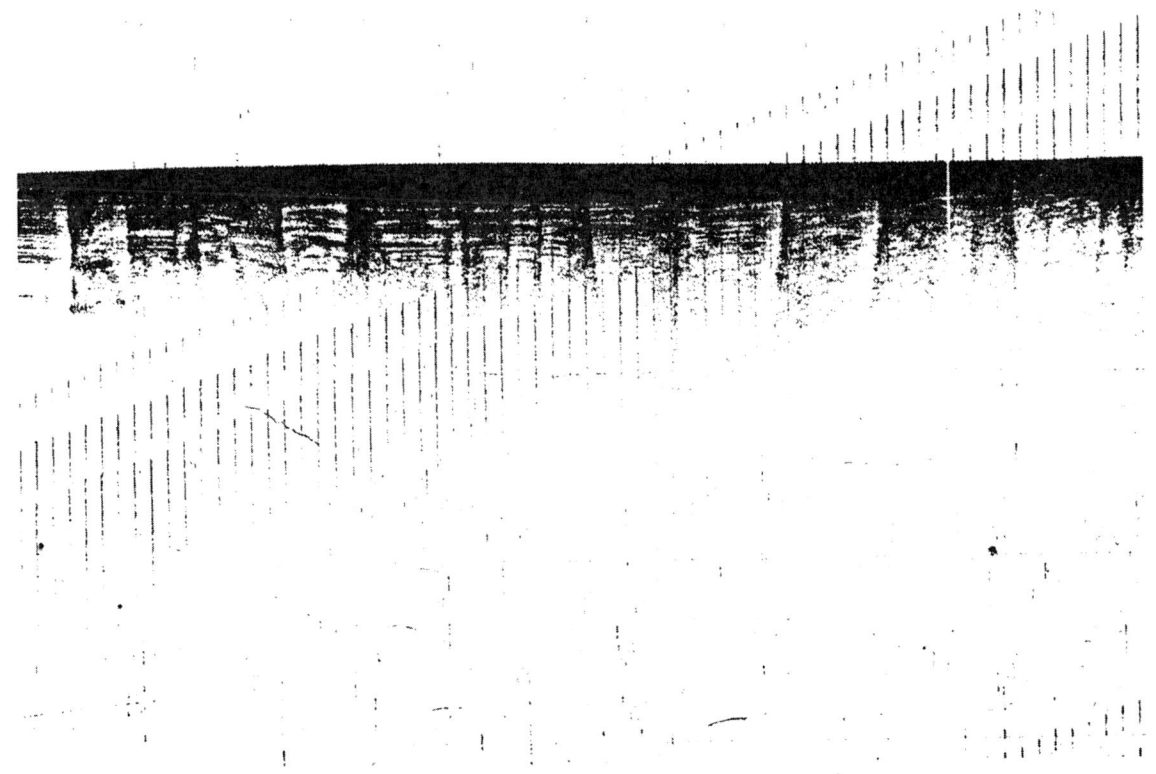

Fig. 10 Example of NAP acoustic sediment type II with characteristic irregular and discontinuous reflectors and with focusing features. Distance between horizontal lines is 100 ms. Profile length is *c.* 6.5 km

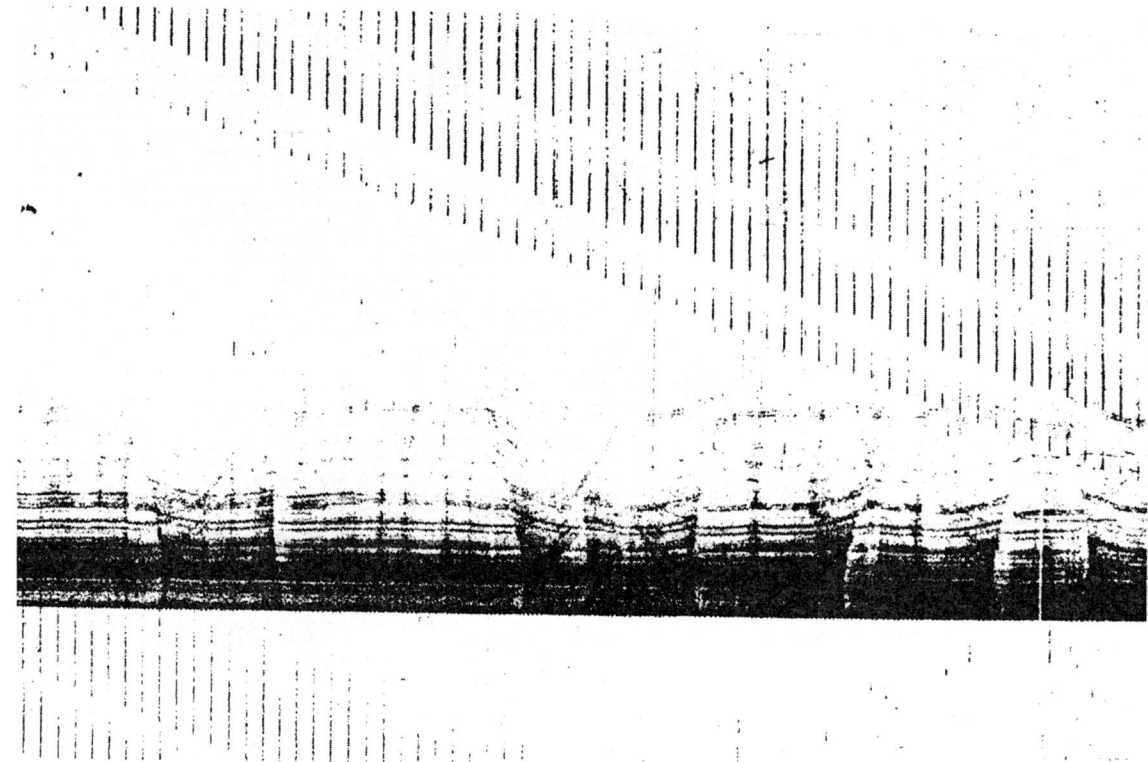

Fig. 11 Example of NAP acoustic sediment type IIIa with parallel and continuous reflectors and some apparent faults and focusing features. Distance between horizontal lines is 100 ms. Profile length is *c.* 6.5 km

of the upper layer varies between 10 and 20 ms; signal penetration is moderate to good (60–90 ms). This clay layer is found in an area with a slightly deeper sea floor, centered near 23°20′N, 63°W and orientated in a WSW to ENE direction.

(5) *Acoustic sediment type IIIc* is representative of continuous sub-bottom reflectors (thickness *c.* 30 m), with a transparent pelagic unit below, in which two distinct reflectors conformable to the oceanic basement occur. Acoustic sediment type IIIc is found in the southeastern sector with its irregular morphology and is most directly affected by the deep contour current.

Not fully understood small faults comparable to those observed in the Madeira Abyssal Plain seem to occur, especially in the central part of the Nares Abyssal Plain, among numerous echo-focusing features

(Fig. 11). These faults are less obvious than those observed in the Madeira Abyssal Plain; but the difference between echo-focusing features and faults is not easy to distinguish (Duin, 1985). It could be that small faults will turn out to be a normal feature of abyssal plains with a turbidite fill, if these abyssal plains receive the same attention as the Madeira and Nares Abyssal Plains.

Stratigraphy

Some 50 piston cores up to 13 m taken from 1981 onwards have shown that the superficial sediments consist of grey clays and silts and brown clays. $CaCO_3$ content in clays is low (5%).

The frequency, thickness and burial depth of silt layers are variable from core to core

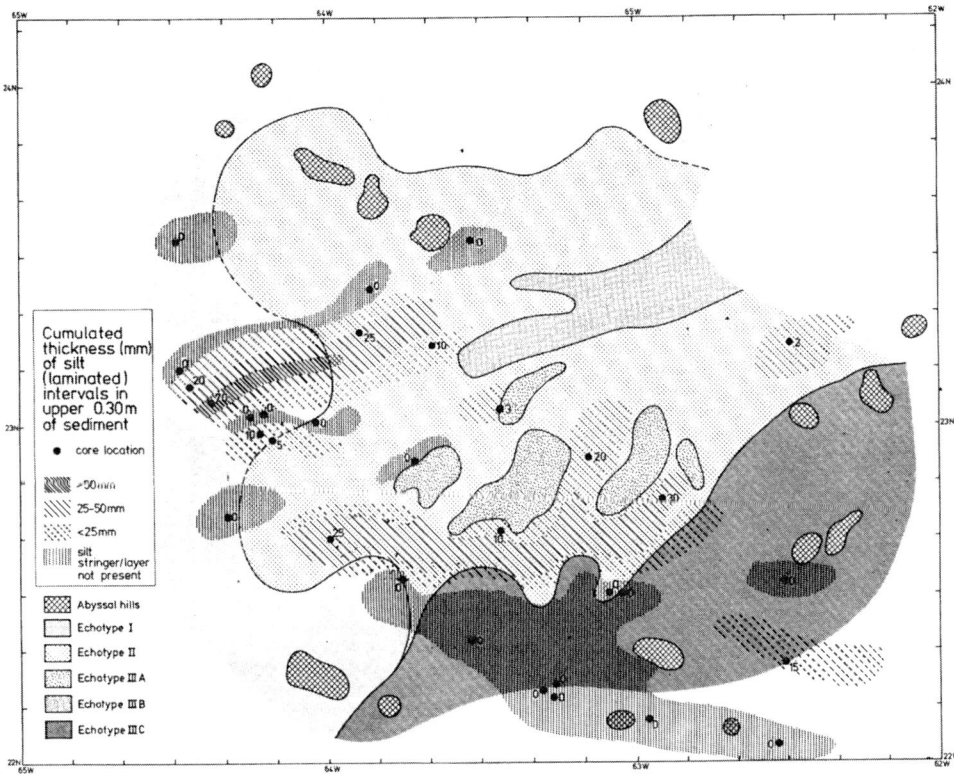

Fig. 12   Nares Abyssal Plain: silt dispersal pattern in uppermost 0.3 m with acoustic sediment types (echotypes). (From Kuijpers and Duin, 1985)

(Kuijpers, 1985). Silt layers containing micaceous fine silt decreasing in grain size to the north-east are generally enriched in organic carbon (maximum 1%) and $CaCO_3$ (up to 17%). Maximum silt-layer thickness (0.2–0.3 m) occurs in the north-west and locally also in the south and to the north-east. The cumulated silt thickness of cores of 0.3 m (Fig. 12) and 6.5 m (Fig. 13) depth intervals shows silt dispersal paths meandering from south-west to north-east. Turbiditic sedimentation rates are difficult to establish but are thought to be of the order of decimetres per 1000 years.

It has been suggested that the silt and turbiditic clay sedimentation pattern is due to the effects of a fluctuating easterly contour current system superimposed on distal turbidity flows entering the Nares Abyssal Plain from the west (Kuijpers and Duin, 1985).

Physical Properties of Sediment

Various physical properties have been investigated in clay-silt cores from the Nares Abyssal Plain (Shephard, 1985). Void ratios range from 2 to 4. Porosities in clay are 75–80%. Water content values of clay layers may be more than 140% and less than 30% in adjacent silt layers. These sharp variations may explain the difficulties so far experienced in obtaining undisturbed cores of more than 10 m length in the Nares Abyssal Plain.

Variations in the shear-strength data from cores also relate to lithological variations. Maximum values occur near the silt base of each turbidite interval. These values are generally twice those for the overlying clayey turbidites. Maximum values are c. 15 kPa at 3.5–6 m below the seabed. Cores rich in brownish clay show a more gradual

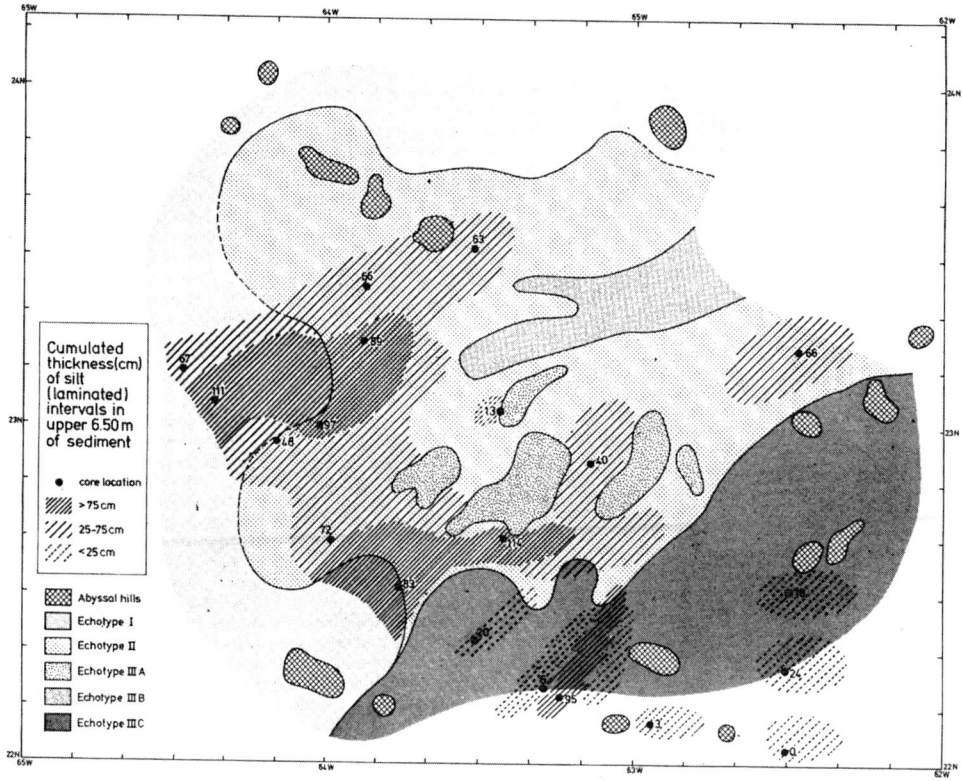

Fig. 13   Nares Abyssal Plain: silt dispersal pattern in uppermost 6.5 m with acoustic sediment
types (echotypes). (From Kuijpers and Duin, 1985)

increase of 2 kPa near the surface to 9 kPa at
8 m down.

Permeability values of the clay layers are
of the order of 10^{-7} to 10^{-8} cm/s, and for silts
they are thought to be 10^{-4} to 10^{-5} cm/s.

COMPARISON OF MAP AND NAP

As outlined above, recent studies have
shown marked differences in transport
mechanisms and sedimentary patterns in
the abyssal environments of the Eastern
and the Western North Atlantic Basin,
which significantly affect sediment properties of the abyssal plain sediments involved.
A summary of these differences was based
on data from the Madeira Abyssal Plain in
the eastern basin and the Nares Abyssal
Plain in the western basin.

The Madeira Abyssal Plain shows a highly
episodic sediment transport mechanism con-

sisting of gravity to turbidity flows; the sediment transport mechanisms in the Nares
Abyssal Plain appear to be much more continuous, consisting of the interaction of turbidity flows and contour-following currents
(Kuijpers and Duin, 1985). Maximum grain
sizes are about 450 m (MAP) and 50 m
(NAP), and maximum $CaCO_3$ contents are
c. 99% and c. 17%. The coarse sediment of the
MAP consists of angular volcanic mineral
grains; however, that of the NAP consists
mainly of well-packed micas, which might
explain the differences in water content.
This appears to influence the performance of
conventional sampling equipment like piston
cores. The great areal sedimentary variability in NAP (Figs 12 and 13) is probably
caused by topographic effects on the flow
regime, and by the fluctuating bottom current strength. In the gravity-flow environment of MAP variability is much lower (Fig.
6), which means that the predictability of a

site in an abyssal plain like the MAP is much higher than in a NAP-type abyssal plain. A sampling station grid of, say, 10/10 miles in MAP would give a degree of reliability similar to or even higher than a 1/10 mile station grid in NAP.

ACKNOWLEDGEMENTS

The present knowledge about the Madeira and Nares Abyssal Plains is to a large extent due to the activities of the national site assessment programmes of Canada, The Netherlands, the United Kingdom and the United States. Permission to publish this paper was granted by the Director of the Geological Survey of the Netherlands. Mr J. Bruinenberg supervised the drawings and Mrs N. Bauer typed the manuscript.

REFERENCES

1. Auffret, G., Buckley, D., Laine, E. P., Schüttenhelm, R. T. E., Searle, R. C. and Shephard, L. E. 1984. NEA Seabed Working Group: status on site qualification for nuclear waste disposal within deep sea sediments. SAND 83-2037, Sandia National Laboratories, Albuquerque, USA, pp. 1–64.
2. Duin, E. J. Th. 1985. Geophysics of the Southern Nares Abyssal Plain, Western North-Atlantic. Geological Studies in the Western North Atlantic. Internal Report, Rijks Geol. Dienst, Haarlem (in press).
3. Duin, E. J. Th., Mesdag, C. S., and Kok, P. T. J. 1984. Faulting in Madeira Abyssal Plain sediments. *Mar. Geol.* **56**, 299–308.
4. Embley, R. W. 1982. Anatomy of some Atlantic margin sediment slides and some comments on ages and mechanisms. In *Marine Slides and other Mass Movements* (Eds S. Saxov and J.K. Nieuwenhuis). Plenum Press, New York–London, pp. 189–213.
5. Kuijpers, A. 1982. Sediment studies in the western Madeira Abyssal Plain. Geological Studies on Abyssal Plains in the North Atlantic, Part II. Internal Report, Rijks Geol. Dienst, Haarlem, pp. 41–107.
6. Kuijpers, A. 1985. Sediments of the Southern Nares Abyssal Plain, Western North Atlantic. Geological studies in the Western North Atlantic. Internal Report, Rijks Geol. Dienst, Haarlem (in press).
7. Kuijpers, A. and Duin, E. J. Th. 1985. Boundary current-controlled turbidite deposition: a sedimentation model for the southern Nares Abyssal Plain, western North Atlantic. *Geomar. Lett.* (submitted).
8. Kuijpers, A., Rispens, F. B. and Burger, A. W. 1984. Late Quaternary sedimentation and sedimentary processes on the Madeira Abyssal Plain, Eastern North Atlantic. *Meded. Rijks Geol. Dienst*, **38**(2), 91–118.
9. Noel, M. 1984. Measurements of sediment temperatures, conductivity and heat flow in the North Atlantic and their relevance to radioactive waste disposal. IOS Report 172, Institute of Oceanographic Sciences.
10. Searle, R. C., Weaver, P. P. E., Schultheiss, P. J., Noel, M., Jacobs, C. L., Huggett, G. J. and McGiveron, S. 1984. Great Meteor East (Distal Madeira Abyssal Plain): geological studies of its suitability for disposal of heat-emitting radioactive wastes. Report Institute of Oceanographic Sciences.
11. Shephard, L. E. 1985. Geotechnical property characteristics of Nares Abyssal Plain sediments: a summary report. Geological Studies in the Western North Atlantic. Internal Report, Rijks Geol. Dienst, Haarlem (in press).
12. Tucholke, B. E. and Ewing, J. I. 1974. Bathymetry and sediment geometry of the Greater Antilles Outer Ridge and vicinity. *Bull. Geol. Soc. Am.* **85**, 1789–1802.

4

The Quaternary Succession on the Northern United Kingdom Continental Shelf and Slope: Implications for Regional Geotechnical Investigations

M. S. Stoker, D. Long, A. C. Skinner and D. Evans, British Geological Survey, Edinburgh, Scotland

ABSTRACT

Regional geological and geophysical surveys have provided an insight into the nature of the Quaternary succession on the Northern UK Continental Shelf and Slope between 56°N and 62°N. In the Central and Northern North Sea the Quaternary sequence exceeds 300 m in thickness, and has accumulated in sedimentary basins which have been the focus of deposition since Permian times. The scale of the sedimentary basins is such that discrete seismostratigraphic boundaries can be mapped over wide areas of the North Sea, and this has led to the establishment of a regional Quaternary stratigraphy for the area. Detailed investigation of the stratigraphic sequence in the North Sea has revealed a more complete sequence of Pleistocene sediments than was previously known. The sequence ranges from Lower Pleistocene to Holocene in age.

On the west and north-west continental margin the Quaternary sequence is at present less well established. A seismic stratigraphy has been set up for the Hebrides, although dating of the sequence remains uncertain. Studies on the north-west shelf and slope are still at an early stage, although reconnaissance mapping data are available. In contrast to the North Sea, the Quaternary sequence in these areas is often restricted to localized basins overlying Mesozoic and Tertiary basins separated by Precambrian basement highs. The thickness of sediment in these basins locally exceeds 250 m although, in general, it is much thinner. Towards the shelf break and down the continental slope the geometry and thickness of the Quaternary sequence is currently being studied.

Geotechnical data are available for much of the northern shelf and slope and have been compiled to produce a summary geotechnical map. This shows the strength of the sediment at 1 m depth below the seabed, although information from deeper in the sequence is also presented. The geotechnical map indicates regional variation in sediment strength, and this is related to the geometry of the Quaternary sequence.

INTRODUCTION

The area of study extends from 56°N to 62°N and covers the Northern UK Continental Shelf and Slope down to about 1100 m water depth (Fig. 1). The British Geological Survey (BGS) has been actively surveying this area since 1968 as part of its regional offshore mapping programme. Consequently an extensive geological and geophysical database has been developed for the offshore area. These data, which total 246 boreholes, about 13 500 shallow sample stations, and in excess of 75 000 km of seismic traverse, have been used to compile a variety of offshore geological maps at 1 : 100 000 and 1 : 250 000 scale (see Fannin, 1980). These maps are especially valuable to the offshore industry as they contain regional background data for more localized site studies. Of particular interest are the

Quaternary maps which present an insight into the geometry of the Quaternary succession and its geotechnical variations, thus providing an essential element of any site investigation project.

This paper presents an overview of the Quaternary succession on the Northern UK Continental Shelf and Slope. It includes a brief reference to the pre-Quaternary geological framework as this has influenced, to some extent, the position of Quaternary basins. Most of the early stratigraphic work in the North Sea and Sea of the Hebrides has been revised since 1980; therefore this study presents an update on the Quaternary geology of these areas. Although work on the West Hebridean and North-West Scottish Shelf is still at an early stage, a brief summary of the available data is included. A large amount of geotechnical data exist for the Quaternary sequence, and it has been possible to produce a summary geotechnical map for most of the North Sea and for local areas in the north-west. It is shown that a knowledge of the Quaternary geology is crucial to the understanding of regional

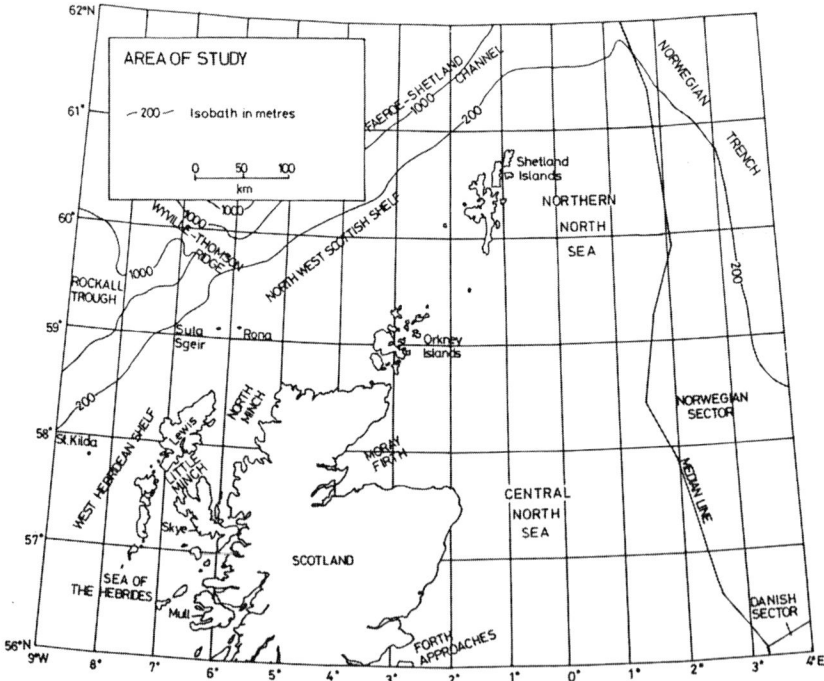

Fig. 1   Location of study area

geotechnical variation on the shelf and slope.

PRE-QUATERNARY GEOLOGICAL FRAMEWORK

The area may be considered as two geological provinces: the North Sea region and the area to the west and north-west of Scotland. The provinces are separated by the predominantly Precambrian basement massif of the Scottish Mainland and its north-northeastward extension to Shetland, which includes Upper Palaeozoic Devonian sediments (Fig. 2.). Mesozoic sediments occur only on the periphery of the landmass on the west coast and around the Moray Firth. These exposures provide a hint of the great change that takes place in the geology off-shore.

In the North Sea, Permian and Mesozoic rocks occur within a few kilometres of the Mainland coast (Evans *et al.*, 1981), except east of Shetland, and thicken eastwards towards the graben system that has been the focus of deposition in the area since Permian times (Glennie, 1984). This graben system, which was formed by extensional faulting during the Mesozoic, consists of the Central and Viking Grabens as well as the westerly branch which forms the Moray Firth Basin. The Mesozoic rocks are overlain by softer Tertiary sediments which again thicken towards a subsiding depositional centre overlying the grabens, although their deposition was not accompanied by faulting.

To the west and north-west of Scotland (Naylor and Mounteney, 1975; Stoker and Fannin, 1984) Permian and Mesozoic rocks are again common offshore, although they generally occur in smaller faulted basins separated by Precambrian basement ridges which often form rugged topography.

To the west of Scotland, early Tertiary igneous activity has produced rocks which tend to form irregular topography, whereas Tertiary sediments on the Atlantic side of Scotland are almost entirely restricted to the outer shelf and slope.

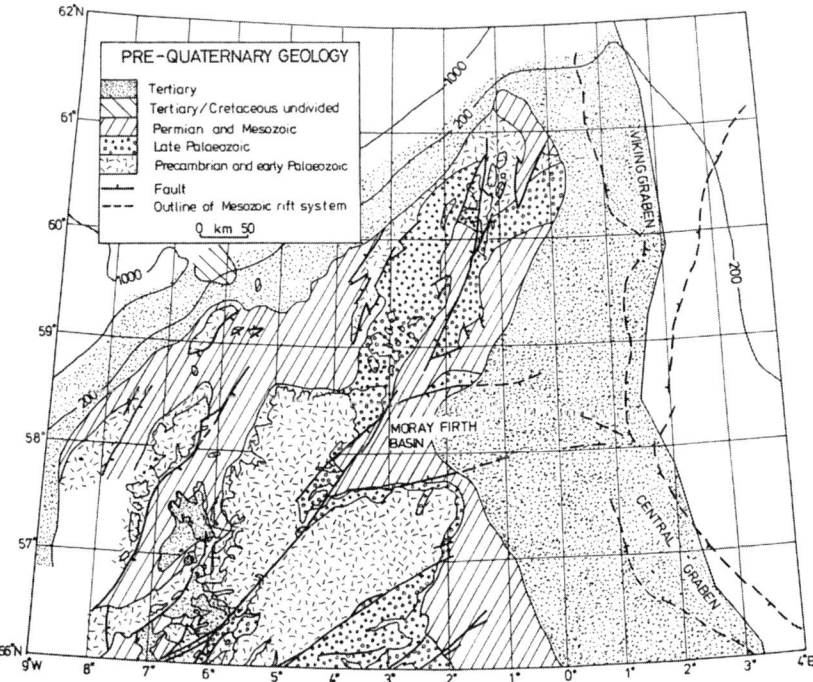

Fig. 2 Pre-Quaternary geological setting

THE QUATERNARY SUCCESSION

Quaternary sediments are widespread over most of the Northern UK Continental Shelf and Slope, although the thickness of the sequence is highly variable (Fig. 3). In the North Sea two distinct basins of deposition are discernible separated by a pre-Quaternary platform area east of Orkney and Shetland. The relatively thin Quaternary sequence overlying this platform is generally less than 150 m in thickness.

To the south, in the Central North Sea, the thickest development of Quaternary sediment occurs in the area of the Central Graben where over 300 m of sediment are preserved. In the eastern part of this area the base of the Quaternary cannot be identified on shallow seismic records due to the interference of multiple reflections. Previous estimates which suggested that the Quaternary sequence may be in excess of 600 m in this area (Caston, 1977, 1979) have recently been questioned (Stoker *et al.*, in press, b), and it is thought that the Quaternary may not be as thick as originally suggested.

Identification of the base of the Pleistocene in this area is currently under review using a widespread coverage of digitally processed sparker data.

In the Northern North Sea the Quaternary sequence thickens to the north-east of Shetland where over 350 m of sediment occurs in the extreme north of the area.

To the west and north-west of Scotland the distribution is much more complex with the greatest thicknesses of sediment generally being preserved in isolated, deep basins overlying Mesozoic and Tertiary sediments. These are bounded by upstanding Precambrian basement ridges which tend to have little or no sediment cover. This is particularly evident in the Sea of the Hebrides where over 250 m of sediment are preserved in deep, narrow, elongate basins or troughs. Discrete basinal areas are similarly noted in the Little Minch and North Minch and in the Lewis–Sula Sgeir area on the North-West Scottish Shelf where over 100 m of sediment occur locally.

Towards the edge of the shelf the base of the Quaternary sequence has yet to be accu-

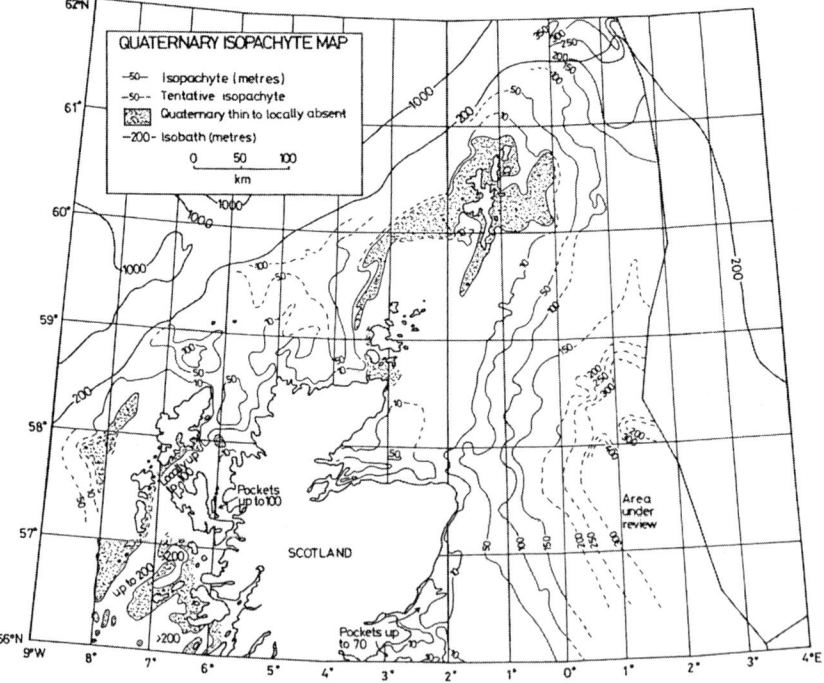

Fig. 3 Distribution and thickness of Quaternary sediments

rately defined. West of Shetland and north-west of Orkney the Quaternary succession appears to thicken towards the shelf break. However, to the south-west of the Wyville–Thomson Ridge the sequence appears much thinner at the shelf break, and the Quaternary cover on the ridge itself is generally thin, and locally absent. The sequence may be thicker again to the south of St Kilda. Down the continental slope the thickness of the Quaternary is presently unknown.

Figure 4 depicts areas of Lower, Middle and Upper Pleistocene as well as Holocene sediments at or near the seabed. It does not include the uppermost 'mobile layer' which reflects present-day seabed processes, although a brief summary of the seabed sediments is included at the end of this section. This Quaternary map is simplified from the 1 : 250 000 map series, although the latter tend to show individual Quaternary formations as the primary mappable unit. Figure 4 is therefore an amalgamation of the stratigraphic data, and each unit on the map

may represent several individual Quaternary formations.

Mapping in the North Sea is almost complete; the blank areas on the map are currently being worked upon. West of Scotland, regional mapping has only been completed in the Sea of the Hebrides, although areas on the north-west shelf are being actively interpreted. Information therefore exists for most of the shelf and slope and this is summarized below.

North Sea

Quaternary stratigraphic sequences have been established in the Central and Northern North Sea. Diagrammatic sections illustrating the stratigraphic successions are shown in Figs 5 and 6, and a basic chronostratigraphic framework which correlates the two areas is present in Fig. 7. This stratigraphic framework is based on the Dutch stratigraphic classification (Zagwijn, 1979) which takes the base of the Quaternary to be

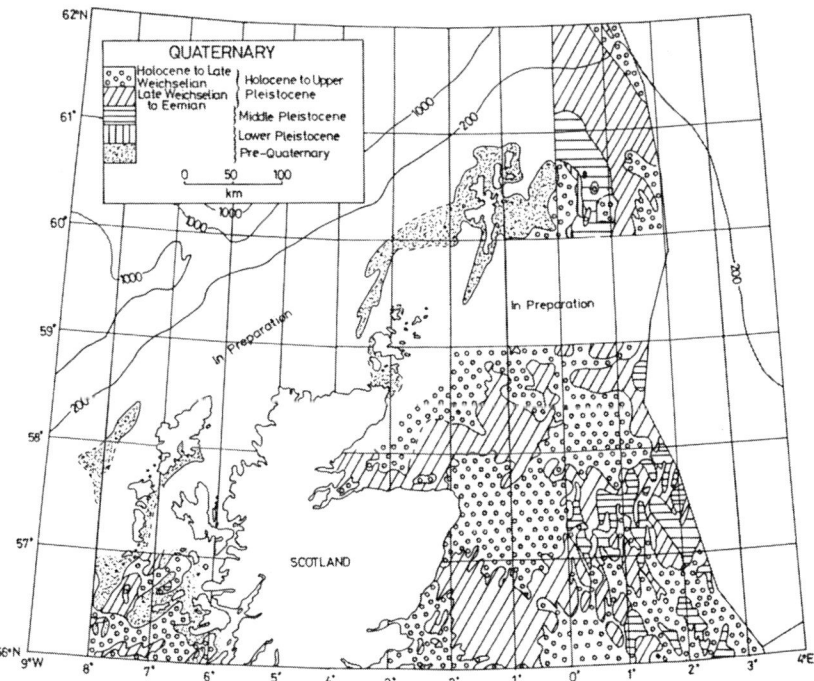

Fig. 4 Simplified Quaternary geology map

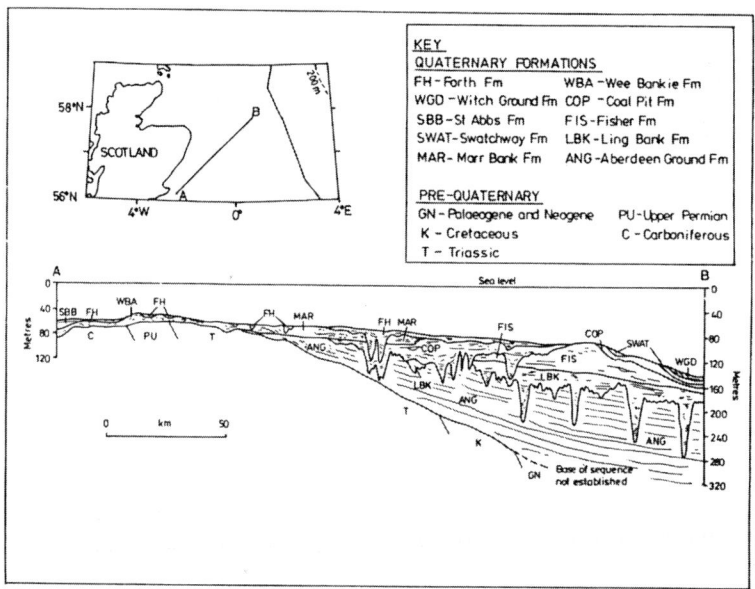

Fig. 5   Cross-section showing the geometry and general seismostratigraphic relationships of the
Quaternary formations in the Central North Sea. (Vertical exaggeration ×234)

at about 2.5 million years BP. The Lower/
Middle Pleistocene boundary is generally
taken at the Brunhes/Matuyama palaeomagnetic boundary (Butzer and Isaac,
1975), which has been variably dated at
730 000 years BP (Mankinen and Dalrymple, 1979) and 790 000 years BP (Johnson,
1982), and is placed towards the base of the
'Cromerian Complex' (Zagwijn, 1979). The
base of the Upper Pleistocene is taken at the
beginning of the Eemian (Butzer and Isaac,
1975), which began at about 128 000 years
BP (Bowen, 1978). The Upper Pleistocene/
Holocene boundary is taken at about 10 000
years BP (Mangerud et al., 1982).

Central North Sea

The Quaternary succession in the Central
North Sea has recently been revised (Stoker
et al., in press, a and b), and a more complete sequence of Pleistocene sediments has
now been recognized. Micropalaeontological
data (Stoker et al., in press, a and b) and
amino acid geochronology (Brigham-Grette
and Sejrup, 1984) support an earlier
palaeomagnetic study (Stoker et al., 1983)
which identified the existence of Lower,

Middle and Upper Pleistocene sediments in
the area. This clearly refutes the earlier
hypothesis that the greater part of the
Quaternary succession is of Upper Pleistocene age (Holmes, 1977; McCave et al.,
1977; Eden et al., 1978).

Ten Quaternary formations have been
established and these can be mapped over
the entire area of the Central North Sea. A
detailed account of the stratigraphy has
been presented elsewhere (Stoker et al., in
press, a) and this is summarized below.

The geometry of the sequence is illustrated in Fig. 5, which shows the succession
increasing in thickness towards the east,
and it is in this area that the most complete
sequence is preserved. In the west, the base
of the succession rests with angular unconformity on rocks of Palaeozoic, Mesozoic and
Tertiary age. To the east, basal Pleistocene
has nowhere been proved and the base of
the succession has still to be defined. The
Aberdeen Ground Formation is the oldest
proven Quaternary unit and ranges from
Lower to Middle Pleistocene (?Tiglian to
Elsterian) in age. With the exception of
minor isolated outliers in the west of the
area (too small to be shown on Fig. 4), this

formation occurs wholly at subcrop beneath younger sediments. The Lower Pleistocene part of the succession consists predominantly of bioturbated marine muds with occasional interbedded sands which were deposited in a warm, temperate, inner to middle shelf environment.

The Middle Pleistocene succession comprises the upper part of the Aberdeen Ground Formation, the Ling Bank and Fisher formations and the basal part of the Coal Pit Formation. On seismic records the sequence is generally much more complex than the Lower Pleistocene and distinct irregular, interformational erosional surfaces characterize the sequence (Fig. 5). A change in environmental conditions is evident since the bulk of the sequence comprises sediments of a predominantly distal glaciomarine nature. These are generally argillaceous sediments with abundant scattered, matrix-supported clasts (dropstones) and shell fragments. Locally in the west, subglacial and proximal glaciomarine sediments have also been identified (Stoker and Bent, in press). One thin (2 m) interglacial horizon has been identified towards the base of the Ling Bank Formation and this has been tentatively assigned a Holsteinian age. Reference to Figs 4 and 5 indicates that Middle Pleistocene strata are at, or near, the surface in the east and north-east of the area.

Upper Pleistocene and Holocene sediments crop out over a wide area of the Central North Sea and comprise most of the Coal Pit Formation, the Wee Bankie, Marr Bank, Swatchway, St Abbs, Forth and Witch Ground Formations. Eemian interglacial strata have been identified near the base of the Coal Pit Formation and up to 30 m of bioturbated marine muds and shelly sands have been recovered locally. The upper part of the Coal Pit Formation consists mainly of proximal and distal glaciomarine sands and muds of uncertain Lower/Middle Weichselian age. The remainder of the Upper Pleistocene sequence is composed of Upper Weichselian sediments, which reflect the last regional glaciation and subsequent deglaciation. The Wee Bankie Formation is a subglacial till, whereas the Marr Bank, Swatchway and St Abbs formations comprise glaciomarine sands and muds. The Forth and Witch Ground formations transgress the Upper Pleistocene/Holocene boundary and are shown on Fig. 4 as the Holocene to late Weichselian subdivision of the Holocene/Upper Pleistocene unit. The Forth and Witch Ground Formations transglacial estuarine and fluviomarine sands and muddy sands overlying late-glacial gravelly muds. The Witch Ground Formation is a predominantly argillaceous, basinal marine sequence with some minor glacial input (e.g. dropstones) only in the lower part of the sequence. Pockmarks are abundant in the Witch Ground Formation. Radiocarbon dating (Holmes, 1977) and pollen analysis (Jansen et al., 1979) of the Holocene sequence suggest that most of it was deposited in the early Holocene between 7000 and 10 000 years BP.

Extension of the Central North Sea stratigraphy into the Moray Firth casts doubt on the suggestion by Chesher and Lawson (1983) that the sequence in this area is strictly Upper Pleistocene to Holocene in age. This assumption is based solely on a radiocarbon age date of 43 500 years BP obtained from shell material in their lowermost unit. However, this basal unit can be correlated seismically with the Aberdeen Ground Formation, which implies that it is Lower or Middle Pleistocene in age.

Northern North Sea

Aspects of the Quaternary sequence in the Northern North Sea have recently been presented by Skinner and Gregory (1983), Stoker et al. (1983) and Rise et al. (1984). Eleven Quaternary formations have been mapped throughout the area, and their generalized stratigraphic relationships are illustrated in Fig. 6.

The basic geometry of the sequence reflects a northeasterly offlapping succession which is thickest in the area of the Viking Graben. It is apparent from cross-

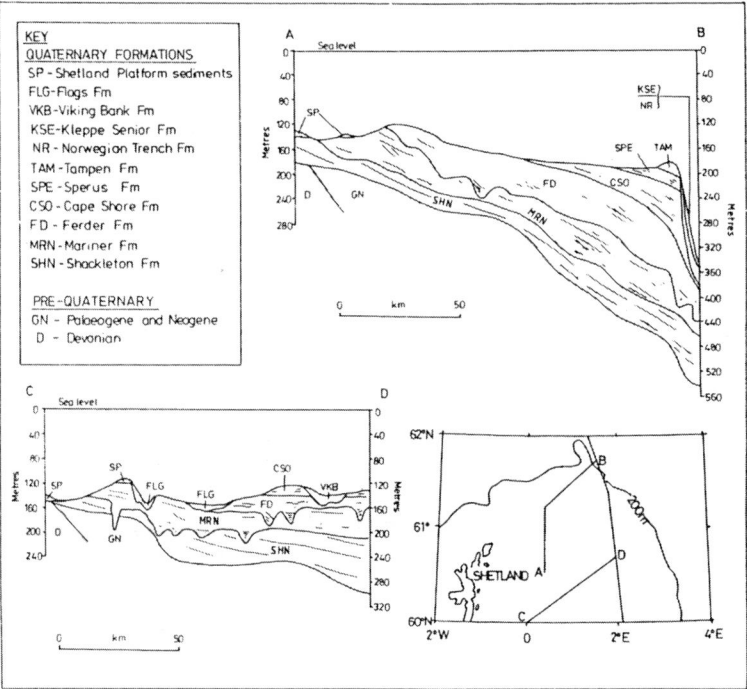

Fig. 6   Cross-section showing the geometry and general seismostratigraphic relationships of the Quaternary formations in the Northern North Sea (Vertical exaggeration ×234)

section A-B in Fig. 6 that although the continental margin is pre-Quaternary in age the present-day profile is clearly due to deposition and selective erosion of Quaternary strata.

In the extreme west of the area the base of the sequence rests with angular unconformity on Tertiary and older strata. To the east, the base appears to rest conformably on Upper Pliocene sediments. The lowest Quaternary unit is the Shackleton Formation, which comprises Lower Pleistocene marine sands and clays that were deposited in a boreal, estuarine-to-offshore basinal environment. These sediments crop out as a narrow strip in the central part of the area (Fig. 4).

The Middle Pleistocene sequence rests unconformably on the Lower Pleistocene sediments and consists of the Mariner Formation and the basal part of the Ferder Formation. Interformational boundaries are markedly irregular. The sediments consist dominantly of glacial or glaciomarine sandy muds, which are very gravelly at the base of the succession in the west although the gravel content decreases to the north and east. These sediments crop out in a broad band to the north and east of the Lower Pleistocene outcrop (Fig. 4).

The Upper Pleistocene/Holocene succession includes the bulk of the Ferder Formation, the Cape Shore, Sperus, Tampen and Norwegian Trench formations. These are of Eemian to late Weichselian age, while the Shetland Platform sediments and the Flags, Viking Bank and Kleppe Senior formations are late Weichselian to Holocene in age. Shelly, sandy, silty clays of Eemian interglacial age have been identified (Skinner and Gregory, 1983) near the base of the Ferder Formation. However, most of the sequence consists of Weichselian sediments, which include glacial gravelly sandy clays, as well as glaciomarine sandy and silty clays, and muddy sands. The late Weichselian to Holocene units are mostly lateral equivalents. The Shetland Platform sediments consist of glacial till overlain by late- to post-glacial clays which form a veneer on the pre-Quaternary bedrock in the area immediately east of Shetland. The Flags Formation and lower part of the Viking Bank Formation are dominantly argillaceous and tend to occupy hollows. Pockmarks occur locally at seabed in the Flags Formation. The upper part of the Viking Bank Formation includes sands and gravels deposited in an environment characterized by shallow mobile barrier islands. The Kleppe Senior Formation crops out in the Norwegian Trench and is characterized by late to post-glacial clays with surface pockmarks and iceberg scour marks in some areas. The Upper Pleistocene and Holocene sediments occur widely at outcrop over most of the Northern North Sea (Fig. 4).

Stratigraphic Correlation

A tentative correlation between the Quaternary formations in the Central and Northern North Sea is shown in Fig. 7. The

stratigraphic sequences have many similarities and both are correlatable with the Quaternary succession established in the Southern North Sea (Cameron *et al.*, 1984). Although much more detailed work remains to be done to better define the individual stratigraphic units, a consistent picture is emerging throughout the North Sea.

West and North-West of Scotland

Regional Quaternary mapping in this area is still at a preliminary stage; however, a great deal of basic data exists and these are briefly summarized below.

Sea of the Hebrides

The Quaternary stratigraphy in this area was originally described by Binns *et al.* (1974) but has recently been revised by Davies *et al.* (1984) whose work is summarized below.

The succession has been divided into eight formations whose relationships are shown schematically in Fig. 8. The pre-Quaternary bedrock surface is characterized by considerable variations in relief in this area and the bulk of the Quaternary sediments is preserved in the basins. There are no direct chronological controls on the Quaternary sediments and their age remains unproven. The basal three units, the Skerryvore, Malin and Canna formations, are considered

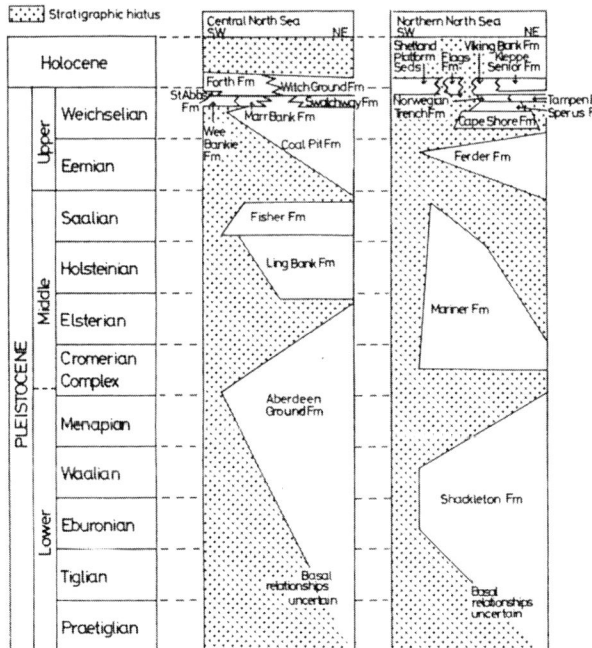

Fig. 7 Correlation of the Quaternary formations in the Central and Northern North Sea with the Pleistocene stages of North-West Europe

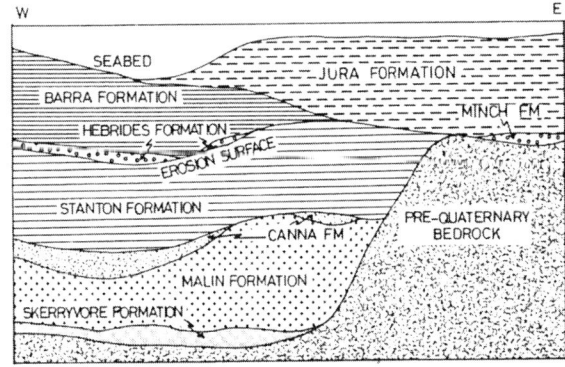

Fig. 8 Schematic cross-section showing the general relationships of the Quaternary formations in the Sea of the Hebrides

to be pre-Weichselian in age. The predominance of glacial and glaciomarine gravelly sand clays and muds throughout these units, excepting the occurrence of temperate marine muds at the base of the Skerryvore Formation, would further suggest at least a pre-Eemian, Middle Pleistocene age. Sediments belonging to this unit occur at or near the seabed only in a small area north-west of Tiree (Fig. 4).

The main part of the succession is of Upper Pleistocene/Holocene age. The Stanton Formation, which is composed of glaciomarine, early/middle Weichselian pebbly, silty clays, is separated from the overlying formations by a major widespread erosion surface which is thought to have been cut at the time of the late Weichselian glacial maximum. A thin diamicton layer overlying this surface, and forming the Hebrides and possibly the Minch formations, may be the remnants of a late Weichselian subglacial till. Late Weichselian gravelly silty clays of glaciomarine origin form the Barra and lower part of the Jura formations. The upper part of the Jura Formation consists of post-glacial (Holocene) sandy and silty clays. The Holocene to late Weichselian unit on the Quaternary map (Fig. 4) records exclusively the distribution of the Jura Formation sediments.

Little Minch and North Minch

A significant thickness of the Quaternary succession in the Little Minch is confined to localized accumulations, but a widespread area of preservation is found in the North Minch. The area has been described by Chesher *et al.* (1981) who related the sequence to that described by Binns *et al.* (1974) in the Sea of the Hebrides. However, as noted above, the stratigraphy erected by Binns *et al.* has recently been revised, and it is presently uncertain how the sequence in the Minches correlates with the revised stratigraphy. Revision of the area is underway.

West Hebridean Shelf

Study of this area is still at a very early stage and it has not yet been fully surveyed. The tentative isopachytes shown in Fig. 3 are based on a preliminary shipboard interpretation of seismic data collected in the St Kilda area. A significant feature is the major rock platform that extends from the Flannan Islands (to the west of Lewis) southwards beyond St Kilda. This is a rugged area of submerged rock platforms and small islands with very localized pockets of Quaternary sediments.

North-West Scottish Shelf and Slope

This is an area of active research and, although a detailed seismostratigraphy has still to be established, aspects of the sequence can be illustrated by using the available geological and geophysical data. A composite lithostratigraphy, based on 36 boreholes widely spread across the shelf, and over 200 cores from the shelf and slope principally west of Shetland and west and north of Sula Sgeir, is shown in Fig. 9. The BGS 6 m vibrocorer had capabilities down to about 500 m water depth. Beyond this depth sediment cores were retrieved using a 3 m gravity

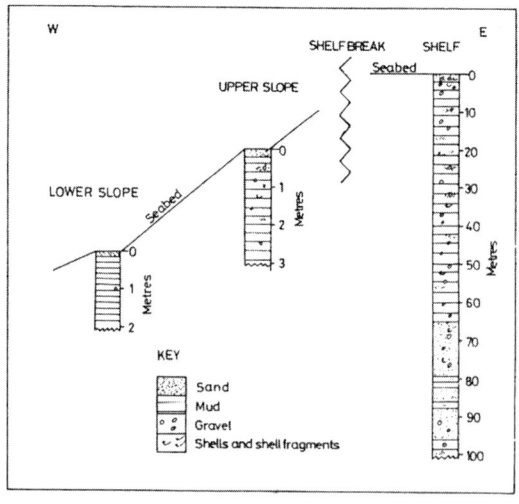

Fig. 9 Composite lithostratigraphy of the upper sediment layers on the North-West Scottish Shelf

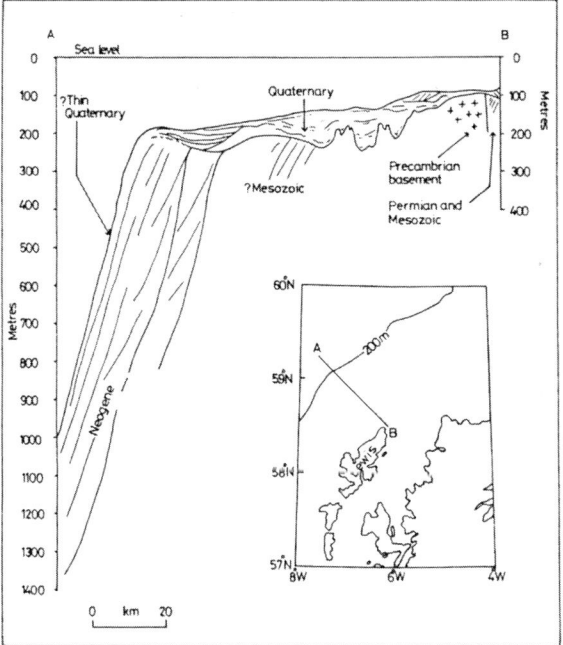

Fig. 10 Cross-section illustrating the nature of the Quaternary succession on the North-West Scottish Shelf. (Vertical exaggeration ×100)

corer. A cross-section across part of the shelf and slope is shown in Fig. 10. The section is located in an area of the shelf where the base of the Quaternary has been identified, although base Quaternary remains undetermined on the slope. The variable thickness of the sequence and its relationship to the underlying pre-Quaternary geology is well illustrated in the section.

On the shelf, the sequence comprises a cover of shelly sand and gravelly sand, mostly less than 2 m thick but locally up to 20 m thick, which overlies predominantly argillaceous sediments characterized by gravelly silty muds with subordinate sandy horizons. Micropalaeontological data suggest that the top sandy unit is Holocene in age and on parts of the shelf is still subject to reworking. The contact between the sand and the underlying argillaceous sediments is usually very sharp and erosional. Locally the top of the argillaceous sequence appears to be weathered, and this weathered zone can occur in present-day water depths of down to 160 m. In some areas the top sand

unit is absent and the argillaceous sediments crop out at the seabed. At present there are no chronological data available for this part of the sequence. Micropalaeontological data have indicated the presence of temperate horizons in several boreholes which suggests the possibility that pre-Upper Pleistocene sediments may be present. However, the subdivision of Quaternary units is still uncertain and reference to Figs 9 and 10 illustrates the lateral and vertical stratigraphic changes. Work in the North Sea has amply demonstrated the variable distribution of Pleistocene and Holocene sediments, and the sequence can vary a great deal in age between adjacent boreholes. Despite uncertainties regarding the dating of the sequence, micropalaeontological data suggest that the bulk of it was deposited under harsh climatic conditions, mainly in a glaciomarine environment. Closer to land areas such as Shetland (Long and Skinner, in press), subglacial sediments have been identified.

Towards the shelfbreak at about 200 m water depth, the top sand unit becomes generally finer, although gravelly sands have been recovered down to 90 m water depth in the Wyville–Thomson Ridge area. Down the slope the sand cover becomes very fine and silty and is generally much thinner (5–20 cm) than on the shelf. As on the shelf, the underlying sediments are argillaceous, while the contact between the two facies varies from sharp to transitional. West of Sula Sgeir, the argillaceous sediments on the upper slope, between 200 m and 600 m water depth, have a distinct gravel fraction and resemble the glaciomarine sediments on the shelf. Below 600 m water depth the argillaceous sediments in this area are gravel free. However, this distribution is not ubiquitous as there is a widespread occurrence of a gravel fraction in the argillaceous facies down to at least 800 m on the south-west flank of the Wyville–Thomson Ridge, and down to 110 m in the area north-west of Shetland.

It is probable that the sediments preserved on the slope represent a more condensed and more complete Quaternary sequence than those preserved on the shelf. Correlation between the shelf and slope has not yet been attempted.

Seabed Sediments

The distribution of seabed sediments shown in Fig. 11 has been simplified from the BGS 1 : 250 000 map series. These maps are based on grain-size analyses of seabed deposits and are classified after Folk (1954).

In the North Sea sand is ubiquitous with gravelly sediments occurring mainly on topographic highs in shallower water, and muddy deposits generally occur in water depths below 120 m (Owens, 1981). On the North-West Scottish Shelf sandy and gravelly sediments are predominant, whereas the sediment distribution in the Hebrides and Minches is highly variable but tends to be depth dependent.

The seabed sediments form the uppermost layer of the Quaternary sequence but their thickness is variable and their age often difficult to determine. In the northern part of the Central North Sea early Holocene sediments occur very close to the seabed and only the top 0.15 m have been subjected to minor reworking (Erlenkeuser 1978). The essentially relict nature of the seabed in this area is confirmed by the localized occurrence on the sea floor of sea-ice scour marks formed about 18 000 years BP (Stoker and Long, 1984). Areas of relict sediment also occur on the Hebridean and North-West Scottish Shelf and in the Northern North Sea where late Weichselian iceberg scour marks have been observed at the seabed (Belderson et al., 1973; Skinner, 1983).

Elsewhere, areas of the shelf are covered by a variety of modern bed forms formed as a result of sand and gravel on the shelf and in the nearshore zone, being transported by storm and tidal currents. In the area around north-east Orkney, for example, mobile sandwaves with a thickness of up to 20 m

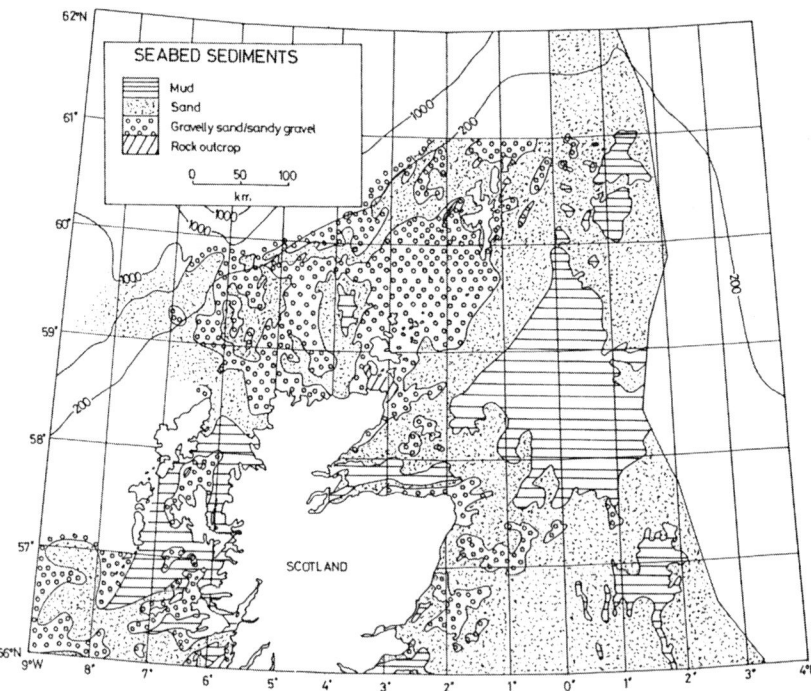

Fig. 11 Simplified map showing the distribution and lithology of the seabed sediments. (Compiled by Colin Graham, BGS)

have recently been described (Farrow *et al.*, 1984). The 'mobile layer' is, however, more related to present-day conditions than to the Quaternary history of the area.

GEOTECHNICAL PROPERTIES

As an integral part of examining cores collected on the UK shelf and slope, simple geotechnical measurements are made using pocket penetrometers and handvanes. Subsamples may also be taken for moisture content and density measurements. Occasionally other geotechnical tests are undertaken, including triaxial studies, consolidation tests, particle-size analyses and Atterberg limits.

Together with data collected by commercial agencies, an attempt has been made to provide ranges of values for various geotechnical parameters for the different seismostratigraphic formations identified in order to produce a simplified geotechnical map of parts of the UK Continental Shelf and Slope (Fig. 12). This was made by

recording the undrained shear-strength value at, or near, one metre below the seabed and noting the lithology of the sample. The values were divided into four groups to indicate the undrained shear strength as measured by a handvane at 1 m depth. One metre was chosen as a useful horizon as it was reached by many of the gravity cores as well as the vibrocores collected, and it is also an important level in studies of a regional nature, such as pipelines. Below 1 m there is a marked reduction in the frequency of readings and their distribution is too spasmodic to attempt to produce a geotechnical map at a lower depth. However, values and comments are recorded on the 1 : 250 000 Quaternary Geology map series.

Geotechnical Map (1 m depth)

The simplified map (Fig. 12) indicates several features which are related to the geological history of the area. The geotechnical properties of the sediments are very variable over this wide area, since the range is dependent on many factors, such as lithol-

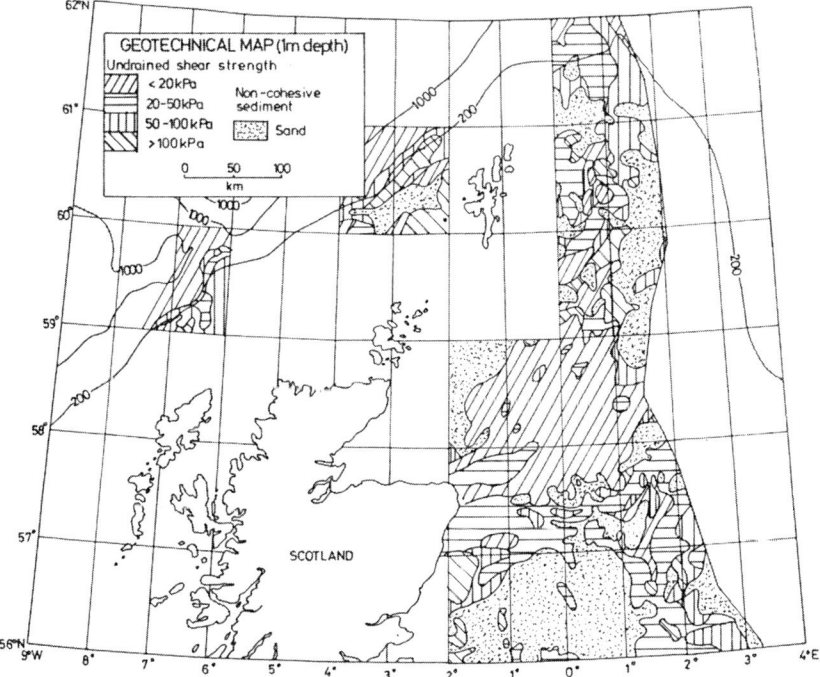

Fig. 12 Simplified geotechnical map (1 m depth)

ogy, clay mineralogy and the stress history of the site. The latter is extremely important when it is considered that a site may have had large thicknesses of ice resting on it, or sediment removed from it. Moreover, large areas of the North Sea were dry land during the Quaternary and dessication and periglacial processes may have affected the sample.

In an arc across the Forth Approaches is an area with shear strengths typically in excess of 100 kPa. This corresponds with the Wee Bankie Formation, believed to represent an end moraine of Late Weichselian age (Stoker *et al.*, in press, a). To the east of this is a large area of sand, which equates with the Marr Bank and Forth formations. In the extreme east lies an area of highly variable shear strength, which reflects the complex distribution of Quaternary units (see Figs 4 and 5) which have been subject to varied stress histories. The pattern is further complicated by the localized occurrence of fluviomarine sediments of Holocene age.

In the northern part of the Central North Sea is a large area of soft sediments, representing a distinct basinal area of Upper Pleistocene to Holocene marine sedimentation. It is known as the Witch Ground Basin. On the southern edge of this basin are areas of sand corresponding to the Swatchway Formation, which lie adjacent to areas of stiff to very hard clays formed by the Fisher Formation. East of Caithness and Orkney is an area of sand and rock outcrop with small patches of mud of various strengths (not shown on the map).

North of the Witch Ground Basin, in the Northern North Sea, is an area of variable lithology and shear strengths similar to the eastern part of the Central North Sea. In the east is a large sandy area with several banks (Rise *et al.*, 1984), which correlates with the Viking Bank Formation and with the sandy southern outcrop of the Cape Shore Formation. East of the Shetlands are several depressions filled with very soft sediments of late Weichselian age (the Flags Formation), overlying Lower to Middle Pleistocene firm to hard clays. North-east of

Shetland the shear strengths are in distinct zones. The ridge at the plateau edge formed by the Tampen Formation (see Fig. 6) forms a well-pronounced area of stiff gravelly and sandy clays thought to represent a lateral moraine of late Weichselian age, while very soft sediments occur in the Norwegian Trench and on the continental slope. The Quaternary geology west of Scotland is complex and no attempt has been made to portray geotechnical data at 1 m depth on the small-scale map. The Sea of the Hebrides has widespread areas of very soft recent clays close to tills and rock outcrops.

The area west of Orkney and Shetland has only recently been examined and so the map only indicates a limited area. However, certain features do show up, such as the sharp break at the shelf edge where stiff glaciomarine clays occur to a water depth of 200 m beyond which very soft muds are recovered on the continental slope. There are two areas of exception, the northern end of the Tampen Formation referred to above and the Wyville–Thomson Ridge where stiff clays are recovered at 1 m depth in water depths up to 350 m beyond which stiff clays are recovered below a slightly thicker (1–2.5 m) covering of very soft muds. It is thought that the ridge may have caused the grounding of large icebergs drifting down the Faeroe–Shetland Channel during periods of lower sea-level, resulting in the overconsolidation of the seabed.

On the north-west shelf the shear strengths are generally high as the near-seabed sediments are possibly tills or glaciomarine sediments which have been subsequently loaded by ice or affected by dessication or periglacial processes. There are small areas of recent soft marine clays and sands on top of these stiff clays. Frequently rock occurs at or close to the surface near the various islands and banks of this area.

Below 1 m depth

The sediments below 1 m depth are extremely variable both vertically and laterally and their shear strength depends on their geological history. The Lower and Middle Pleistocene sediments (the Shackleton, Mariner and Aberdeen Ground formations of the North Sea and the Skerryvore and Malin formations of the Sea of the Hebrides) vary from normally consolidated to heavily overconsolidated with undrained shear strengths locally in excess of 1500 kPa, although figures of 500–800 kPa are typical.

Some Upper Pleistocene sediments exhibit high strengths near the seabed (approximately 5 m depth) for various reasons. In the Northern North Sea, where strengths of 800 kPa have been recorded in the upper 10 m, some cementation has been noted (M. J. Edge, 1984, personal communication). In the UK block 16 similar high values have been recorded (Marsland et al., 1982), probably due to permafrost, evidence for which has been identified in several parts of the North Sea (Derbyshire et al., in press).

Elsewhere low shear strengths have been recorded well below seabed from within channels or troughs extending to 150 m below the shoulder of the channel. In the North Sea these channels are recorded at several horizons within the Quaternary sequence and they are an important aspect in the evaluation of a site for foundation conditions. Other phenomena, which can be assessed on a regional scale, have been described in detail elsewhere (Fannin, 1980). They include gassified sediments, pockmarks and slope stability. Pockmarks, for example, occur predominantly in the Witch Ground, Flags and Kleppe Senior formations; therefore their distribution can be mapped over a wide area of the shelf and slope. Areas of slumping and mass movement have been identified over most of the Northern UK Continental Slope, although, in contrast to similar features recorded within the Norwegian Sector (Fannin, 1980), they are relatively small-scale structures up to 25 m depth and 2–5 km in length. Areas of slumped sediment have also been identified within open channels in the North Sea.

Lithological variation is also important, particularly the recognition of buried gravelly sequences which often cause problems for offshore drilling companies when they attempt to get holes 'spudded-in'.

CONCLUSIONS

Regional offshore mapping has provided a basis for detailed geological studies. The development of a regional Quaternary framework has led to a greater understanding of how and why the geotechnical properties of sediments vary on the shelf and slope. The occurrence of Quaternary units of different ages and with different stress histories, at or near the seabed, is a major factor governing the regional geotechnical variations. Therefore an appreciation of the regional geological setting is a vital part of any site-investigation project.

ACKNOWLEDGEMENTS

We would like to thank all of our colleagues in the Marine Geology and Marine Geophysics Research Programmes who have assisted in the collection of the data. This paper is published with the permission of the Director, British Geological Survey, NERC.

REFERENCES

1. Belderson, R. H., Kenyon, N. H. and Wilson, J. B. 1973. Iceberg plough marks in the Northeast Atlantic. *Palaeogeog., Palaeoclim., Palaeoecol.* 13, 215–224.
2. Binns, P. E., Harland, R. and Hughes, M. J. 1974. Glacial and post-glacial sedimentation in the Sea of the Hebrides. *Nature* 248, 751–754.
3. Bowen, D. Q. 1978. *Quaternary Geology: a stratigraphical framework for multidisciplinary work.* Pergamon Press, Oxford. 221 pp.
4. Brigham-Grette, J. and Sejrup, H. P. 1984. Stratigraphic resolution of amino-acid geochronology in North Sea Quaternary sedi-

ments (Abstract). In *Quaternary Stratigraphy of the North Sea* (Eds I. Aarseth and H. P. Sejrup). Symp. Univ. of Bergen, pp. 30–32.

5. Butzer, K. W. and Isaac, G. L. 1975. *After the Australopithecines*. The Hague.

6. Cameron, T. D. J., Laban, C. and Schüttenhelm, R. T. E. 1984. Flemish Bight: Sheet 52°N 02°E. Institute of Geological Sciences and Rijks Geologische Dienst. 1 : 250 000 Series: Quaternary Geology.

7. Caston, V. N. D. 1977. A new isopachyte map of the Quaternary of the North Sea, 1. *Rep. Inst. Geol. Sci.* No. 77/11, 1–8.

8. Caston, V. N. D. 1979. A new isopachyte map of the Quaternary of the North Sea. In *The Quaternary History of the North Sea* (Eds E. Oele, R. T. E. Schüttenhelm and A. J. Wiggers). Symp. Univ. Uppsala, pp. 23–28.

9. Chesher, J. A. and Lawson, D. 1983. The geology of the Moray Firth. *Rep. Inst. Geol. Sci.* No. 83/5.

10. Chesher, J. A., Smythe, D. K. and Bishop, P. 1981. The geology of the Minches, Inner Sound and Sound of Raasay. *Rep. Inst. Geol. Sci.* No. 83/6.

11. Davies, H. C., Dobson, M. T. and Whittinghton, R. J. 1984. A revised seismic stratigraphy for Quaternary deposits on the inner continental shelf west of Scotland between 55°30'N and 57°30'N. *Boreas* 13, 49–66.

12. Derbyshire, E., Love, M. A. and Edge, M. J. In press. Fabrics of probable segregated ground-ice origin in some sediment cores from the North Sea basin. In *Soils and Quaternary Landscape Evolution* (Ed. J. Boardman). John Wiley.

13. Eden, R. A., Holmes, R. and Fannin, N. G. T. 1978. Quaternary deposits of the central North Sea. 6. Depositional environment of offshore Quaternary deposits of the Continental Shelf around Scotland. *Rep. Inst. Geol. Sci.* No. 77/15.

14. Erlenkeuser, H. 1978. The use of radiocarbon in estuarine research. In *Biogeochemistry of Estuarine Sediments*. Proc. UNESCO/SCOR Workshop, Melreux, Belgium, pp. 140–153.

15. Evans, D., Chesher, J. A., Deegan, C. E. and Fannin, N. G. T. 1981. The offshore geology of Scotland in relation to the IGS shallow drilling programme, 1970–1978. *Rep. Inst. Geol. Sci.* No. 81/12.

16. Fannin, N. G. T. 1980. The use of regional geological surveys in the North Sea and adjacent areas in the recognition of offshore hazards. In *Offshore Site Investigation* (Ed. D. Ardus). Graham & Trotman, London, pp. 5–22.

17. Farrow, G. E., Allen, N. H. and Akpan, E. B. 1984. Bioclastic carbonate sedimentation on a high-latitude, tide-dominated shelf: Northeast Orkney Islands, Scotland. *J. Sed. Pet.* 54, 373–393.

18. Folk, R. L. 1954. Sedimentary rock nomenclature. *J. Geol.* 62, 345–351.

19. Glennie, K. W. 1984. The structural framework and the pre-Permian history of the North Sea area. In *Introduction to the Petroleum Geology of the North Sea* (Ed. K. W. Glennie). Blackwell Scientific Publications, Oxford, pp. 17–39.

20. Holmes, R. 1977. Quaternary deposits of the central North Sea. 5. The Quaternary geology of the UK sector of the North Sea between 56°N and 58°N. *Rep. Inst. Geol. Sci.* No. 77/14.

21. Jansen, J. H. F., Doppert, J. W. C., Hoogendoorn-Toering, K., De Jong, J. and Spaink, G. 1979. Later Pleistocene and Holocene deposits in the Witch and Fladen Ground area, northern North Sea. *Neth. J. Sea Res.* 13, 1–39.

22. Johnson, R. G. 1982. Brunhes-Matuyama magnetic reversal dated at 790,000 yr BP by marine-astronomical correlations. *Quat. Res.* 17, 135–147.

23. Long, D. and Skinner, A. C. In press. Glacial meltwater channels in the northern isles of Shetland. *Scott. J. Geol.*

24. Mangerud, J., Birks, H. J. B. and Jäger, K. D. 1982. Chronostratigraphical subdivisions of the Holocene: a review. *Striae* 16, 1–6.

25. Mankinen, E. A. and Dalrymple, G. B. 1979. Revised geomagnetic timescale for interval 0–5 my BP. *J. Geophys. Res.* 84, 615–626.

26. Marsland, A., Prince, A. and Love, M. A. 1982. Role of soil fabric studies in the evaluation of the engineering parameters of offshore sediments. In *Proc. 3rd Int. Conf. on Behaviour of Offshore Structures*. MIT, Boston, Vol. 1, pp. 181–202.

27. McCave, I. N., Caston, V. N. D. and Fannin, N. G. T. 1977. The Quaternary of the North Sea. In *British Quaternary Studies — Recent Advances* (Ed. F. W. Shotton). Clarendon Press, Oxford, pp. 187–204.

28. Naylor, D. and Mounteney, S. N. 1975. *Geology of the Northwest European Continental Shelf*, Vol. 1. Graham Trotman Dudley, London.

29. Owens, R. 1981. Holocene sedimentation in the North-western North Sea. *Spec. Publs Int. Ass. Sediment.* **5**, 303–322.

30. Rise, L., Rokoengen, K., Skinner, A. C. and Long, D. 1984. Norlige Nordsjø. Kvartaergeologisk kart mellom 60°30′ og 62°N, og øst for 1°Ø. (Northern North Sea. Quaternary geology map between 60°30′ and 62°N, and east of 1°E.) M. 1 : 500 000. Institutt for Kontinentalsokkelundersøkelser (IKU), Norway.

31. Skinner, A. C. 1983. Cormorant: Sheet 61°N–0°. Institute of Geological Sciences and Continental Shelf Institute, Norway. 1 : 250 000 Series: Sea Bed sediments.

32. Skinner, A. C. and Gregory, D. M. 1983. Quaternary stratigraphy in the northern North Sea. *Boreas* **12**, 145–152.

33. Stoker, M. S. and Bent, A. In press. Middle Pleistocene glacial and glaciomarine sediments in the west central North Sea. *Boreas*.

34. Stoker, M. S. and Fannin, N. G. T. 1984. A geological framework for the North West United Kingdom continental shelf and slope. Internal Rep. No 84/21, BGS Marine Geology Research Programme.

35. Stoker, M. S. and Long, D. 1984. A relict ice-scoured erosion surface in the central North Sea. *Mar. Geol.* **61**, 85–93.

36. Stoker, M. S., Long, D. and Fyfe, J. A. In press (a). A revised Quaternary stratigraphy for the central North Sea. *Rep. Brit. Geol. Surv.*

37. Stoker, M. S., Long, D. and Fyfe, J. A. In press (b). The Quaternary succession in the central North Sea. *Newsletters on Stratigraphy*.

38. Stoker, M. S., Skinner, A. C., Fyfe, J. A. and Long, D. 1983. Palaeomagnetic evidence for early Pleistocene in the central and northern North Sea. *Nature* **304**, 332–334.

39. Zagwijn, W. H. 1979. Early and Middle Pleistocene coastlines in the Southern North Sea basin. In *The Quaternary History of the North Sea* (Eds E. Oele, R. T. E. Schüttenhelm and A. J. Wiggers). Symp. Univ. Uppsala. pp. 31–42.

5

The Benigraph: A Sea-floor Mapping System

A. Løvik, Bentech A/S, Tromsö, Norway

ABSTRACT

The Benigraph sea-floor mapping system is an integrated system for pipeline surveys. It consists of the following sub-units: a surface positioning system, a tow-fish positioning system, a high-frequency scanning sonar, facilities for data storage and presentation and a tow fish with cable and handling equipment. The specifications of the Benigraph enable it to measure the bottom topography with a depth resolution of 5 cm and a resolution of 20 cm in the horizontal plane.

INTRODUCTION

In the field of seabed mapping for offshore engineering applications, information on the topography is traditionally acquired by line surveys. The instrumentation used is normally a combination of a side-scan sonar and a topographic echo-sounder, together with the required positioning systems and systems for topside storage and presentation of the retrieved data. The Benigraph sea-floor mapping system has been developed to overcome the many shortcomings and uncertainties of the traditional systems. It is a fully integrated modular quantitative topographic system that provides a real-time 3D colour display of the sea floor. It continuously records and displays the seabed in a format suitable for interactive processing and map-making.

The Benigraph has been designed to measure the bottom topography with a depth resolution of 5 cm and a horizontal resolution approaching 20 cm in both directions. In this paper, first of all an overall description of the system will be given. This will be followed by a detailed outline of the subsystems used and a description of their main specifications.

THE BENIGRAPH CONCEPT

In developing the Benigraph the object was

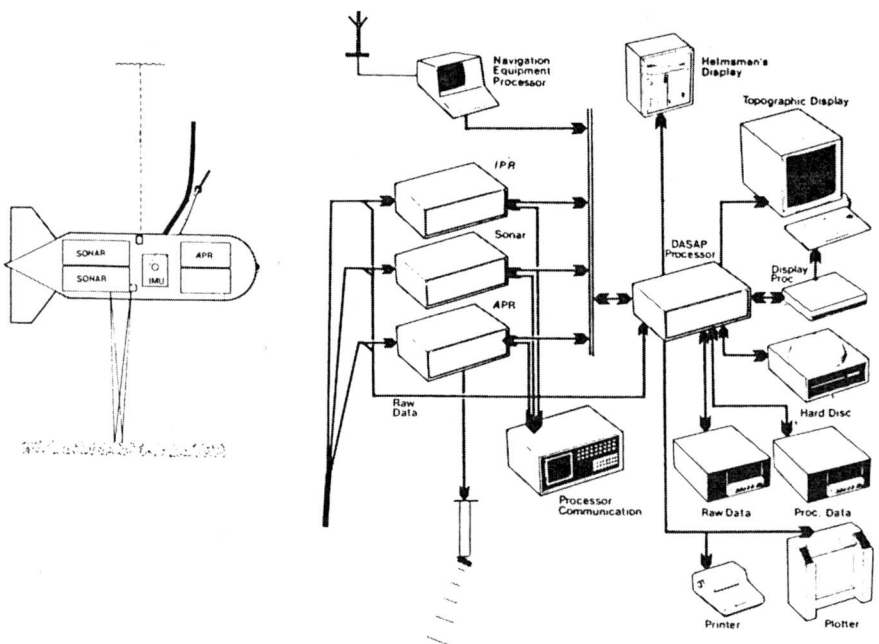

Fig. 1 Block diagram of the Benigraph system

to produce a high-resolution real-time topographic sea-floor mapping system which would allow quantitative assessment of the spans and the coverage of a pipeline. It was intended that, for pre-route pipeline surveys, the system would be able safely to provide the information needed for the determination of the pipeline route. The system has been designed to map the sea floor at water depths down to 300 m without loss of accuracy.

Figure 1 shows an overall block diagram of the Benigraph. The system is divided into the following subsystems:

- surface positioning system;
- tow-fish positioning system;
- scanning sonar;
- system for data storage and presentation;
- tow fish and handling equipment.

The tow fish houses parts of the sonar and its own positioning system. The data are transmitted up to the surface units by means of three digital links. The sonar processor performs bottom detection and geometrical correction on the data, based on information from the tow-fish positioning

system (FIPOS). The FIPOS consists of two units:

- the acoustic positioning reference system (APR);
- the inertial positioning reference system (IPR).

The APR measures the ship's position in relation to the tow fish. This information and the ship's position from SYLEDIS, together with measurement on the inertial platform, is used to give the absolute position of the tow fish. The output of the IPR processor also contains measurements of the attitude of the tow fish. The three parallel processors for the sonar, APR and IPR systems communicate with the data storage, acquisition and presentation system (DASAP) on an IEEE 488 bus. The DASAP stores both raw and processed data and presents the sea-floor on a topographic display. The helmsman's display provides information on the desired track and the position of both the tow fish and the ship.

DASAP is the master system on the IEEE 488 bus and distributes the real-time clock information needed to time-tag the data in all the other subsystems. The hardware in the DASAP consists of an HP1000 A900 computer, an ICAN graphic terminal and the necessary peripheral equipment to store and present the data. Among the many software modules in the DASAP are:

- data reduction algorithms;
- survey journal;
- survey statistics;
- redisplay of covered area;
- graphics.

A description of two of the subsystems, the sonar and tow fish positioning system, is given in the following section.

THE TOW-FISH POSITIONING SYSTEM

The tow-fish positioning system is composed of two units, an inertial and an acoustic positioning system (IPR and APR).

The APR is designed to measure the position of the vessel in relation to the tow fish with either no installations on the ship or only minor ones. The system is thus an inverted one, having the receiving array mounted on the tow fish. There are two modes of operation, passive and active. In the passive mode the position is measured by locking onto the noise generated by the propeller. The active mode uses a frequency-coded signal transmitted from a hull-mounted or towed transducer. The system may use either or both modes.

Figure 2 shows a block diagram of the signal processing in the APR. As can be seen from this figure, the measurement of the two angles is carried out by a correlation method.

The required time difference, and thus the angle, is found by cross-correlating the hydrophone signals. This is accomplished in the frequency domain by the use of an array processor, which performs the Fourier transforms needed. This processor is a co-processor with a VME bus to a MC 68000 processor. By also measuring temperature and depth the APR gives an angular resolution of $0.1°$ and a slant range resolution of $0.1°$. The position update rate is 2 Hz.

In order to estimate positions, a reference frame has to be established in the tow fish. This is done by using the inertial system, the IPR. The IPR system also combines the surface position with the relative position and estimates the errors in the inertial platform by the use of a Kalman filter. Figure 3 shows the functions of the system.

Both the heading reference unit (HRU) and the vertical reference unit (VRU) are gimballed systems. The total IPR performance gives a heading accuracy of $0.5°$ at $60°N$ and an accuracy in roll and pitch of $0.05°$. The total MINAV package is placed in the tow fish and the IPR on-board computer is a VME based MC 68000 microcomputer.

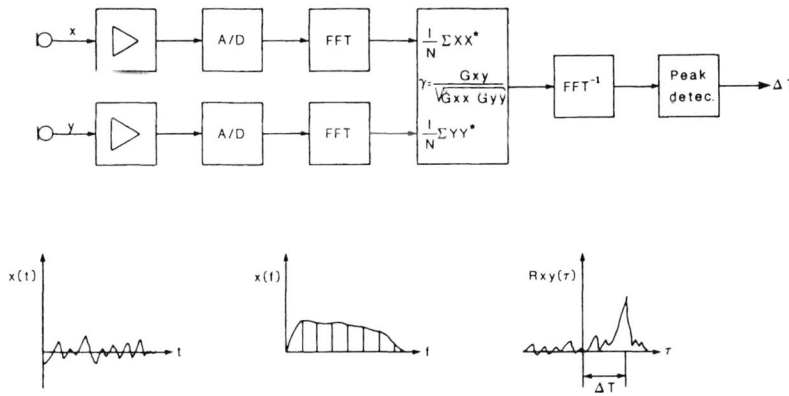

Fig. 2 APR correlation scheme

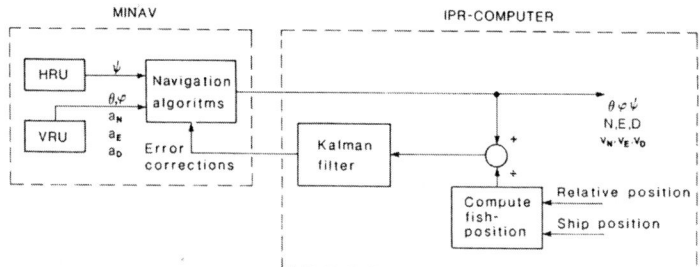

Fig. 3   IPR functional diagram

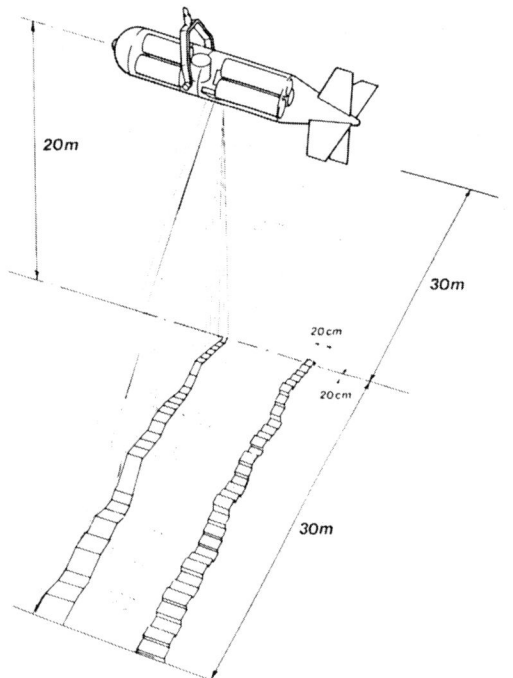

Fig. 4.   Sonar operation

THE BENIGRAPH SONAR

The purpose of the Benigraph sonar is to measure the sea floor with the spatial resolution necessary to allow a quantitative assessment of the condition of the pipeline. To make this possible, the angular resolution was set at 0.5° and the range at less than 10 cm. The cross-track coverage is of the order of 100°. This gives the operation shown in Fig. 4. The fish is towed approximately 20 m above the seabed, giving a coverage of 60 m cross-track on the bottom. The horizontal ensonified area in a beam underneath the sonar is 0.20 × 0.20 m. The sonar uses a separate transmitter.

When the fish is towed close to the bottom the sonar may operate at a fairly high frequency, thus making it possible to use smaller transducers than would otherwise be necessary. In order to operate at greater ranges the sonar is designed to work at 500 kHz and 740 kHz, in addition to the nominal frequency of 1 MHZ.

Figure 5 shows a block diagram of the

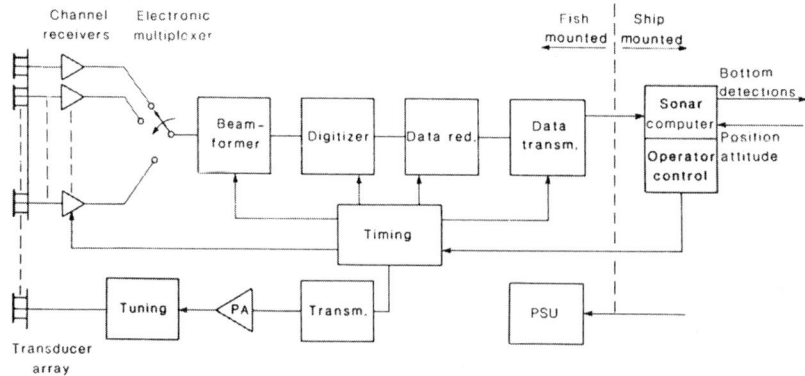

sonar. Each of the 256 elements has a separate channel receiver with a digitally controlled TVG and a quadrature detector. The outputs are multiplexed into the beamformer. In order to form all the 200 beams within a range cell of less than 5 cm, the beam-former is realized as a Chirp–Fourier transformer. The technology chosen is one which uses surface acoustic waves (SAW). The output is digitized and a preliminary data reduction is performed before the data are transmitted on a 3.2 Mbit/s PCM link to the surface processor. Here bottom detection is performed in a specially built bit-slice processor. This processor works as a co-processor on a VME bus to a MC 68000 processor. The MC 68000 performs geometrical corrections and tags the sonar hit-points with absolute position and mean reflectivity. These data are sent to the DASAP for storage and presentation.

OPERATION

All three special processors, the sonar, APR and IPR, may be operated from one menu-driven console. The operator can set all parameters and detection thresholds. Once the set-up is completed, the DASAP will read and store the parameters and master the operation.

The survey set-up done by a surveyor is completed on the ship's positioning system and read and stored by the DASAP. Thus, during normal operation, the Benigraph operator has merely to use the topographic display. All operator interactions in the DASAP are menu driven. Figure 6 shows an example of the simulated display when the Benigraph is towed over a pipeline. The topographic information is shown in the largest square in Fig. 6. Below this picture information about position and time are given. To the right a selection of input modes are shown. Other areas are for system warnings and the listing of new and current input.

CONCLUDING REMARKS

The Benigraph sea-floor mapping system is designed especially for pipeline surveys. The system uses state-of-the-art technology to provide an integrated real-time high-resolution survey tool. In contrast to conventional systems, the retrieved information is quantitative, allowing assessment of pipeline conditions such as the measurement

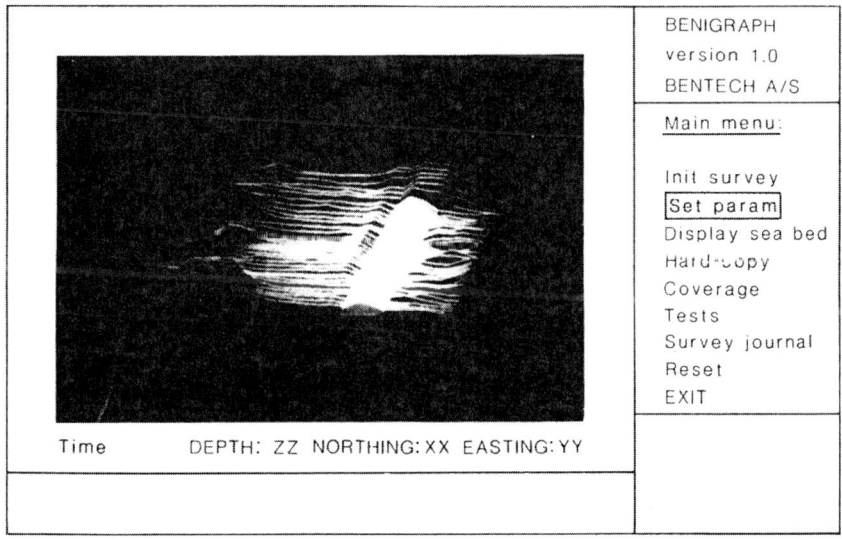

Fig. 6 Benigraph display

of free spans. Therefore we believe that the system will have a major impact both on the quality of pipeline surveys and the ease with which they can be carried out. Other applications will follow in areas such as hydrography and defence.

In a separate but linked project a scanning shallow seismic source is being developed for use in conjunction with the Benigraph. This source, which is capable of resolution at a penetration depth of 60 m or more, will utimately provide the means for achieving a fully integrated remote sensing survey package.

ACKNOWLEDGEMENTS

This work has been carried out under a contract from A/S Norske Shell. The advice and guidance from the technical staff at Shell Internationale Petroleum Maatchappij, The Netherlands, has been of invaluable assistance to the project. Special thanks are due to Dr C. D. Green and Ir. J. G. Riemersma for their enthusiastic support.

6

Seabed II: High-resolution Acoustic Seabed Surveys of the Deep Ocean

R. W. Hutchins, *J. Dodds*, Huntec (70) Limited, Toronto, Canada, *and G. Fader*, Bedford Institute of Oceanography, Dartmouth, Nova Scotia, Canada

ABSTRACT

Recent developments in seabed survey technology for the deep ocean have utilized a tethered deep tow vehicle with side-scan sonar, high-resolution seismic reflection profiler, motion and position sensing and real-time digital processing of data for regional and site-specific geological mapping. Among the major design considerations were the bandwidth and penetration of the profiler system; the range of the side-scan sonar; the system's accuracy of position and attitude measurements; towed body stability; and ease of deployment and recovery. The 1000 J boomer sound source, operating at a depth of 2000 m, provided over 150 m of penetration through upper continental slope sediments with a resolution of 2 m, prior to deconvolution processing. This represents a major advance in resolution and penetration over the performance of conventional single-channel airgun and sparker seismic reflection systems and 3.5 kHz profilers.

A geological interpretation of data col-lected in the summer of 1984 demonstrates the system's capabilities. The continental slope east of Sable Island, adjacent to the Scotian Shelf, is underlain by a thick sequence of glacial and post-glacial sediments extending to water depths greater than 2000 m. From 200 to 640 m water depth the seabed is covered with relict iceberg furrows. Extending from 500 to 1100 m water depth, a large field of pockmarks (gas escape craters) occurs. Locally, slumped zones indicate recent instabilities that could affect present and planned hydrocarbon drilling activity.

INTRODUCTION

This paper describes a new integrated sea-floor mapping system, 'Seamor', developed under the Seabed II research programme, and presents some results from sea trials during July 1984. Seamor uses a high-performance seismic profiler and a side-scan sonar as the primary sensor systems for quantitative near-bottom acoustic surveys of the sea floor in water depths beyond 500 m Additional sensors for determining attitude, geographic position, and the velocity vectors of the sensor platform provide the information required to produce scale-rectified motion-compensated geographically positioned and oriented mosaics of the sea floor, and corresponding seismic profiles and 'feature metrics' of the scattered and reflected energy, which are used for characterizing the bottom and sub-bottom units (Hutchins *et al.*, 1982).

Before describing the system, it will be useful to review briefly the evolution of developments in quantitative acoustic survey methods as applied to systematic lithostratigraphic mapping. The precision of these methods has steadily improved over the years and they now provide quantitative estimates of the surface roughness, the correlation area, and the hardness and variability of the sea floor, in addition to stratigraphic and bathymetric profiles of greatly improved image quality. This information is relevant to sea-floor engineering studies and is particularly useful in the design of cost-effective sampling and testing programmes (Hutchins, 1980, pp. 92–97; King, 1980).

Improved sound sources with increased frequency bandwidth and multichannel receivers now provide the means for measuring acoustic attenuation, acoustic impedance and sound speed as a function of depth at a horizontal scale of a few metres and a vertical scale of better than a metre (Dodds, 1980; Fagot, 1983).

BACKGROUND

A great deal of work has been done since the early 1940s concerning the relationships between the acoustic and physical properties of the sea floor. The propagation of sound in the ocean, of interest to the military sonar systems specialist, is strongly dependent upon the reflectivity, scattering and absorption properties of the bottom and sub-bottom to depths of several wavelengths at the frequencies of interest.

A knowledge of the bottom geology, therefore, is necessary for the prediction of acoustic propagation in the ocean.

In more recent times, geologists and geophysicists have been using acoustic reflectivity measurements to characterize geology, and this is one of the principal missions of 'Seamor'. Quantitative estimates of the acoustic reflectivity of the sea floor using a shipboard echo-sounder were carried out by Breslau (1967). King (1967) used textural analysis for systematic regional mapping of the Canadian shelf. Real-time quantitative estimates of the acoustic reflectivity of the deep sea floor using a near sea-floor deployed 4 kHz seismic profiler have been carried out on a systematic basis by the Scripps Institution of Oceanography since 1974 (Tyce, 1976). Prediction of attenuation from a wedge of sediments using a sparker sound source deployed on the surface of the sea was reported by Li and Taylor Smith (1969) but met with little success due to scatter in the data. Tyce (1976, pp. 692–695), using this approach and near sea floor measurements of reflectivity from the Carnegie Ridge, obtained attenuation measurements in general agreement with those obtained from core samples taken in the same area.

This early work was frustrated by high variability in the acoustic reflectivity measurements. Although much of this variation may be due to lithologic changes, a great deal is due to motion of the sensor platforms, variation in elevation and orientation of the source and receiver, focusing and defocusing caused by large-scale topographic features and scattering by small-scale surface roughness.

In general, we may consider the horizontal resolution of a normally incident acoustic pulse as being equivalent to the diameter of the first Fresnel zone, the area which lies within one quarter of a wavelength of the minimum distance of the reflecting surface to the source-receiver. High-resolution surveys, therefore, require that the source-receiver system be deployed close to the sea floor (Fagot, 1983).

THE SEABED PROJECTS

During the last decade, the authors have been associated with the seabed research projects. The object of these projects has been to improve methodology for high-resolution lithostratigraphic mapping of the Canadian continental shelf.* Research activity under the Seabed I project had three main thrusts. First, the development of high-resolution quantitative seismic profiling technology having improved layer resolution at the seabed, greater penetration in hard bottom areas; useability in the weather and sea conditions generally encountered on the eastern Canadian shelf, and reduced sensitivity to variables which contribute to reflection amplitude variations (Hutchins, 1974; Hutchins et al., 1976; Simpkin, 1978; Hutchins, 1980, pp. 92–95). Second, to acquire data using the technology. Third, to develop appropriate geoacoustic models for inversion of the data to derive measures of features of the bottom and sub-bottom units as aids to lithologic classification (Cochrane and Dunsiger, 1979; Dodds, 1980; Ellis and Chapman, 1980; Hutchins, 1980; Parrott et al., 1980; Dunsiger et al., 1981; Hutchins et al., 1982).

During the last few years some 50 000 line kilometres of acoustic reflectivity and sub-bottom profile data have been acquired (King and Fader, 1967; King and Fader, 1976; King, 1980; Fader and King, 1981; King and Fader, 1984), with successive generations of the original Deep Towed Seismic (DTS) system (Simpkin, 1978). The current version of the DTS has a tow depth capability of 500 m and includes a side-scan sonar in addition to the quantitative seismic profiler and motion sensors for heave and depth correction included in the earlier version.

In 1982 the scope of the Seabed project was enlarged to extend the technology to deep-water surveys and to develop an integrated real-time data processing system to provide a true on-line quantitative mapping capability. This development is identified as the Seabed II project and the deep-towed survey system as 'Seamor'.

OVERVIEW OF THE SYSTEM

Seamor is a near sea floor quantitative acoustic survey system for operation in water depths from 500 to 6000 m employing a ship-deployed tethered towed body (fish) as a vehicle for the in-water sensors. System outputs include scale-rectified geographically oriented and positioned areal maps of the sea floor from side-scan sonar sensor data and bathymetric and sub-bottom profiles from the seismic profiler sensor data.

The processing for the side-scan data assumes a flat sea floor so there will be some geographic distortion of features for a non-flat sea floor. In addition, the system computes and displays the acoustic reflectivity, 'reflectivity metrics' (Parrott et al., 1980), along the tow fish track. The horizontal scales of the side-scan and seismic profiler displays are identical. The vertical scale of the seismic profiler display assumes a constant sound speed and may be exaggerated. In addition to the hard copy graphic recordings, all unprocessed sensor data are recorded using a high-density digital tape recorder (HDDR). The simplified block diagram is shown in Fig. 1. The tow fish position, velocity and attitude are computed in real time by the Tow Track Recovery System (TTRS) from information provided by sensors in the tow fish, an acoustic Fish Positioning System (FPS) and from the ship's positioning and navigation system. All data, including the output data from the TTRS, are processed digitally in real time through the Information Processing and Display Subsystem (IPDS) which produces the hard copy system outputs identified earlier. In addition to operating in real time, the IPDS may be operated in playback mode

---

* The seabed projects have been a joint research and development undertaking involving several agencies of the Federal Government of Canada, universities, and Huntec (70) Limited as prime contractor. The goal is improved methodology for marine geological mapping through acoustic remote sensing.

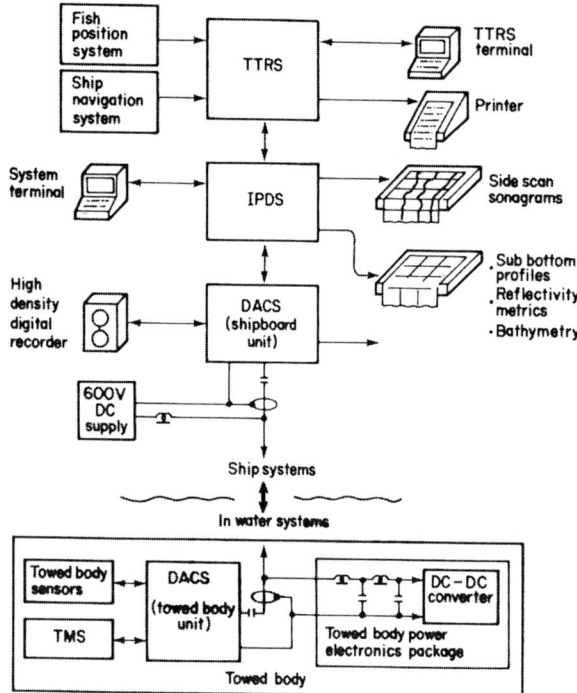

Fig. 1   System block diagram

and routinely executes internal performance verification checks. Diagnostic modes of operation are provided for identification and localization of faults.

The electrical core of the main tow cable is a single, RG8U 50 Ω coaxial cable which carries both power and information between the ship and towed body. Power for the system elements located in the towed body is provided by a 600 V DC supply on the ship. The Data Acquisition and Control System (DACS) receives commands through the IPDS and transmits them via the downlink to the towed body where they are used for control of the sensors and towed body systems. Data from the towed body sensors is transmitted via the DACS uplink where they are processed by the IPDS and TTRS to produce the required outputs. Analogue outputs of all tow fish sensor data are available at the DACS analogue output interface.

to produce the system outputs from the tape-recorded digital data.

All communications between the operators and the system are by means of standard ASCII keyboards and screens. System status, as well as the TTRS outputs, are automatically logged and printed out in hard copy. The system is self-calibrating

DEPLOYMENT AND HANDLING

The configuration and general operating principles of the towed body ('in-water' assemblies) of Seamor are shown in Figs 2 and 3 respectively. Referring to Fig. 2, the towed body consists of two mating parts, the sensor vehicle and the depressor. The

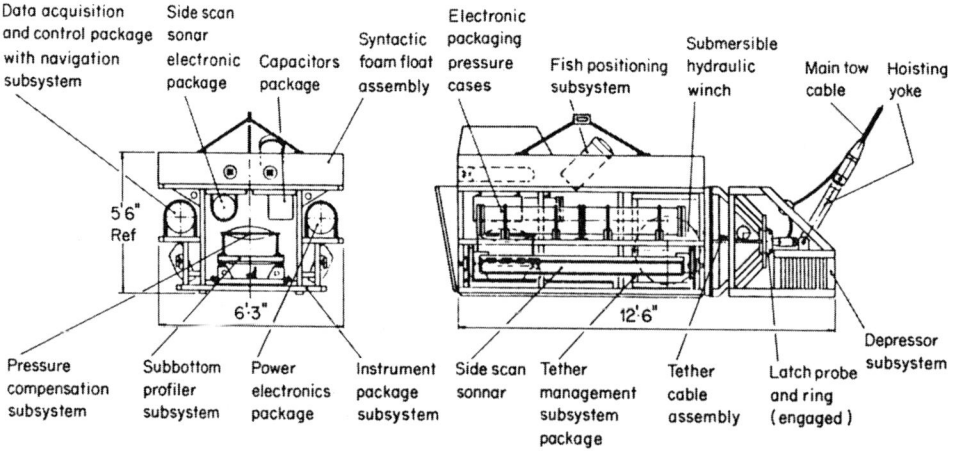

Fig. 2   Towed body arrangement showing the sensor vehicle and depressor in their coupled configuration

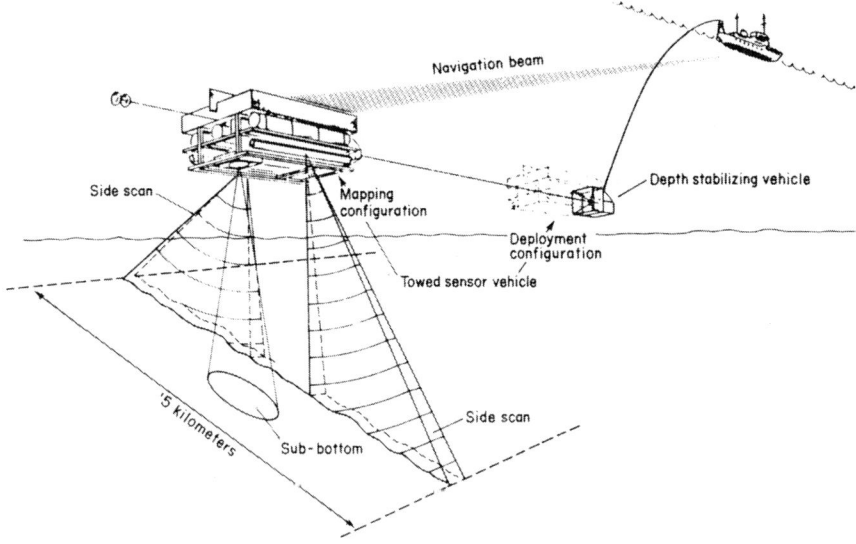

Fig. 3 Operating principle of the two-stage towed body. Note the hydrophone streamer for the seismic profiler trailing behind the sensor vehicle

depressor is attached to the free end of 5000 m of the main tow cable loaded onto the drum of the main winch, which is located on the ship. The sensor vehicle is fitted with a tether winch on which is spooled 100 m of neutrally buoyant cable, the tether cable. The free end of the kevlar stress member of the tether cable is secured to the depressor. The sensor vehicle is firmly secured to the depressor by a remotely operated latching arrangement so that the towed body may be handled during launch and recovery operations as a single unit. To launch the towed body, it is hoisted off the deck by a crane and lowered over the stern of the ship into the water. Following release of the latches which secure the sensor vehicle to the depressor, the winch operates to pay out the tether cable until the sensor vehicle is approximately 100 m behind the depressor, its normal position for acquiring data. With the ship underway the main tow cable is paid out until the towed body has reached its desired operating depth. The sensor vehicle has slight positive buoyancy and tracks behind the depressor, effectively decoupled from heave and other motions. The depressor is of the dead-weight type, and its depth is determined by the cable scope and ship's speed.

To recover the towed body, the main winch on the ship is operated to bring the towed body assembly to the surface. The tether winch now operates to heave in the tether cable, bringing the sensor vehicle up to the depressor where the latches automatically engage, locking the two assemblies together. It is then recovered by the crane as a single unit. The towed body suspended by the main tow cable is shown in Fig. 4 and in its stowage position on the deck of CSS *Hudson* in Fig. 5. General specifications for the handling system are shown in Table 1.

TOWED BODY SYSTEMS

The sensor vehicle provides a stable platform for the following system elements identified in Fig. 2:

- The sub-bottom profiler sensor systems which include the sound-source transducer; the pressure compensator; air bottles for the pressure compensator (not labelled); energy storage capacitors in an oil-filled pressure-balanced package; high-voltage DC-DC inverter supply for capacitor charging included in the power electronics package; hydrophone

Fig. 4 Towed body being deployed from CSS *Hudson*

streamer (not shown) and a near-field calibration hydrophone (not labelled).

- The side-scan sonar sensor system which includes the port and starboard tranducers in oil-filled pressure-balanced enclosures and the power amplifiers and trans-

Fig. 5 Towed body in its deck stowage position on CSS *Hudson*

TABLE 1
General specifications—handling systems

Deployment depth*	2000 m
Tow speed at 2000 m	1 m/s (2 kN)
Cable scope	4500 m
Cable tension	61.6 kN
Towed body: coupled configuration	
Length	3.81 m
Height	2.29 m
Width	1.90 m
Weight in air	2900 kg
Depressor weight	900 kg (air) 600 kg (water)
Sensor vehicle length	3.2 m
Tow cable: double armour contrahelically wound stress member	
Diameter	1.73 cm
Length	5000 m
Breaking strength	180 kN
Electrical core single coax RG8U 50 Ω coaxial	
Winch: electrohydraulic	
Length	3.5 m
Height	2 m
Width	2.54 m
Line speed at 44 kN	1.5 m/s max.
Weight with 5000 m cable	95 kN
Horsepower	150 max.
Winch power pack	
Length	1.83 m
Height	1.5 m
Width	2.3 m
Weight	2245 kg

*May be extended to 6000 m with change of syntatic foam and additional cable

ceiver units housed in the side-scan sonar electronic package pressure case.

- The DACS package which transmits sensor data to the ship on the uplink and decodes the downlink command and control data from the ship through the RG8U coaxial core of the main tow cable.
- The navigation subsystem sensor package which consists of acceleration and angle-rate sensors along each of the three orthogonal vehicle axes, using accelerometers and rate gyros respectively; a three-component fluxgate magnetometer and a pressure depth sensor.

The DACS and navigation sensor packages are housed in a single pressure case.

- The power electronics package which consists of DC-DC inverters which provide regulated output voltages for the various towed-body systems, using power from the 600 V DC supply on the ship transmitted through the same RG8U cable which serves the DACS.
- The Fish Positioning Subsystem, a pinger, or ultra-short baseline receiver, depending upon the particular configuration, for determining the position of the sensor vehicle relative to the ship or, in the inverted configuration, the ship relative to the sensor vehicle.
- The Tether Management Subsystem described earlier, which consists of a submersible hydraulic winch, slip rings for electrical interfacing between the tow vehicle systems and the tether and main tow cables and the latching mechanism for locking the tow vehicle and depressor together into a single unit.

DESIGN CONSIDERATIONS

Some of the considerations involved in the design of the system are discussed in the following subsections.

Data and Power Transmission

Tow speed and depth performance are major considerations. Because the size and power of the handling system depends upon the volume of the cable, it is necessary to minimize the electrical cross-section of the cable since it decreases the cable density, does not contribute to strength and adds to the cable volume. The most efficient type of cable from the point of view of information carrying capacity is a coaxial cable. The information carrying capacity of a particular cable varies inversely with its length squared and directly with its diameter squared for constant input signal power. Some form of data multiplexing is therefore required. The side-scan sonar and seismic

profiler sensor signals have dynamic ranges in excess of 100 dB and frequency bandwidth for the seismic profiler hydrophone exceeds 8 kHz. Low noise, wide dynamic range, freedom from cross-talk, adequate bandwidth, and linearity and transmission over long cables without amplification are the requirements which must be met.

These requirements favour digital time division multiplexing for the DACS. The uplink uses alternate marks inverted at 10^6 bits per second and achieves a bit error rate of one part in a hundred million. The uplink modems can easily be converted for future use with a fibre optic cable core. The downlink uses frequency shift keying at 13 and 18 kHz. Specifications for the DACS are given in Table 2.

Tow Track Recovery System (TTRS)

The purpose of the TTRS is to provide data for mapping the returns from the side-scan sonar and sub-bottom profiler, and for compensation of these returns for tow fish motion. To accomplish this purpose, the TTRS has two functions — absolute and relative positioning. In absolute positioning, it outputs the tow fish position and

TABLE 2
Specifications for digital acquisition and control system

Uplink
SDLC protocol 10^6 bits per second
Alternate marks inverted
Bit error rate   1 in 3×10^8 bits
Word length   4 gain bits plus 12 data bits

Side-scan sonar and seismic profiling
2 channels each, 16 kHz sampling rate, binary gain ranging for 132 dB dynamic range

Auxiliary sensors
18 Channels, 80 Hz sampling rate, binary gain ranging for 108 dB dynamic range

Downlink
13–18 kHZ frequency shift keyed for command and control of tow-fish systems

motion dynamics information, which is processed in the IPDS to position patches on
the seabed with respect to some local or
global datum to an accuracy consistent with
the position of the ship. In relative positioning, it outputs tow fish attitude and motion
dynamics information, which is processed in
the IPDS to position successive acoustic
returns with respect to each other to an
accuracy and repeatability consistent with
the acoustic sensor resolution and the
required image quality.

To determine the attitude and motion
dynamics of the tow fish, components of the
TTRS in the tow fish measure the water
pressure (tow fish depth), acceleration,
angular rates and local magnetic field vectors. All this data is output in the form of
analogue signals to the DACS uplink, since
the TTRS processing components are
located on the ship rather than in the tow
fish with the sensors.

To determine the position of the tow fish,
an appropriate fish positioning system and
ship navigation system are used. The fish
positioning system determines the position
of the tow fish with respect to the ship. The
ship navigation system determines the position of the ship. TTRS uses this information,
together with information about the velocity
of the tow fish, to estimate the position of
the tow fish. The structure of the TTRS is
shown in Fig. 6.

The shipboard components of the TTRS
processes the data in real time and outputs
the tow fish position, attitude and motion
dynamics to the IPDS. TTRS also displays
the data on a CRT and printer. Any of the
data (statistical summaries where the data
are dynamic) can be displayed to the
operator on a CRT at a rate selected by the
operator (with a maximum 0.1 Hz update
rate). Any of the data can also be logged on a
printer at an operator specified rate (with a
maximum 0.01 Hz update rate). Lastly, the
operator may assign any of the data, with
appropriate upper and lower bounds, to 8
analogue (±10 V) output channels. These

Fig. 7 Shipboard systems. The Information
Processing and Display System is shown on the
right. The Honeywell digital tape recorder and
the Numerex array processor are part of the
IPDS on-line processing system. The Data
Acquisition and Control System and its 22
analogue output interfaces is located in the left-
hand rack

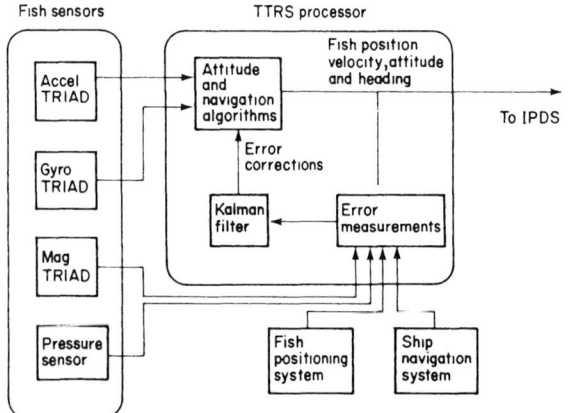

Fig. 6   Block diagram showing structure of the
Tow Track Recovery System

TABLE 3
Specifications for Tow Track Recovery System

Capabilities
Full six degree-of-freedom orientation of tow fish and determination of its motion dynamics. 0 to 6000 m depth, single or two-stage bodies can be interfaced to other mapping and survey systems independently of the rest of the Seabed II ship electronics. Can be reconfigured with different sensors to suit different user requirements and accuracies. Provides a number of reversionary modes of operation through sensor redundancy

Tow fish attitude and position measurements
Pitch error   1.0° variation in any 3 s period
Roll error   0.5° variation in any 3 s period
Yaw error   0.2° variation in any 3 s period

Fish positioning subsystem
Slant range error   2% slant range
Bearing error   0.25°

Ship positioning system
Positioning error   150 m

Sensors
Columbia triaxial accelerometer   SA-307TX
Northrop triaxial rate gyro   50162
Develco triaxial magnetometer   9200C
Bell & Howell pressure sensor   BHL4101-18

Shipboard processor
Intel ISBC 86/30 with 8087 floating point
    co-processor
Intel ISBC 464 EPROM memory board

Options
Data translation DT1843 analogue output board
Intel ISBX 311 analogue input board

channels are updated at the same rate as the variables are internally updated and can be used to observe the dynamic behaviour of the data. A strip chart recorder, an analogue $X-Y$ plotter, and/or an oscilloscope may be attached to the analogue interface. Specifications for the TTRS are given in Table 3.

The Seismic Profiler

The sound source requirements for an effective quantitative high-resolution seismic profiler are a wide-frequency bandwidth for resolution of thin layers, sufficient energy at low frequencies for penetration, well-defined directional characteristics with a narrow acoustic beam capability for good horizontal resolution and high repeatability and durability. This combination of characteristics is difficult to achieve at reasonable cost using piezoelectric ceramic transducer technology (e.g. pingers and echo sounders) which operate most efficiently at a single design frequency. Surface-deployed seismic sources, such as sparkers and airguns, are not suitable for deep deployment because of decreased efficiency at increased ambient pressures, energy supply difficulties in the case of airguns, variability of the output pulse characteristics with time and depth and non-directionality, which increases reverberation levels and multipath echos.

An appropriate solution is the impulsively driven plane piston. By impulsively driven we mean that a plane circular piston in contact with the water accelerates rapidly outwards, then gradually comes to rest and slowly retracts to its original position. The acoustic pressure signature measured at a point in the far field along the acoustic axis is a replica of the piston acceleration history. A suitable technology for producing this kind of motion is the 'boomer' sound source (Edgerton, 1964). Through appropriate design (Hutchins, 1974), and with the addition of pressure compensation to permit operation at depth, the boomer produces a highly repeatable intense positive pressure pulse free of secondary oscillations (McKeown, 1975).

The sound source used in Seamor has a piston diameter of 0.5 m and is driven by a 60 μF capacitor discharge system which operates at a maximum energy level of 1000 J per shot and a maximum average power of 500 W. The high-voltage DC supply and energy storage capacitors are housed in a pressure case on the sensor vehicle as noted earlier. Along the main beam axis the peak acoustic intensity is 220 dB referenced to 1 μPa at 1 m, the width of the positive pressure excursion is 200 ms and the peak amplitude of the negative pressure excursion is less than 25% of the positive peak. The energy in this pulse is distributed over a broad frequency band with a gentle peak at 2.5 kHz extending down to 500 Hz and upwards to a roll-off at 6 kHz. The low-frequency components are distributed over a broad beam, while the higher frequency components are concentrated around the main beam axis. The beam width at 2.5 kHz is approximately 60°. This spatial distribution of energy is particularly suitable for near sea-floor profiling using a dual channel processor. The received signal is passed through a high-pass filter, the output of which represents the energy reflected or scattered from a small portion of the sea floor due to the small effective beamwidth of the incident pulse, thus providing precise bathymetric profiles but little or no penetration. The same signal, but without filtering, is then used for generating the sub-bottom profile on a second graphic recorder.

In quantitative terms, the wide-frequency bandwidth and the spatial distribution of the energy of the incident pulse is equivalent to a multifrequency echo-sounder. It provides additional information about frequency-dependent processes, such as scattering from a rough bottom (surface scattering) and scattering from within the bottom (volume scattering). By use of appropriate geoacoustic scattering models, the reflection data may be inverted to yield quantitative estimates of bottom and sub-bottom reflectivity, surface roughness and correlation distance, attenuation in the sediments and volume scattering coefficients (Cochrane and Dunsiger, 1979; Dodds, 1980; Hutchins *et al.*, 1980; Parrott *et al.*, 1980; Dunsiger *et al.*, 1981).

In Seamor, two receiving hydrophones are used. One of these hydrophones is located in the near field of the sound source and is used to calibrate each shot. The other hydrophone is a three-meter long array of ten elements connected series parallel, streamed behind the sensor vehicle. The analogue signals from each hydrophone are fed into independent input channels of the

TABLE 4
Specifications for seismic profiler

Sound source: boomer with active pressure compensation

Circular piston	0.50 m diameter
Peak acoustic power (1000 J)	220 dB related to 1 μPa at 1 m
Frequency bandwidth	300 Hz to 10 kHz
Maximum operating depth*	2000 m

Power system: 60 μF at 1–6 kV adjustable in 1 kV increments

Maximum energy	1000 J at 6 kV
Average power	500 W
Input	300–600 V DC at 3 A
Repitition rate limited by average power	

Receiving system: streamer hydrophone length 3 m, 10 elements

Sensitivity	203 dB related to 1 V/μPa
Bandwidth	300 Hz to 6 kHz

Calibration system: single LC10 hydrophone in near field of source

*May be extended to 6000 m with minor changes to presure compensation system

DACS where they are passed through anti-aliasing filters, amplified in a programmed binary gain amplifier, sampled at 16 kHz and encoded for time division multiplexing along with the other sensor signals to the surface. Specifications for the seismic profiler are given in Table 4.

The Side-scan Sonar

The side-scan sonar has a range of 3 km each side when towed at an elevation of 300 m. The transducers have a horizontal beamwidth of 1.7°. Each transducer consists of a double bank of individual motors operating at 27 kHz on the left side and 30 kHz on the right side and provides a vertical beam of 50°. Provision is made for adjusting the tilt angle of the transducers to accommodate different operational conditions. The double bank of elements allows for measurement of the arrival angle of the back-scattered signal, if means are provided for measurement of phase in the sensor vehicle electron-

TABLE 5
Specifications for side-scan sonar

Frequency
  Left side   27 kHz   Right side   30 kHz

Source level
  228 dB related to 1 μPa

Beamwidth
  Horizontal   1.7°   Vertical   50°

Pulse duration
  0.3–3.2 × 10^{-3} s

Repitition rate
  0.4–5.0 s

Maximum operating depth
  6000 m   Swath-width at 300 m elevation is 6 km

ics package. At present, the summed output from the two banks of elements are amplified in a time-varied gain amplifier which applies a gain function. The gain function consists of a time-squared factor designed to correct for two-way spreading losses and an exponential time factor to correct for sound absorption in the water. After the gain stages the signals are frequency translated

down to 3 kHz for input to the DACS uplink.
  Calibration of the voltage gain of the system via its TVG is required for quantified acoustic back-scatter measurements. The calibration sequence may be remotely initiated at any time during system operation. Condensed specifications for the side-scan sonar are given in Table 5.

RESULTS OF THE 1984 SEA TRIALS

The 1984 sea trials of the Seabed II system were conducted on the Scotian Shelf and Scotian Slope off southeastern Canada aboard the Canadian research vessel CSS *Hudson* of the Bedford Institute of Oceanography. Although the IPDS processing software for the side-scan sonar and seismic profiler were incomplete, unprocessed sensor data available at the DACS analogue output interface was fed to graphic recorders to display the side-scan sonar and seismic profiler sensor data. The trials were intended for verification of performance of the hardware as a condition of acceptance by the Crown. The system performed beyond expectations and geological survey data were collected in the Verrill Canyon area identified on the location map (Fig. 8).

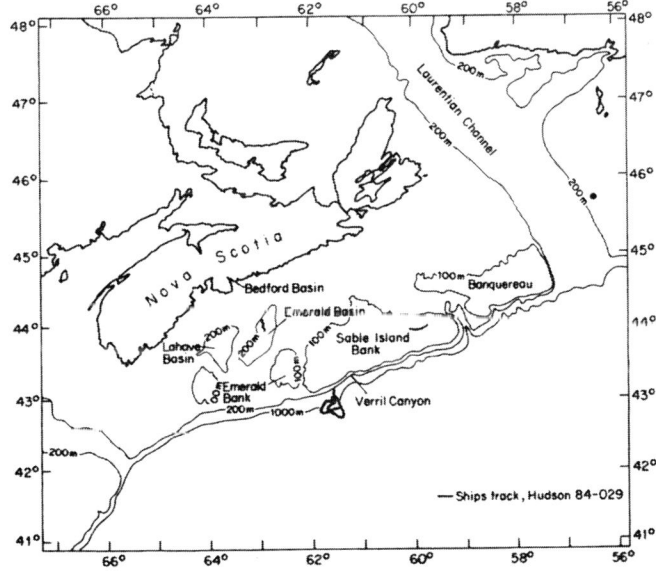

Fig. 8 Location map showing the track of CSS *Hudson* near the Verrill Canyon area during the test cruise in July 1984

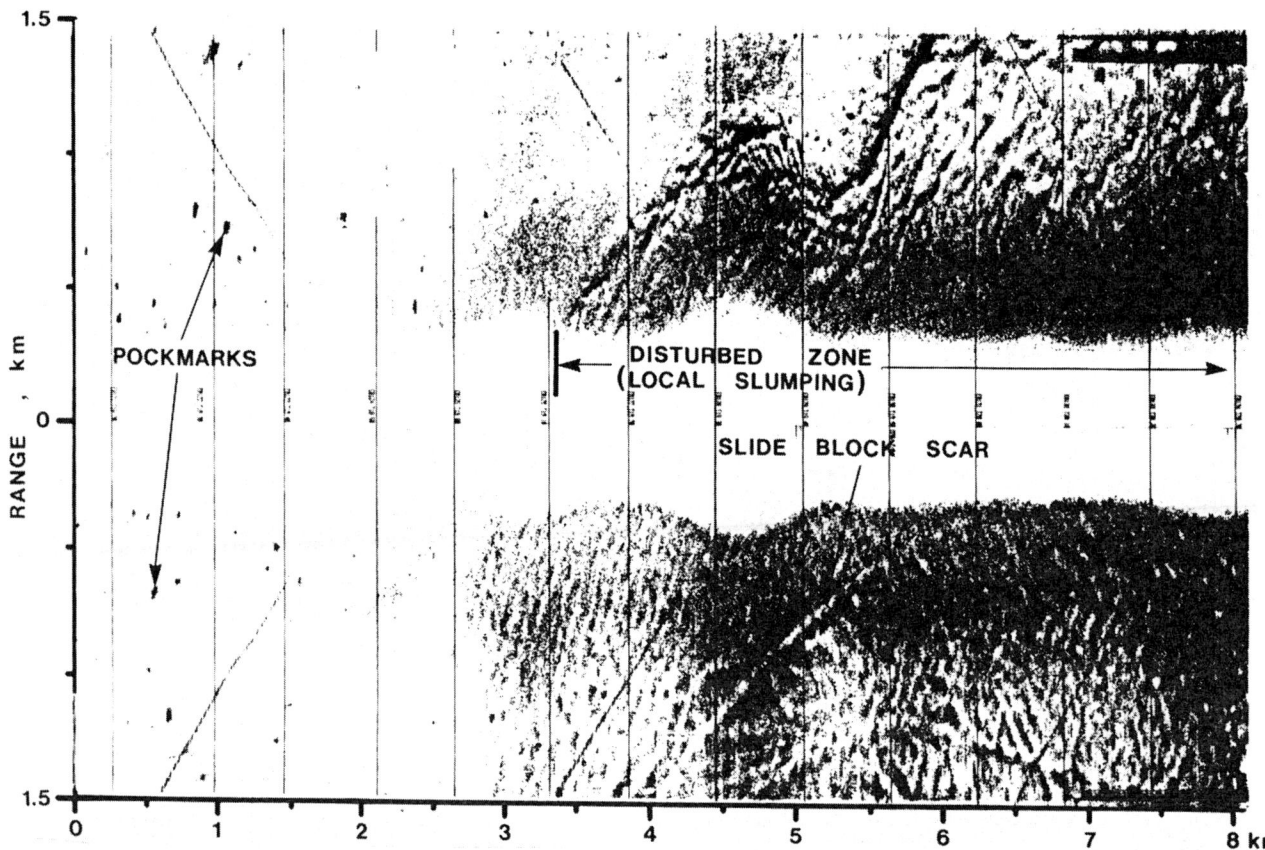

Fig. 9 Sidescan sonogram* from the upper continental slope off Nova Scotia, near Verrill Canyon, in 100 m water depth, collected with the Seabed II (Seamor) system. The aspect ratio is approximately 1 : 1 and the range is 1.5 km across track. From the 0–3 km marks across the profile, the seabed is covered with pockmarks (gas escape craters). Extending from the 3–8 km marks is a large zone of disturbed sediments with a subparallel hummocky morphology resulting from local slumping. The curvilinear features across the disturbed zone are interpreted as 'slide block scars' or impressions

Analysis of these side-scan sonograms and high-resolution seismic reflection profiles showed previously unknown geological features at the seabed and in the subsurface.

Essential to the interpretation of seabed geology is the need to obtain both side-scan sonar and seismic reflection data from a study area, in order to understand a three-dimensional configuration of structure and morphology. To illustrate this system interdependence, Fig. 9 (side-scan sonogram) and Fig. 10 (high-resolution seismic reflection boomer profile) are presented. They

* This sonogram has not been processed, scale rectified or geographically positioned.

were collected concurrently from the same area of the upper slope extending from 500 to 1300 m water depth.

From 200 to 600 m water depth over the edge of the shelf the seabed appears to be covered with iceberg furrows which are both degraded and modified through erosion, and partially buried, as evidenced by the occurrence of discontinuous iceberg furrow rims. These iceberg furrows were previously unknown from the upper slope of this area. Downslope, from 500 to 1100 m water depth, the seabed is covered with pockmarks (gas escape craters) which range in depth to 4 m and cover 5–10% of the sea-

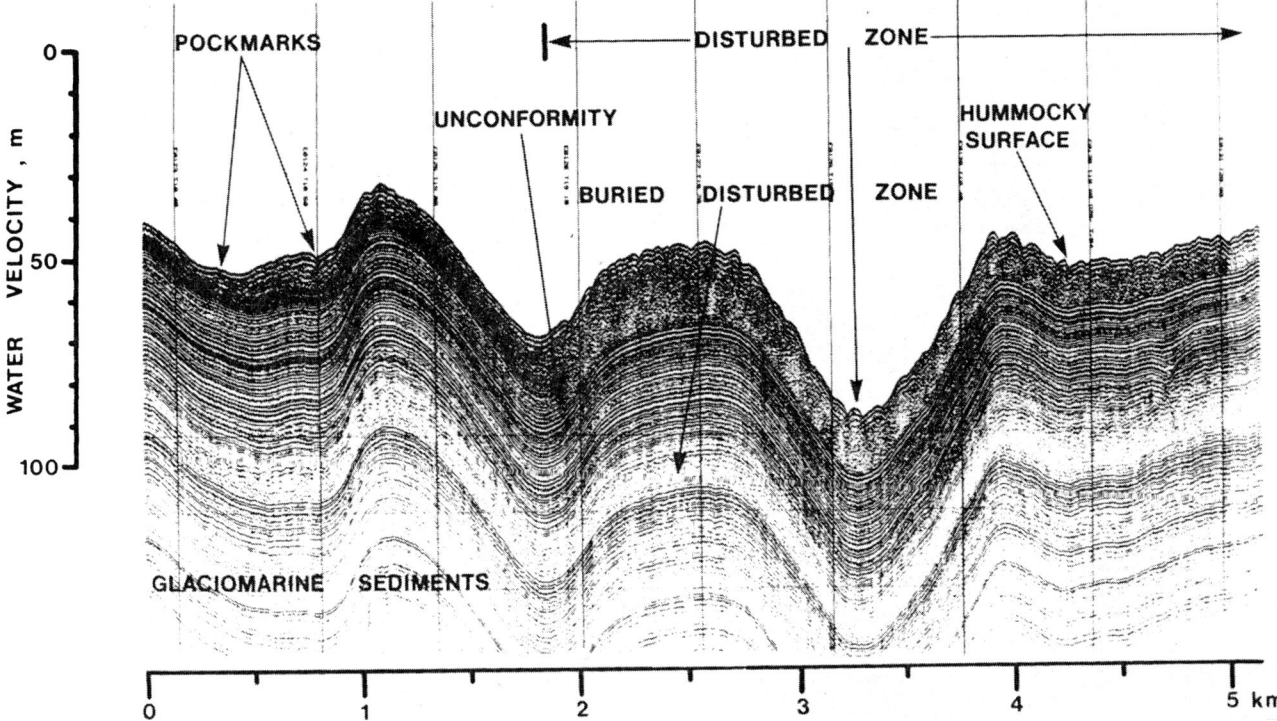

Fig. 10 High-resolution seismic reflection boomer profile* from the upper continental slope off Nova Scotia, Canada, near Verrill Canyon, in 1000 m water depth, collected with the Seabed II (Seamor) system. Over 200 m penetration is obtained with a layer resolution of approximately 2 m. The seabed, to a depth of 20 m, consists of a zone of sediments characterized by incoherent reflections. These sediments have pockmarks developed on their surface from the 0–2 km marks, and are hummocky, resulting from local slumping and 'slide block scarring' from the 2–5 km marks. The continuous coherent reflections are interpreted as representing glaciomarine sediments. The presence of a zone of buried incoherent reflections at a sediment depth of 50 m may represent an instability event which occurred at 50 000 years BP. These profiles will enable the time intervals for major sedimentational and erosional events to be determined, which is of prime importance for the safety of hydrocarbon drilling operations

bed. Pockmarks also had not been previously detected in these water depths or environments on the Canadian continental slope. The survey continued downslope across a zone of locally slumped sediments (Fig. 9) from 2 to 5 km marks. These sediments exhibit a subparallel hummocky morphology with relief of up to 5 m and occasional pockmarks developed on their surface. Long curvillinear depressions interpreted as 'slide block scars' occur across the surface of these disturbed

This recording has not been processed, scaled or motion compensated.

sediments. On the seismic reflection profiles, the disturbed sediment is up to 20 m in thickness and internally characterized by incoherent reflections. It would have been difficult to penetrate and resolve the subsurface stratigraphy of these sediments with conventional 3.5 kHz profilers.

Of significant importance is the occurrence of a buried zone of incoherent reflections at a depth of approximately 50 m (Fig. 10), similar to the zone of disturbed, locally slumped sediments at the seabed. This buried zone is of regional occurrence and, from information based on limited biostratigraphic and chronostratigraphic control from nearby wells and piston cores, is estimated to be approximately 50 000 years BP in age (D. Piper, personal communication). This buried zone possibly correlates with the maximum extent of ice-sheet advance during the Wisconsonian glaciation of the southeastern Canadian continental shelf (Kin and Fader, 1984). The new information provided by the Seabed II system provides a framework for understanding the time interval for instabilities on the continental slope, an essential parameter in safe hydrocarbon exploration and development.

ACKNOWLEDGEMENTS

The Seabed II project was supported with funds from the following agencies of the Government of Canada: the National Research Council; the Defence Research Establishment Atlantic; the Department of Energy, Mines and Resources; and the Department of Fisheries and Oceans. The user of the system is the Atlantic Geoscience Centre at the Bedford Institute of Oceanography.

We wish also to acknowledge the many contributions of our colleagues in government and academe, in particular, L. Collett, M. Keen, K. Manchester, L. H. King, D. McKeown, D. Chapman, J. Ross, N. Cochrane, R. Peters, P. Simpkin, R. Parrott, Dave Dunsiger (deceased), Dave Ross and, as well, our friends at Huntec.

REFERENCES

1. Breslau, L. R. 1967. The normally incident reflectivity of the seafloor at 12 kHz and its correlation with physical and geological properties of naturally occurring sediments. Woods Hole Oceanogr. Inst., Ref. 67–16.
2. Cochrane, N. A. and Dunsiger, A. D. 1979. Seabed roughness characterization by broadband acoustic echo sounding. In *Proc. Fifth International Conference on Port and Ocean Engineering under Arctic Conditions*, Trondheim, Norway, August 13–17.
3. Dodds, D. J. 1980. Attenuation estimates from high resolution sub-bottom profiler echoes. In *Bottom-Interacting Ocean Acoustics* (Eds W. A. Kuperman and F. B. Jensen). Plenum Press, New York.
4. Dunsiger, D. A., Cochrane, N. A. and Vetter, W.J. 1981. Seabed characterization from broadband echosounding with scattering models. *IEEE J. Oceanic Engng* OE-6(3).
5. Edgerton, H. F. and Hayward, E. C. 1964. The Boomer sonar source for seismic profiling. *J. Geophys. Res.* 69, 3033–3042.
6. Ellis, D. D. and Chapman, D. M. F. 1980. Propagation loss modelling on the Scotian Shelf: comparison of model predictions with measurements. In *Bottom-Interacting Ocean Acoustics* (Eds W.A. Kuperman and F. B. Jensen). Plenum Press, New York.
7. Fader, G. B. and King, L. H. 1981. A reconnaissance study of the surficial geology of the Grand Banks of Newfoundland. Paper 81-1A, Current Research Part A, Geological Survey of Canada, pp. 45–56.
8. Fagot, M. G. 1983. A deep towed sound source and hydrophone array system: performance analysis and hardware description. In *Acoustics and the Seabed* (Ed. N. G. Pace). Bath University Press, Bath, pp. 369–377.
9. Hutchins, R. W. 1974. Computer simulation of a transiently excited underwater sound projector. In *Proc. Oceans 74, IEEE International Conference on Engineering in the Ocean Environment*.
10. Hutchins, R. W. 1980. Development of new geophysical methods for site investigation studies. In *Offshore Site Investigation* (Ed. D. A. Ardus). Graham & Trotman, London, pp. 76–102.
11. Hutchins, R. W., McKeown, D. and King, L. H. 1976. A deep tow high resolution seismic system for continental shelf mapping. *Geosci. Can.* 3(2), 95–100.
12. Hutchins, R. W., Dodds, D. J., Parrott, D. R. and King, L. H. 1982. Characterization of seafloor sediments by geo-acoustic scattering models using high resolution seismic data. In *Proc. Oceanology International*, Brighton, March 1982.
13. King, L. H. 1967. Use of a conventional echo sounder and textural analysis in delineating sedimentary facies: Scotia Shelf. *Can. J. Earth Sci.* 4, 691–708.
14. King, L. H. 1980. Aspects of regional surfi-

cial geology related to site investigation requirements on eastern Canadian shelf. In *Offshore Site Investigation* (Ed. D. A. Ardus). Graham & Trotman, London, pp. 37–59.

15. King, L. H. and Fader, G. B. 1976. Application of Huntec Deep Tow high resolution seismic system to surficial and bedrock studies — Grand Banks of Newfoundland. Paper 76-1C, Report of Activities Part C, Geological Survey of Canada.

16. King, L. H. and Fader, G. B. In press. Wisconsinan glaciation of the Continental shelf — southeast Atlantic Canada. Geological Survey of Canada.

17. Li, W. N. and Taylor Smith, D. 1969. Identification of seafloor sediments using underway acoustics. *Geophys. Prospect.* **17**, 231–247.

18. McKeown, D. L. 1975. Evaluation of the Huntec (70) Hydrosonde Deep Tow Seismic System. Report B1-R-75-4, Bedford Institute of Oceanography, Dartmouth, Nova Scotia, Canada.

19. Parrott, D. R., Dodds, D. J., King, L. H. and Simpkin, P. G. 1980. Measurement and evaluation of the acoustic reflectivity of the seafloor. *Can. J. Earth Sci.* **17**(6) 722–737.

20. Simpkin, P. G. 1978. Evaluation of broadband high resolution seismic data for seafloor sediment classification. In *Proc. Oceanology 78*, Brighton, England. Society for Underwater Technology, Birdcage Walk, London, England.

21. Tyce, R. C. 1976. Near bottom observations of 4 kHz acoustic reflectivity and attenuation. *Geophysics* **41**(4), 673–699.

7

High-resolution Geophysical
Surveys for Engineering Purposes

K. P. Games, Gardline Survey Ltd, UK

Abstract

In recent years it has become increasingly
evident that there is a need to bridge the
gap that existed between the ultra-high
resolution (but with limited penetration)
provided by analogue equipment and the
so-called high-resolution shallow seismics.
The latter generally provides resolution of
about 10 m, with limited data in the top few
hundred milliseconds, but gives satisfactory
information at greater depths. To some
extent, vertical resolution can be improved
by paying more attention to the deconvolution techniques at the processing stage, and
lateral resolution can be improved by migration of the data. However, these options —
particularly that of migration which is very
costly and therefore depends on the client's
willingness to pay — are seldom within the
control of the survey company. Hence, two
years ago, Gardline's Geophysical Research
and Development Department undertook
the design of a new acquisition system which
would do much towards achieving such
objectives. The approach to this project is
outlined, and an account of the system is a
high-resolution multichannel hydrophone
array (the mini-streamer). The technical
specifications of this mini-streamer are
described, together with the theoretical
characteristics and responses of the array
design.

The mini-streamer, moreover, is part of
an integrated digital package consisting of
source (mini-sleeve exploder), array (the
mini-streamer), recording system (an
upgraded DFS V) and a processing capability developed in close co-operation with a
specialist high-resolution processing house.

The new system was tested over three
separate locations, and the results of these
test lines are discussed. As a result of these
tests — all of which were very successful —
it has been possible to assess the optimum
operational capabilities of the system, as
well as its geophysical potential.

Since then, nine site surveys have been
successfully completed with the system, and
a case study of some of these sites is pre-

sented along with illustrations of the quality of data obtained.

INTRODUCTION

At the first OSSI conference, held in 1979, Arthur (1980) gave a comprehensive and convincing argument in favour of high-resolution multichannel seismic techniques to bridge the gap between high-frequency profilers and deep seismic data. Since then, in site survey work, it has become the accepted norm that such high-resolution data are acquired by means of a 300 m or 600 m streamer, usually with 12 or 24 channels. An example of the specifications set out by clients is given in Table 1.

Various companies with requirements of this kind have developed or introduced systems which are usually termed high resolution — e.g. Moore and Roy (1981) — and which are aimed at acquiring data with a resolution of the order of 10 m (i.e. 8–10 m in the strata just below the seabed). This still leaves a very large gap between standard digital data and the ultra-high resolution obtainable with single-channel analogue equipment, which can often give resolution of the order of 30 cm. Especially in areas like the North Sea where the penetration of such analogue profilers is limited — sometimes down to just one or two metres below seabed, and at best down to 50–60 m — there is a clear need for the ability to improve the resolution of the multichannel data, especially for engineering purposes.

TABLE 1
Typical specifications for a North Sea site survey

Sampling rate	1 ms
Record length	1.5 s
High-cut filter	248 Hz/72 dB/octave
Low-cut filter	8 Hz/18 dB/octave
Shot interval	12.5 m
Number of groups	24
Group length	25 m
Group interval	25 m
Cable depth	3 m

GENERAL CONSIDERATIONS

It was with this need for improved resolution in mind that a digital ultra-high resolution package, specifically intended for engineering considerations, was developed.

One immediate improvement which can be made is to increase the sampling rate to either 0.5 or 0.25 ms, thus ensuring that higher frequencies are included in the data set.

However, this is not of much benefit if other factors limit the high frequency content, such as the response of the streamer or ghost effects due to the tow depth of the streamer and/or source. In general, with a good set of data, resolution can be improved by various techniques at the processing stage. The most effective way of doing this is by careful attention to the deconvolution parameters in order to improve vertical resolution, and by migration of the data to improve lateral resolution. However, these options are ways of maximizing the information which can be obtained from the data, and can only improve the resolution within the limits imposed by the acquisition process. Also, to a large extent the time and money spent on the processing are often outside the control of the contractor. So the obvious question is: How can we improve the *potential* resolution of our data by changing the acquisition parameters? These obviously include improvement in the seismic source and recording, but our main efforts were concentrated on the receiver system.

A NEW MULTICHANNEL HYDROPHONE ARRAY

The ideas for the design of this new array — referred to as the 'mini-streamer' — came about gradually in the practical atmosphere of site survey work on board ships! One disadvantage of multichannel systems as opposed to single hydrophones is the time taken for a signal to arrive at all the hydrophones in a given group. This means there

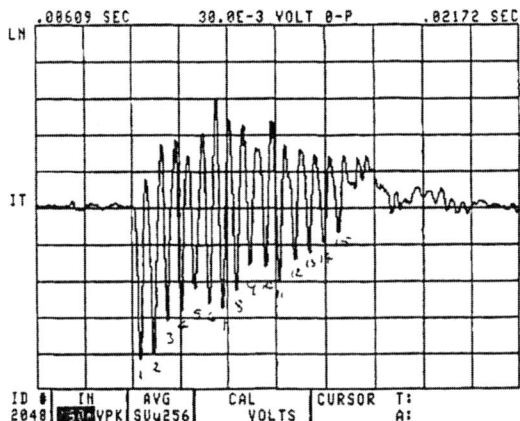

Fig. 1 A record of the direct arrival at the first trace of a standard streamer. Note the response of each of the 15 hydrophones

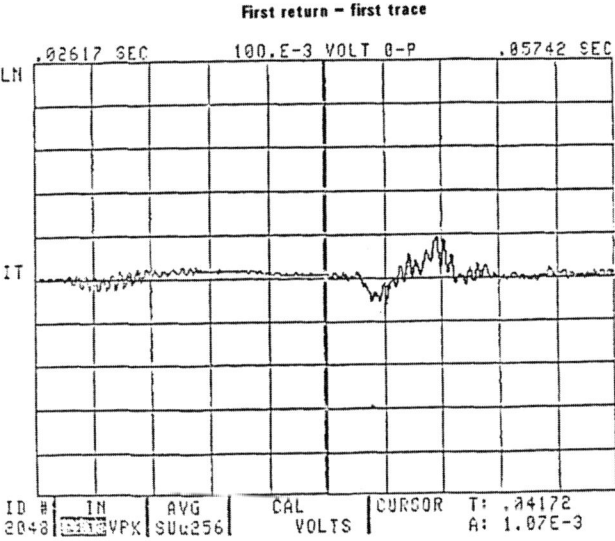

Fig. 2 A record of the first return at the first trace of a standard streamer. The response of each hydrophone, while still distinguishable, is somewhat smoothed out compared with those in Fig. 1

is a time difference between the first and last hydrophone in the groups, which has the effect of broadening any incoming pulse.

Figure 1 shows the direct arrival at the first trace of a standard streamer, and the response of all 15 hydrophones present can be clearly seen. Figure 2 shows how this effect is somewhat smoothed out for the first return at the same first trace, but clearly this effect is going to influence the resolution of the data. So ideally we would like to shorten the group length, while retaining as many hydrophones as is physically possible. Figure 3 shows the time difference in milliseconds between the arrival of a signal at the start of a group and at the end of a group for various group lengths and at selected target depths. Clearly, as the group length gets shorter, the change in the time delay becomes less significant, and we opted in our design for an array length of 2.5 m. With such an array, the time difference for a target at 50 m is 0.04 ms compared with a time difference of 1.03 ms for a standard 12.5 m group length.

Our initial intention was to try to fit 11 hydrophones per group, but as this proved impracticable we decided on 7 hydrophones per group. The sensitivity is 5.24 V/bar, and the outside diameter of the streamer is 4.763 cm. A group interval of 12.5 m was chosen so as to make the system compatible with a

standard streamer with group length 12.5 m, but mainly it was chosen to retain moveout within the confines of a six-channel system. The benefits of doing this have become apparent as will be explained in the section on site surveys performed so far with the mini-streamer.

When the design parameters had been chosen, the next step was to do some computer modelling based on this design to confirm that, at least in theory, we would be achieving our objectives. Figures 4 and 5 show the pattern response for a standard streamer (12.5 m group length) and for the mini-streamer. The improvement is striking, with the first notch for the mini-streamer occurring at a K value of 340 cycles/km compared with a value of 75 cycles/km for the standard streamer. Figures 6 and 7 show a combined response diagram for source and streamer. The parameters used in this particular example were: zero reflection time 80 ms; stacking velocity 1500 m/s; a mini-sleeve exploder source consisting of three elements at a depth of 0.6 m; and a streamer depth of 3.0 m. There is a clear improvement in the case of the mini-

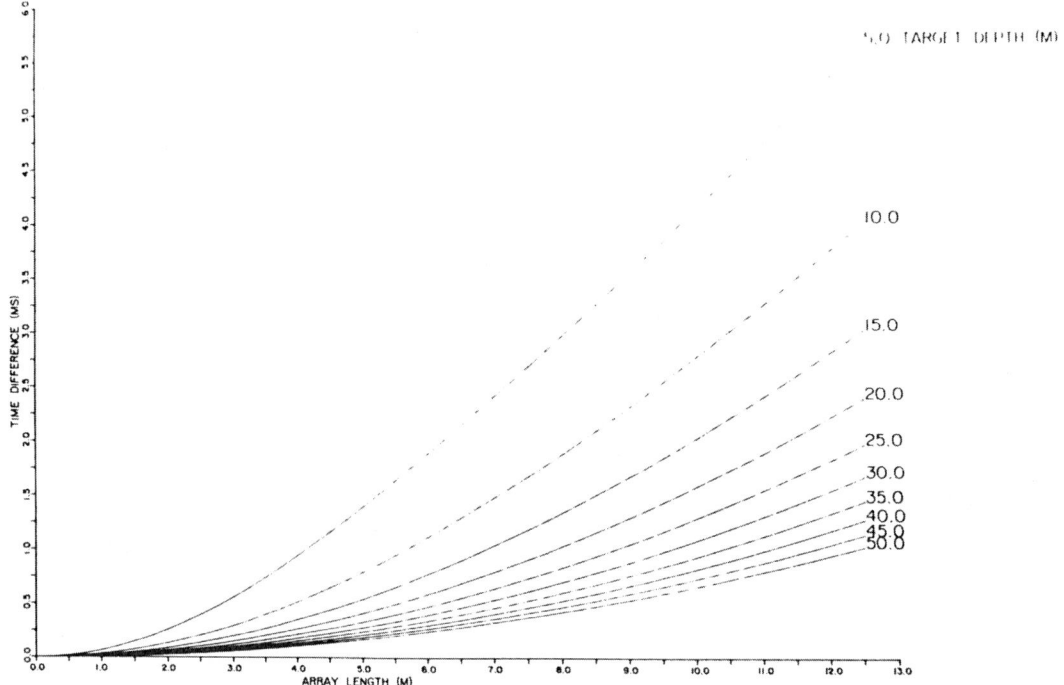

Fig. 3 A geophysical representation of the difference in time between a signal reaching the start of an array and the end of the array for various array lengths and for different target depths

streamer, particularly since its optimum tow depth is not 3 m, but 0.75 m, at which latter depth there is yet further improvement. Note especially the very low attenuation up to 1000 Hz for offsets less than 50 m.

THE TOTAL HIGH-RESOLUTION PACKAGE

Before describing the mini-streamer sea trials, there follows a brief description of the

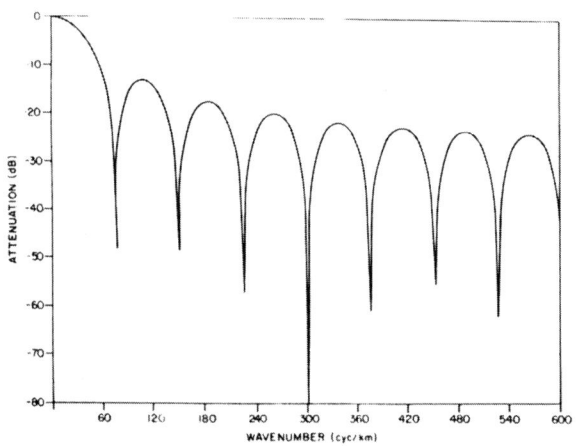

Fig. 4 Pattern response (attenuation versus wave number) for a standard streamer of group length 12.5 m

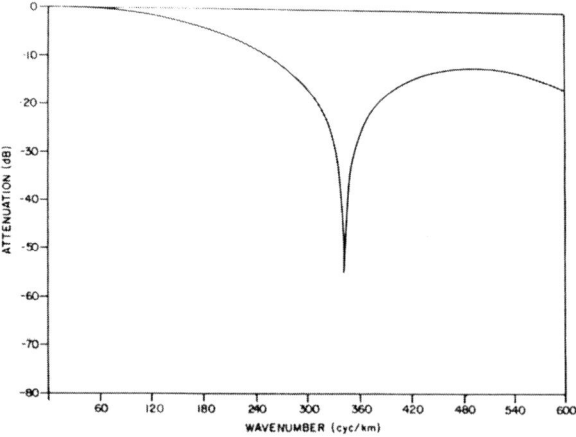

Fig. 5 Pattern response as in Fig. 4 — this time for the mini-streamer with group length 2.5 m

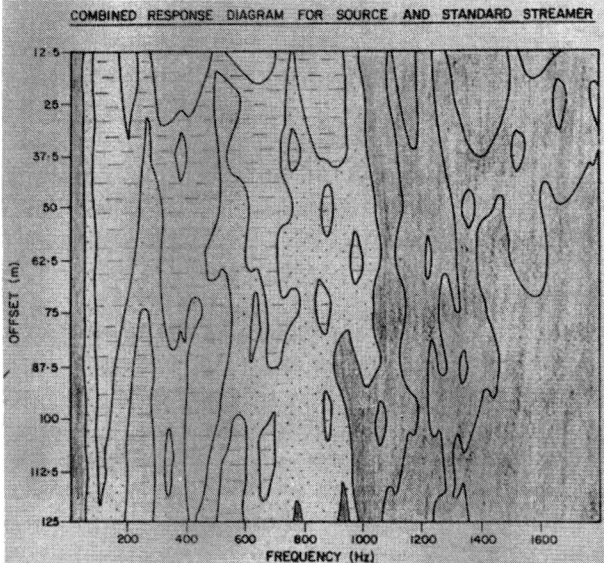
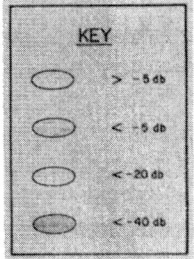

Fig. 6 Combined response diagram for a source at 0.6 m, and standard 12.5 m group length streamer at 3.0 m depth. Zero reflection time is 80 ms, stacking velocity 1500 m/s

complete system used both in these trials and in the site survey work to date. The only source we have used has been a mini-sleeve exploder — by now well known and accepted in the industry as a reliable high-frequency source. It produces a 1 ms wide initial pulse, followed by a very small bubble oscillation 20 ms later. The Fourier transform of the pulse reveals a flat spectrum from 60 to 1000 Hz.

In principle, of course, any source can be used which provides a broad-band acoustic pulse with a very narrow pulse width.

The recorder used was a modified Texas

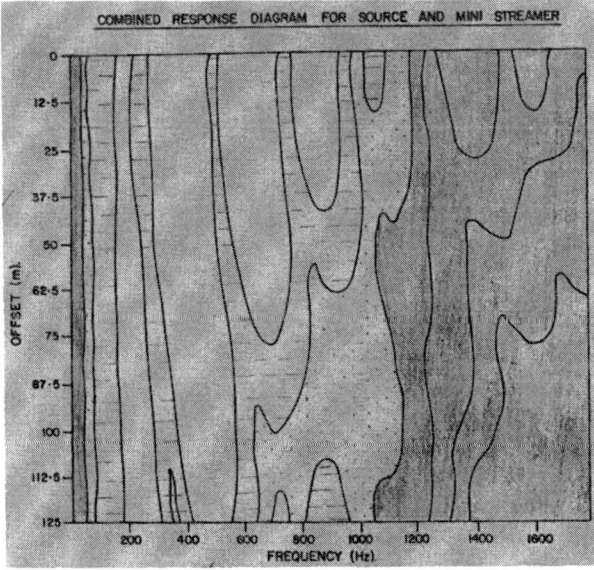
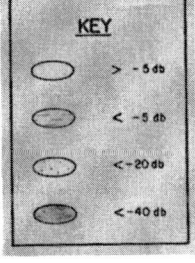

Fig. 7 Combined response diagram for a source at 0.6 m, and the mini-streamer at 3.0 m depth. Zero reflection time is 80 ms, stacking velocity 1500 m/s

Instruments DFS V recorder, which could sample simultaneously 6 channels at 0.25 m and 24 channels at 1.0 m.

SEA TRIALS

In November 1983 the first mini-streamer was ready for testing, and so a test-line was run in the Central North Sea. Figure 8 shows a section of this test-line recorded at a sampling rate of 0.25 m. The recorder filter settings were 27–1024 Hz, while at the processing stage the high-cut filter was set at 800 Hz, as there was information present up to this frequency. There are several interesting features which were encouraging on this first trial: the sharpness of the reflectors, especially at around 890 m (13 m

below seabed), which we believe to relate to real changes in sediment type; the bubble effect (seen at 100 m is not too serious and has been fairly well removed by the processing; finally, the pinchout seen towards the left-hand side of the section is extremely well resolved. The lower reflector is so strong that it almost certainly represents a hard layer, and we can follow it right up to the pinchout at which there is a vertical separation between the layers of 3.0 m.

Moreover, the vertical displacement of about 2 m at the left of the pinchout, which would not have been apparent if we had used a standard streamer, is not a fault but a buried relief feature, believed to have been caused by the sub-cropping hard layer. Such an interpretation on seismic data alone owes much to the very high resolution achieved.

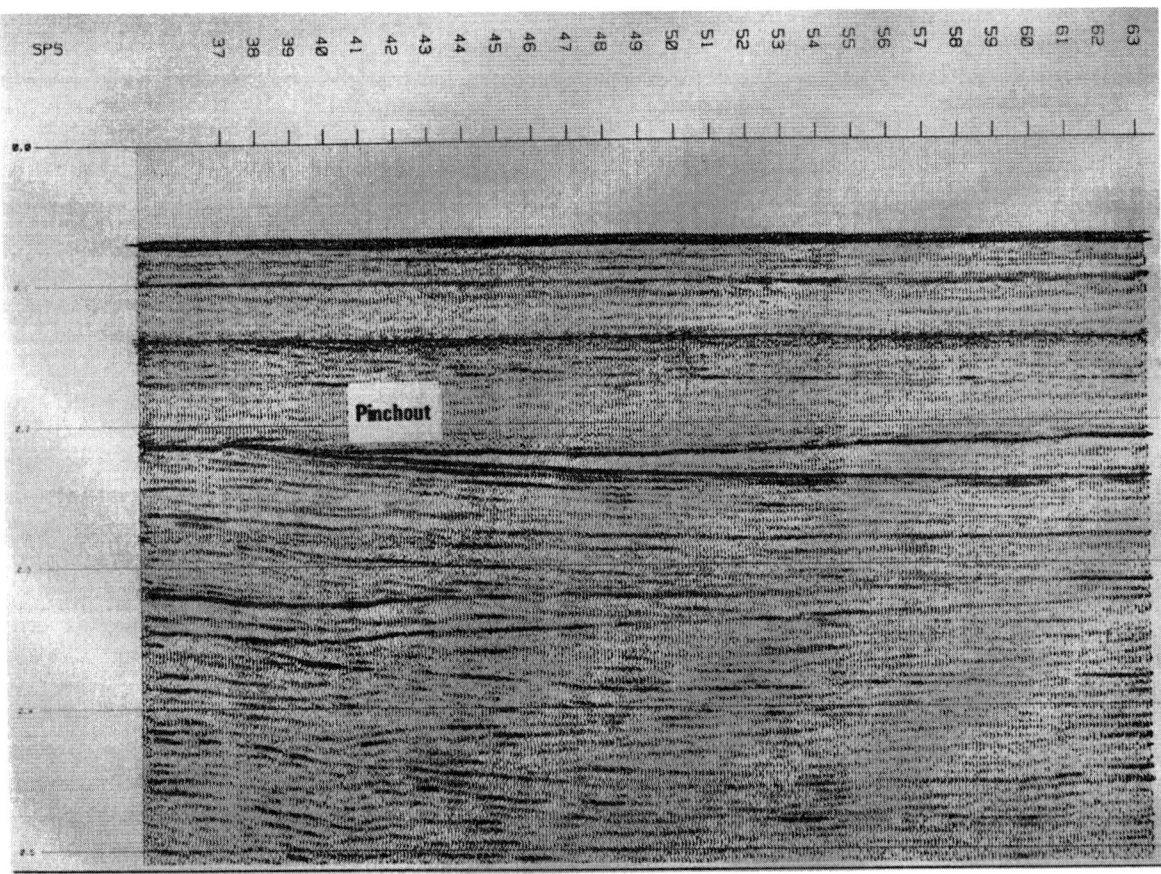

Fig. 8 A section from the Central North Sea using the mini-streamer with a sampling rate of 0.25 m. The high-cut filter at processing was set at 800 Hz to include all significant frequencies present

The system was then used for a more extensive series of trials in the sheltered waters of Spey Bay. In these trials we set out to obtain direct comparisons between ultra-high resolution analogue data, very high resolution 0.25 m mini-streamer data, and high resolution 24 channel, 600 m streamer data. To do this the same 5 km stretch was used for running separate lines with a sub-tow boomer, the mini-streamer and a 600 m streamer.

Figure 9 shows a section taken from the analogue data. Three reflectors, A, B and C, can be clearly identified at depths of 8 m, 18 m and 24 m below seabed respectively, before the multiple obscures any further penetration.

Figure 10 shows the same section at approximately the same horizontal scale taken from the mini-streamer data. Again, the same three reflectors are clearly identifiable. As would be expected, we obtained penetration below the multiple (which has

been well suppressed by the processing) compared with analogue data, but the main point of interest is that we were able to obtain such good correlation between analogue and digital data.

This gave us confidence in identifying real events on the mini-streamer data, and suggested that in certain cases the analogue data may have been superfluous, though in general they served as both a profiling technique and a sort of calibration tool for the digital data. Note also that the bubble effect of the mini-sleeve has been effectively removed by the processing.

Figures 11 and 12 show the same sections on the mini-streamer and standard streamer data. The horizontal scales are the same, but note the vertical scales. There was little penetration achieved below 0.5 s, even with the 600 m streamer, but the data present offer a good comparison between the two systems. In particular, the reflectors A, B and C cannot be identified on the 600 m

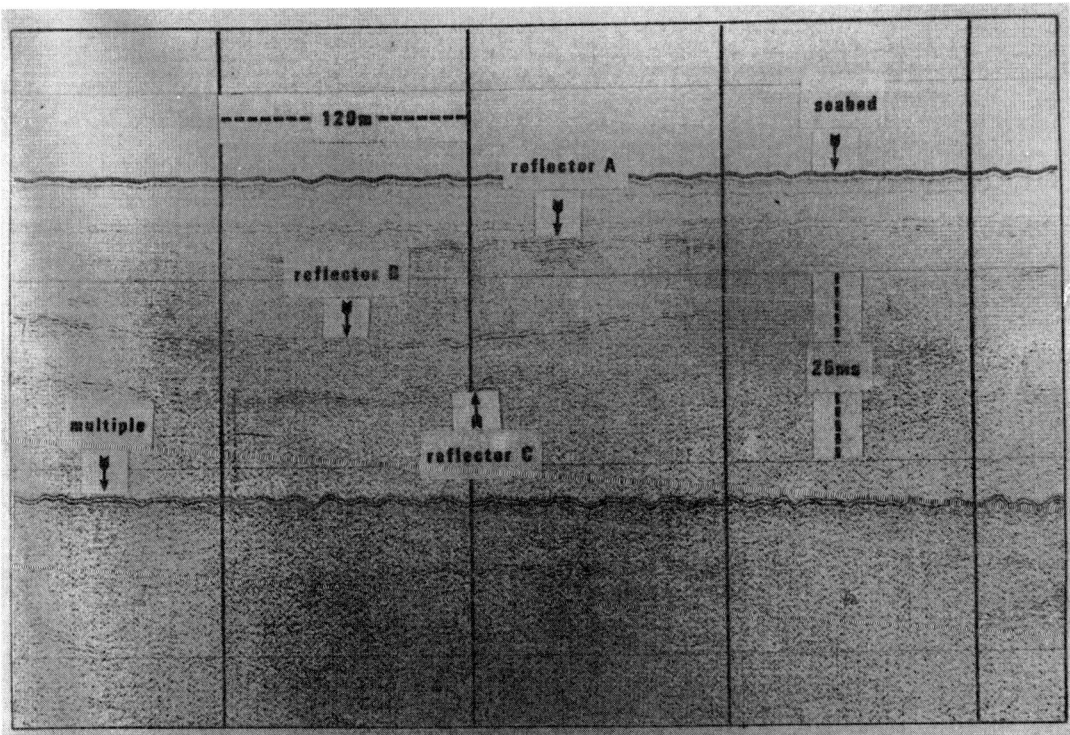

Fig. 9 Section of a sub-tow boomer record from trials in Spey Bay. Note reflectors A, B and C, which occur at depths of 8, 18 and 24 m below seabed

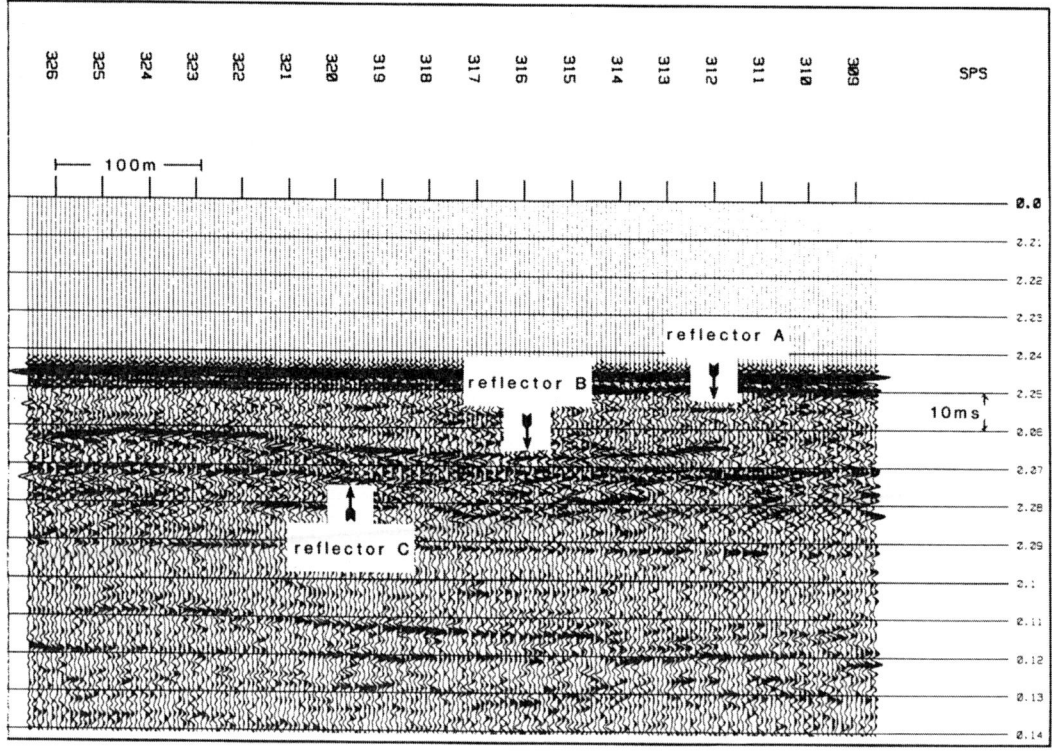

Fig. 10 The same section as Fig. 9 but recorded on mini-streamer, sampled at 0.25 m. The same three reflectors, A, B and C, can be clearly identified showing the very high resolution obtainable

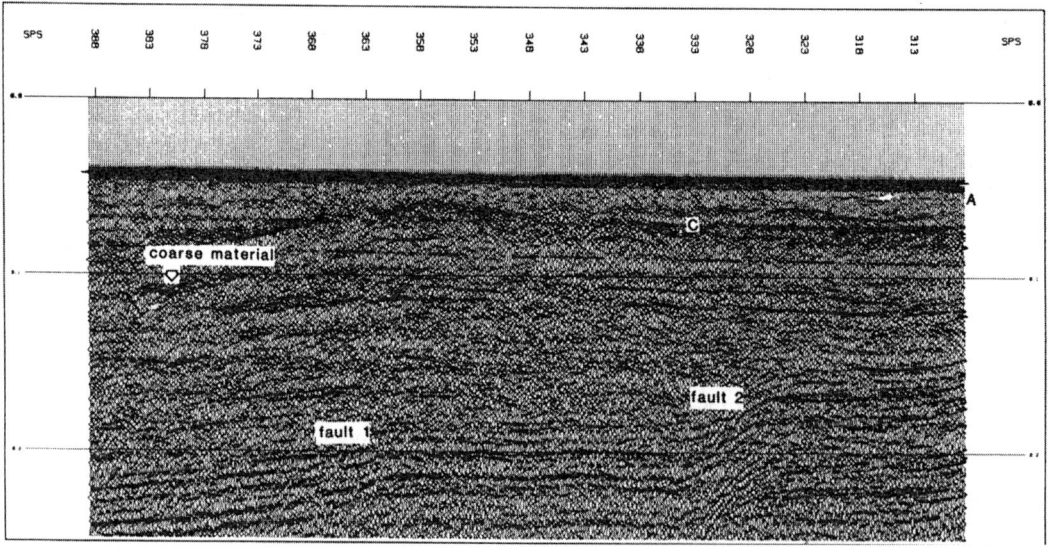

Fig. 11 The same line as Fig. 10 obtained using the mini-streamer, but reproduced at a scale chosen to display a larger section with penetration of 250 m. Note the two faults, and the high-amplitude reflector which probably represents a deposit of coarse material

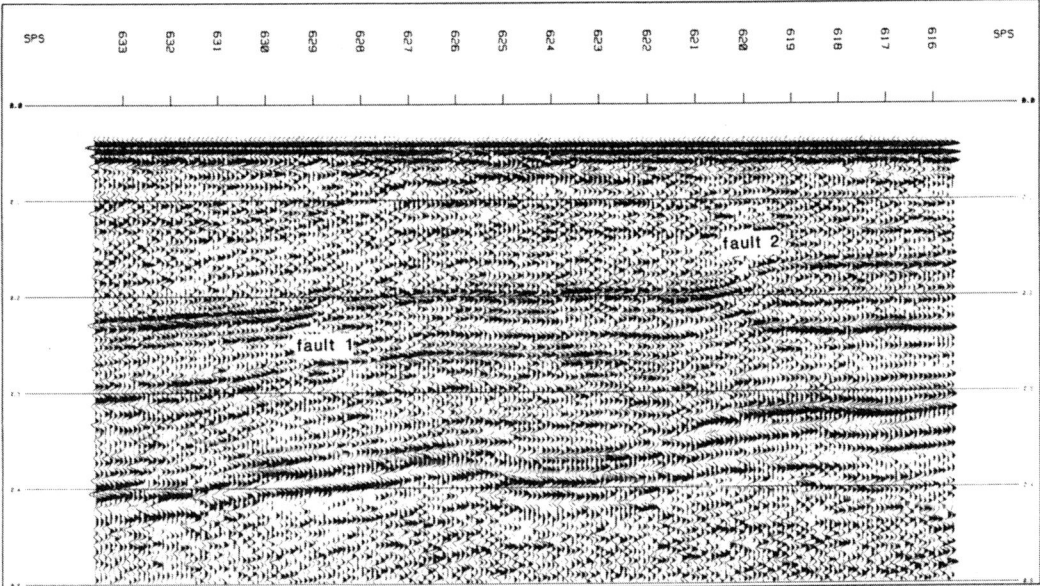

Fig. 12   A section covering the same line as Fig. 11, but this time obtained using a standard 24-channel 600 m streamer. Note the same faults identifiable on both sections, but the difference in resolution as highlighted by the absence of any real detail, particularly in the top 100 ms

streamer, and the two faults, although clearly visible on the 600 m streamer data, appear in much greater detail on the mini-streamer section.

One more line trial was run in a totally different area — the results again being very satisfactory — before it was felt that the system was sufficiently proven to be offered to clients on a commercial basis.

SITE SURVEYS PERFORMED WITH THE MINI-STREAMER

Since the mini-streamer came into operation for site survey work in the summer of 1984, nine site surveys have been completed using the new high-resolution package in one form or another.

In all of these data quality has exceeded expectations. Five of these site surveys were conducted using the mini-streamer as the first active section of a conventional 12.5 group length streamer. The data consisted of 6 channels sampled at 0.25 ms and 12 or 18 channels sampled at 1.0 ms. The success of this system of dual sampling and recording

has justified the choice of a 12.5 m group length for the mini-streamer. To limit the number of illustrations, only two of these sites will be shown here, but the same general conclusions and quality of results apply to all five sites. The other four sites were platform site surveys and ideally suited to the resolution capability of the mini-streamer. These four sites were surveyed using the mini-streamer as the sole digital acquisition system. Again for brevity, only one of these sites is described in detail.

JOINT MINI-STREAMER AND STANDARD STREAMER SURVEYS

Of the five site surveys carried out using the mini-streamer as the first active section of a longer streamer, four were located in the Southern North Sea and one north of the Shetlands. In the latter case, the water depth of 165 m is too deep to expect optimum results with the mini-streamer. Despite this there was a marked improvement in resolution as far down the section as the multiple, probably mainly due to the 0.25 m sampling rate.

In one of the Southern North Sea sites surveyed — which was otherwise not spectacular geologically — the most prominent event was a fault at the top of the Upper Cretaceous chalk with a diffraction pattern at c. 700 m. At this horizon the diffraction patterns from the edge of the fault plane could be seen on the 1.0 m (standard streamer) section. However, on the 0.25 ms data these diffractions did not appear — almost certainly due to the short group length of the mini-streamer — thus the sharpness of the reflectors at this level was increased. Another interesting feature was the degree of penetration achieved with the six-fold 0.25 m data — down to 700 m, and in fact with some data present down to the end of the record at 1.0 s. In the region 0–400 m, the mini-streamer section was clearly superior in terms of resolution and detailed information obtainable. The top 300 m of this consisted of flat-lying Quaternary sands and clays with some very minor channelling. Immediately below this a sequence of Tertiary shales and sandstone was cut by channelling which could be more clearly defined on the mini-streamer section. However, below this channelling and down to the top of the chalk very little could be seen on the 0.25 m data compared with the 1.0 m data. Hence for this site the ideal use of the two systems would be to use the mini-streamer data down to 400 m and the standard streamer data for the rest of the record, while also using the better definition present around the top of the chalk at c. 700 m from the mini-streamer section.

Another rather geologically uninteresting site surveyed in the Southern North Sea consisted of dense sands with occasional thin clay layers down to an erosion surface at about 180 m which was possibly base Tertiary. Beneath this our records showed a thick sequence of mainly discontinuous horizontal reflectors with a few depositional features such as onlap being visible.

On the mini-streamer record there were some minor reflectors, e.g. at 20 m and 70 m below seabed, which probably corresponded to the clay layers referred to above. In fact the reflector at c. 20 m could be identified on the analogue records. There was useful data present on this section down to about 600 m, which is more than adequate for engineering investigations. By comparison, on a conventional streamer record of the same section the shallow reflectors could not be resolved, and the erosion surface was not as clearly delineated. However, we did achieve better penetration on this section, so if we had been interested in both the shallow features and the deeper structures the ideal section would have been a combination of previously recorded data. Such a section was produced by merging 0.25 m data down to 500 m with the 1.0 m data below 500 m. The merging parameters were determined by routine testing during processing. Hence, if dual sampling is performed, a final section can be produced to order with the proportion of very high resolution data varied to suit the geology of the area.

MINI-STREAMER ONLY SURVEYS

Four platform site surveys were successfully completed using the mini-streamer as the sole digital system; one of these carried out in the Southern North Sea will now be presented in detail.

Figure 13 is a typical east–west section across this site, and the record highlights the fact that the geology of the area is dominated by a graben. This feature trends broadly north–south through the site and at its widest part is over 1500 m across. It is interpreted as a typical example of a collapse structure caused by salt dissolution (Jenyon 1984). The strong reflector at c. 350 m is the top of the Upper Cretaceous chalk. This chalk, particularly on its eastern side, dips towards the graben, but then shallows along the edge of the graben to produce a rim syncline on many lines. On the western side of the graben the chalk is shallower than to the east, due to the irregular move-

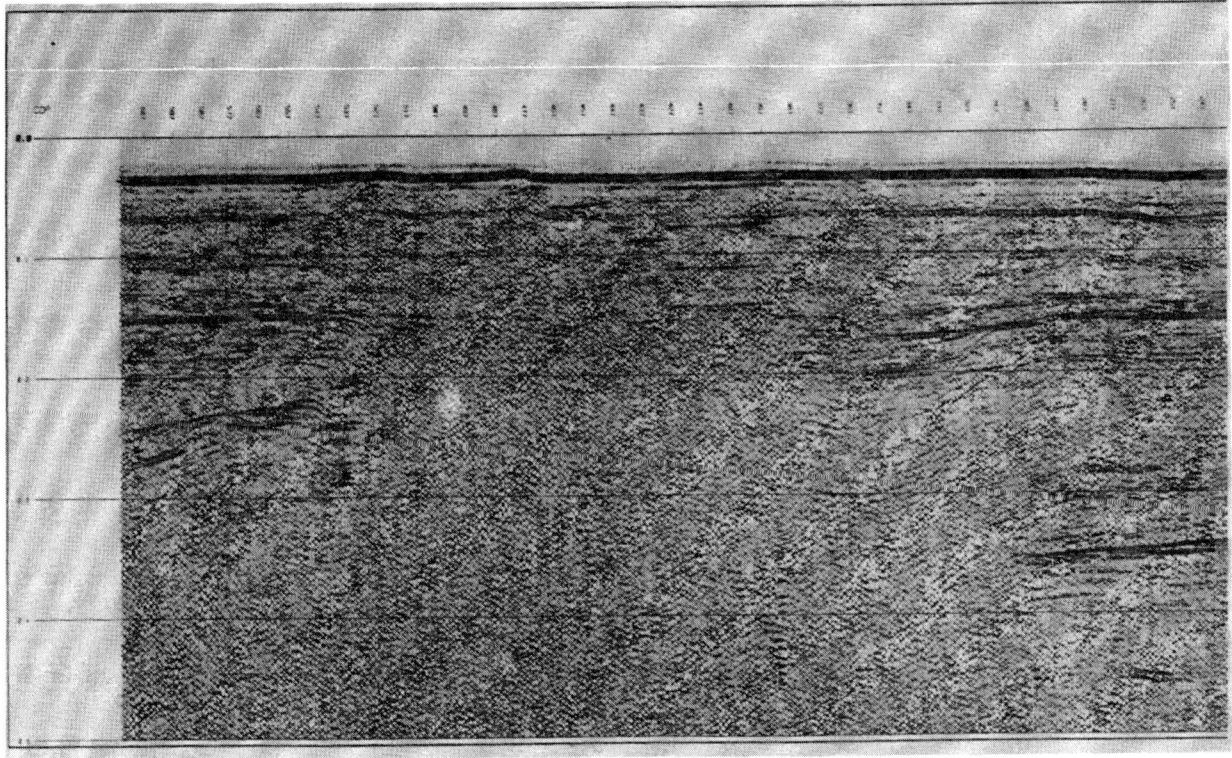

Fig. 13   A section from a platform site survey in the Southern North Sea performed using the mini-streamer as the only digital system. See text for explanation of the geology

ments of the underlying salt. This has been called the 'trapdoor effect' (Trusheim, 1960) and is clearly demonstrated in Fig. 13.

The origin of the graben lies in the history of the underlying salt movement in the area. It is believed that, under the effect of a considerable overburden, salt beds may be triggered into movement, probably by fault activity. Various stages in the growth of the resultant diapirs are recognized. The morphology of the salt on this site is that of a 'pillow'. The updoming of the salt produces extensional faulting in the overlying strata, particularly the brittle chalk. According to Jenyon, the resulting fracture system will allow access of ground water to the salt, thus leading to its solution. This in turn causes collapse of the overlying strata into the incipient void, thus forming the graben seen in Fig. 13. The graben remained active well into Tertiary times.

Sometime during the mid-Tertiary, an important erosive phase developed. This is seen as a strong sub-horizontal reflector at c. 150 m cutting across underlying deposits.

Figures 14 to 16 are also taken from this site, and exemplify some of the more interesting features seen on the mini-streamer sections.

Figure 14 shows a direct comparison between the analogue and digital sections for the same line. It is interesting to see how easily the sand waves can be identified on the digital data: the maximum height of these sand waves is about 4 m.

Below the seabed on this site we see some strongly channelled sequences to depths of 30–40 m, and in places up to 60 m. The base of such channels is well defined and consistently outlined, even where it passes below the water bottom multiple. This is good evidence that the weathering has been essentially mechanical rather than chemical.

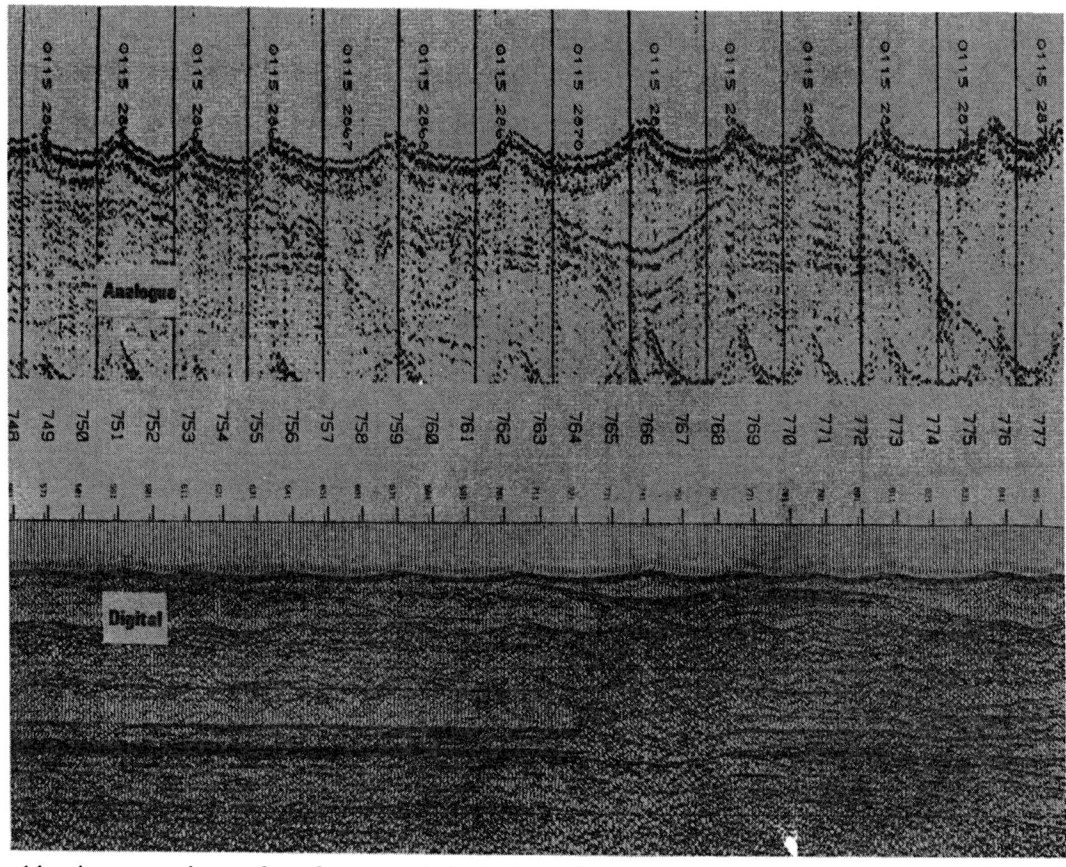

Fig. 14 A comparison of analogue and digital sections from the same site as Fig. 13. Note the definition of the sand waves on the digital section, and the channelling in the centre and on the right-hand side of the sections just below seabed.

Furthermore, the infill is structureless with no evidence of slumping. The inference is a homogeneous sand-filled channel, cut into firm clay or perhaps a dense granular sediment. It can be seen from Fig. 14 that such features can be clearly followed on both analogue and digital lines. The latter often shows the channels to be associated with very strong energy scattering and a velocity anomaly indicating lower velocities within the channels. Since there is little evidence to suggest organic deposits, the alternative is an interpretation of loose gravel and pebble-bearing granular sediment. This would then contrast with a more compact clay or dense sand in the channel sides.

Along the axis of the graben is a zone of 'bright' spots. These are local high-amplitude reflectors which are clearly phase reversed compared with the seabed pulse: an example is given in Fig. 15 where the bright spot can be identified on both analogue and digital lines. In the environment visualized they are almost certainly caused by organic-rich deposits. This may mean anything from a small organic fraction, through peat beds, to tiny gas pockets, although peaty clay or peat is the most likely.

There is a strong correlation between these peat deposits and acoustic masking associated with the graben axis. This masking appears to originate at very shallow levels, and it may be that continued subsidence of the graben could have determined the site of low-lying marshy ground suitable

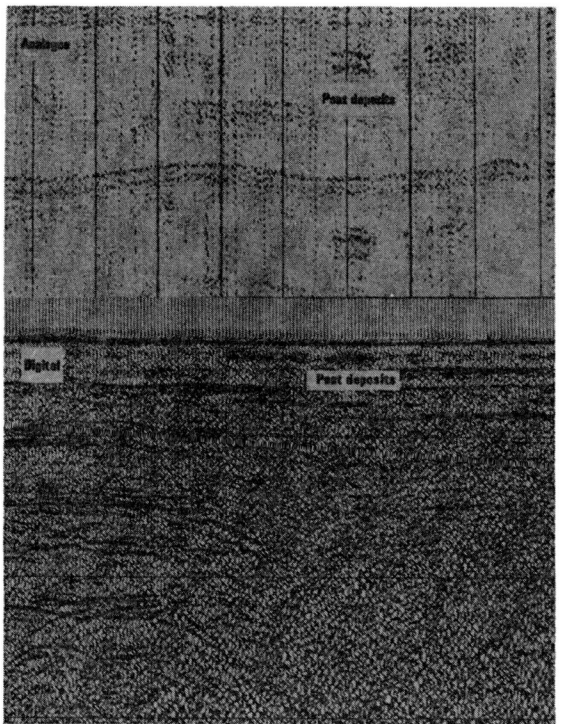

Fig. 15 An example of phase-reversed 'bright' spots on analogue and digital lines from the same site as Fig. 13

for peat formation. Nevertheless, there remains some possibility that the bright spots are gas pockets, originating from upward gas migration along the graben axis.

One other feature of interest is a possible Danian outlier present on top of the chalk in the rim syncline east of the graben (see Fig. 16). Due to the high resolution of these records, the outlier can be distinguished from the Upper Cretaceous chalk to a separation of less than 4 ms.

From the wealth of mini-streamer data at this site, we were able to make specific engineering recommendations to the client. In particular we advised that features such as the channels, peat and sand-wave areas and the major graben structure be avoided. We then outlined areas within the site which would be more favourable for platform siting than the original location proposed. The recommendations were acted upon and the location was changed.

Fig. 16 A section from the same site as Fig. 13 showing the presence of a Danian outlier in the rim syncline of the chalk

CONCLUSIONS

The first season with the new high-resolution system is complete. The main conclusion is that we believe it has achieved its original objectives and done much to bridge the gap known as the 'twilight zone'. Some useful lessons have been learned too: in particular that the shots must be accurately timed to ensure proper stacking. After processing we must re-educate ourselves to interpret data of such high frequency content and high resolution. Ideally there should be correlation with analogue data and/or borehole information, if we wish to achieve the maximum benefit from the data.

For the future we can offer 12 channels (i.e. two active sections) of the mini-streamer, giving an increased CDP fold and hence improving the signal-to-noise ratio. It will also increase the far offset distance, thus achieving more accurate velocity calculations and making multiple attenuation more effective.

It has become apparent during the course of these site surveys that there are many areas where the mini-streamer would be a most useful tool. Primarily, of course, we have demonstrated its value as an engineering aid in platform site surveys, but another use, highlighted by the last site discussed, is the detection of shallow bright spots, particularly where there is a risk of gas.

Depths of around 50 m below seabed are often too deep for analogue equipment, and yet standard digital seismics would most likely fail to resolve features at such depths.

The mini-streamer can also be used in areas where shallow channelling or deposits of coarse material are expected to cause problems: the results shown here clearly demonstrate that such features can be identified with the new high-resolution system.

ACKNOWLEDGEMENTS

Work on the design and development of the mini-streamer was carried out by A. J. S. Ogilvie who has also provided much help in the preparation of this paper. The data illustrated here were all processed by DPTS Ltd of Sevenoaks to whom we are indebted for their considerable effort and expertise. Finally, thanks are due to all those colleagues at Gardline who have been involved with this work and without whose skill and dedication such high-quality results could not have been achieved.

REFERENCES

1. Arthur, J. C. R. 1980. *Offshore Site Investigation* (Ed. D. A. Ardus). Graham & Trotman, London, pp. 77–86.
2. Jenyon, M. K. 1984. Seismic response to collapse structures in the Southern North Sea. *Mar. Petrol. Geol.* 1(1), 27–36.
3. Moore, N. A. and Roy, C. H. 1981. A high resolution marine data recovery technique for stratigraphic exploration. (Abs) *Geophysics* 46(4), 449.
4. Trusheim, F. 1960. Mechanism of salt migration in Northern Germany. *AAPG Bull.* 44(9), 1519–1540.

An Integrated Approach to the Investigation of New Development Areas

C. D. Green, B. Heijna and P. Walker, Shell Internationale Petroleum Maatschappij BV, The Hague, The Netherlands

SUMMARY

1. Geophysical, geotechnical and petrophysical data have been synthesized into a workable geological history of the Troll Field, Block 31/2 area.
2. Well-marked soil units exist within the study area with no large differences in lithology or soil properties. The integration of geophysical and geotechnical data permits lateral interpolation of soil type. This enables geotechnical predictions to be made within the resolution of the database.
3. Seabed pockmarks have been mapped and analysed and a procedure to quantify their importance in engineering terms has been proposed. This procedure will express 'pockmarks' in terms of intensity versus age and depth, and pockmark return period. This will express pockmark risk as design environmental forces for incorporation into design procedures.

INTRODUCTION

The development of the Troll Field in the Norwegian Trench, Block 31/2 (see Fig. 1), would require the installation of a production platform at a water depth in excess of 300 m with associated templates and flowlines. Such a development would require extensive knowledge of the sub-seabed soils in order to provide reliable predictions of geotechnical parameters for design and development.

With this objective, a study integrating high-resolution seismic soil investigations and geological data was carried out. Several sources of geophysical and geotechnical data are available in the area, but this paper deals primarily with data obtained from the seismic survey and soil investigations undertaken for the purposes of this study.

By using engineering seismic and soil boring correlation techniques the integration of both geophysical data and soil investigation

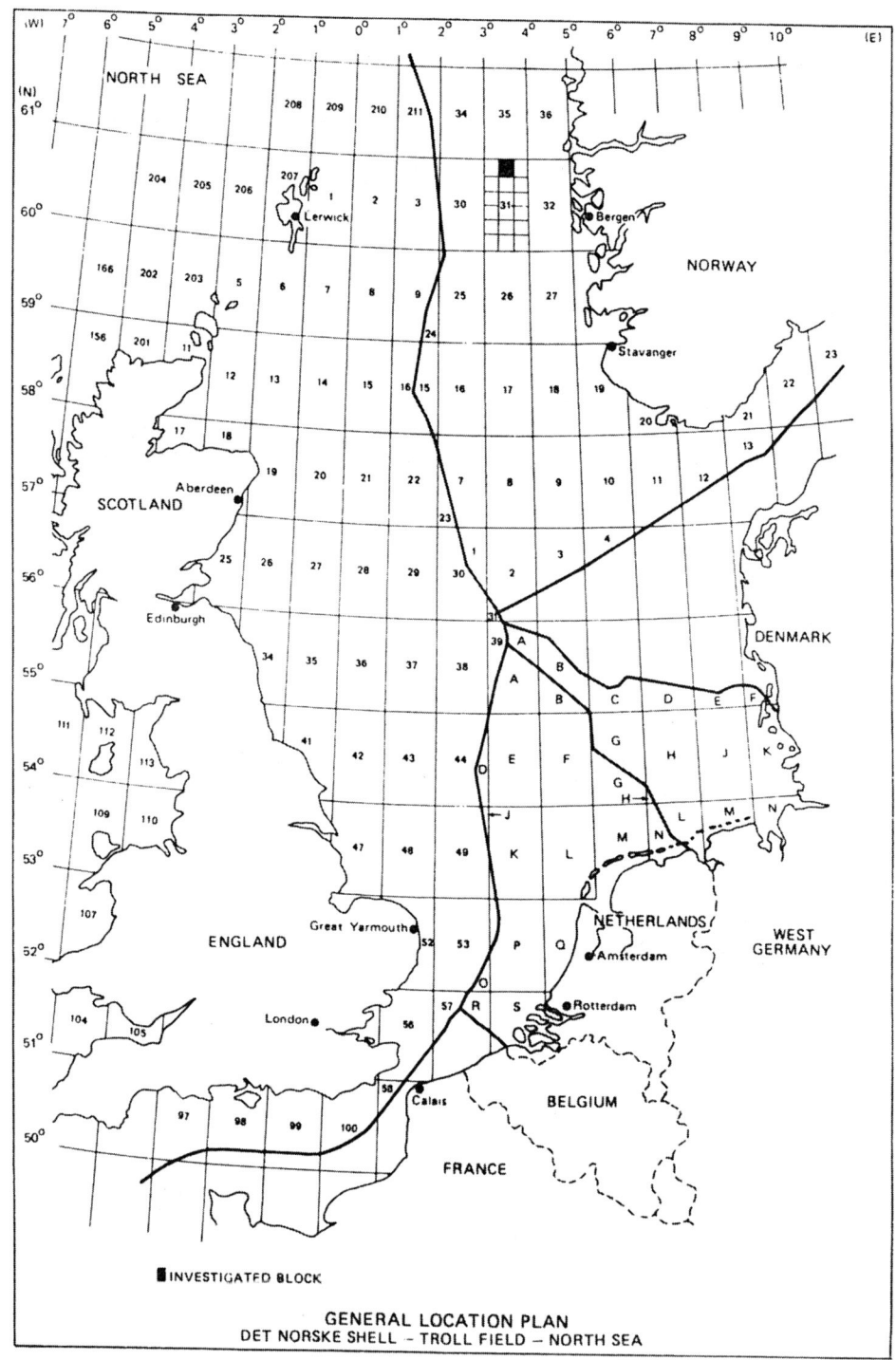

INVESTIGATED BLOCK

GENERAL LOCATION PLAN
DET NORSKE SHELL – TROLL FIELD – NORTH SEA

Fig. 1

results has been accomplished. Further integration with available geological data and results from earlier seismic and soil surveys has resulted in a workable geological model for the identification of seabed and sub-seabed soil lithologies, together with the soil properties required for general foundation engineering. In this paper a description of soil conditions over the area surveyed is presented. Furthermore, zones with potential problems are identified, mapped and discussed.

The objectives of the study were to integrate the available geophysical, geotechnical and geological data in order to determine the composition, thickness, lateral extent and geotechnical conditions of the sub-seabed soils, and therefore to aid in the selection of optimum foundation parameters by the accurate prediction of soil profiles. Such predictions should be possible not only at the specific location of the platform site but also at any development locations within the Troll Field covered by the data grid.

GEOPHYSICAL, GEOTECHNICAL AND GEOLOGICAL DATA

Geophysical Data

Geophysical data were obtained using the mini-sleeve exploder seismic source. Eight channels of data were recorded through a 10 m group length/interval streamer onto a DFS-V recorder with a sampling rate of 0.25 m/s and analogue peripherals (see data sample, Fig. 2). In addition, a second pass over the grid was performed using a Huntec deep tow boomer (see data sample, Fig. 7) and a side-scan sonar.

Horizontal control was achieved with a Syledis integrated with a Simrad HPR acoustic system for both the side-scan sonar and Huntec tow fishes. Data obtained from rig-site surveys and a 1981 survey by Gardline Surveys (MSE) have also been included in the study.

The objectives of the high-resolution engineering seismic survey were:

(a) to provide site-specific geophysical information at the selected platform location;
(b) to provide a regional reconnaissance and appraisal of seabed and sub-seabed soil distribution over the likely development area in the Troll field;
(c) to provide an optimum selection of site investigations for soil calibration;
(d) to provide information on problems and obstructions within the area;
(e) to provide a geophysical data-base outlining major physical features and environments, major soil types and their distribution;

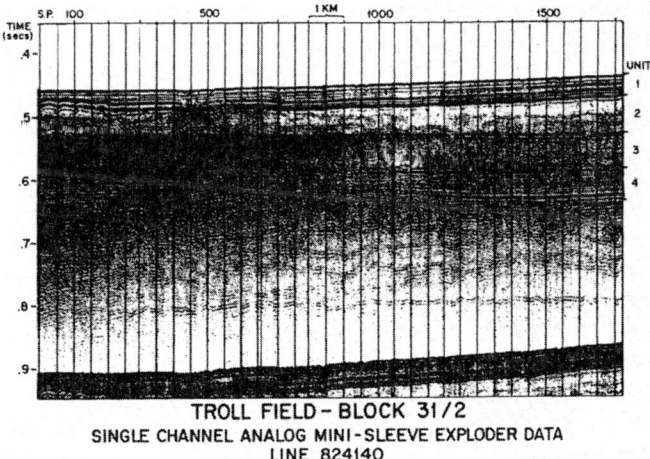

TROLL FIELD – BLOCK 31/2
SINGLE CHANNEL ANALOG MINI-SLEEVE EXPLODER DATA
LINE 824140

Fig. 2

(f) subsequently to integrate the results of the soil investigation so that the data-base would provide detailed information for use in both the conceptual and design phases of seabed production facilities and flowlines.

Geotechnical Data

The 'primary' soil investigation was undertaken by Fugro in the summer of 1983 from the drilling vessel MV *Bucentaur*.

The investigation comprised five borings in the site-specific area and a further four borings in the general area at sites selected from the seismic data. The depth attained varied from 29 to 149 m below the mudline. Samples were recovered after each boring, and restricted laboratory testing, X-ray analysis and geological logging was carried out on board. Extensive laboratory work was performed later by NGI and Fugro on land. At selected intervals within the boreholes, quasi-static cone penetration tests were undertaken, along with borehole vane tests, pressuremeter tests and ambient pressure sampling. In addition, continuous quasi-static cone penetration tests and vane tests were made from the mudline in the area of the platform location, and a separate borehole was used for hydraulic fracture testing.

In addition, 15 continuous seabed cone penetration tests from the mudline to depths of up to 40 m below the seabed and two seabed vane tests from the mudline to depths of 22 m were carried out as the site-specific area.

The objectives of the final 'primary' soil investigations can be summarized as follows:

(a) to obtain site-specific geotechnical information at the platform location;
(b) to obtain regional information on soil conditions at selected locations before examining the stratigraphy and geotechnical properties of the soils in these locations;
(c) to analyse the results from laboratory

and *in situ* tests in order to determine the engineering properties of the soils;
(d) to use the soil data obtained to calibrate the detailed engineering seismic data;
(e) to investigate the features of geotechnical and geological significance identified from the seismic survey, such as pockmarks and seismic 'masking';
(f) to compute all the properties and provide reliable characteristic geotechnical properties for the soil units identified.

Data from previous exploratory soil borings completed in 1981 were also included in the study.

Geological Data

The available petrophysical log data can be divided into:

(1) data obtained in two soil borings during the primary geotechnical survey;
(2) data obtained when exploration wells were drilled in the area.

In two of the soil borings, SB3 and GBA, downhole logging was undertaken. The items measured were:

(a) gamma-radiation (gamma-ray log);
(b) hole diameter (caliper log);
(c) acoustic velocity in the nearby soil (velocity log).

All logs were run from the mudline to the maximum depth of each hole. Downhole logging was performed in all the exploration wells. The items measured were:

(a) gamma-radiation;
(b) spontaneous potential;
(c) resistivity;
(d) velocity.

The exploration well logs commenced at the end of the 30 in casing; therefore they were only useful for the correlation of lower soil units in the area.

The geological data consisted of analyses undertaken on samples obtained from the soil borings at the site-specific platform location. Unfortunately, the testing prog-ramme was restricted to those samples not required for further geotechnical analysis. This imposed constraints on both sample size and sub-seabed depth.

The samples were analysed for:

(1) dinoflagellate cysts and foraminifera by the British Geological Survey, Keyworth, United Kingdom;

(2) carbon-14 dating by the Low Level Measurement Laboratory, Harwell, United Kingdom;

(3) amino acid dating by the University College of Wales, Aberystwyth, United Kingdom.

Geological input from earlier investigations has also been included in the study.

GEOLOGICAL REGIONAL OVERVIEW

Introduction

In this section a brief overview of the geological history and Quaternary sedimen-tation of the near-seabed strata in the Nor-wegian Trench is presented. There are some discrepancies within the literature concern-ing these topics and these will be discussed in the following paragraphs.

Emphasis has been placed on the litera-ture available on the Norwegian Trench area. However, considerably more data are available on the English part of the North Sea (see Caston, 1977; Eden et al., 1978; Fannin, 1979, etc.). It is of interest that good correlation is possible between the English and Norwegian stratigraphy, despite the considerable distances separat-ing the areas. The reader's attention is drawn to Fig. 3 which is a time stratigraphic table including the major dates and geologi-cal periods and stratigraphy discussed below.

The Norwegian Trench forms a prominent morphological feature in the North Sea. Including the extension of the Skagerrak, it is about 900 km long, 80–90 km wide, with a maximum depth below sea-level of 700 m in the Skagerrak.

The origin of the Norwegian Trench has been a subject for much discussion. Accord-ing to Rokoengen and Rønningsland (1983) the formation of the Norwegian Trench was probably initiated by tectonic movements combined with fluvial errosion in the Pliocene. During the Quarternary period, glacial processes modified and accentuated the channel-like morphology. However, according to van Weering (1983), Sellevol and Sundvor (1973) and Floden (1972) the Trench was primarily formed by glacial ero-sion during the Quaternary ice ages and tec-tonics were of little importance.

Within the near-seabed strata of the Norwegian Trench area described in the lit-erature there is a well-marked angular unconformity. There is, however, some con-flict within the literature concerning the age of this unconformity. According to van Weering (1983), Floden (1972) and others it marks the base of the Quaternary sequence, but Rokoengen and Rønningsland (1983) have described it as being of mid-Pliocene age.

Since this paper primarily concerns near-seabed soils, this unconformity marks the depth at which the strata are discussed in detail. The depth of the unconformity varies from a few tens of metres sub-seabed near the coast and western flank of the trench to 220–240 m in the centre of the Norwegian Trench off Bergen and about 500 m at the shelf edge (Sellevol and Sundvor, 1973; van Weering, 1983). Below the unconformity the strata dips gently towards the west with an angle of dip of less than 5°. The stratification of the Quaternary sediments clearly indi-cates two main directions of transport, one across the trench and another parallel to the Trench. The transport across the Trench was glacial, while the transport parallel to it was more likely fluvial or glaciofluvial.

The greater the depth below the seabed, the less information there is in the literature on the seabed strata.

In general, most authors agree on the stratigraphy of the youngest sediments;

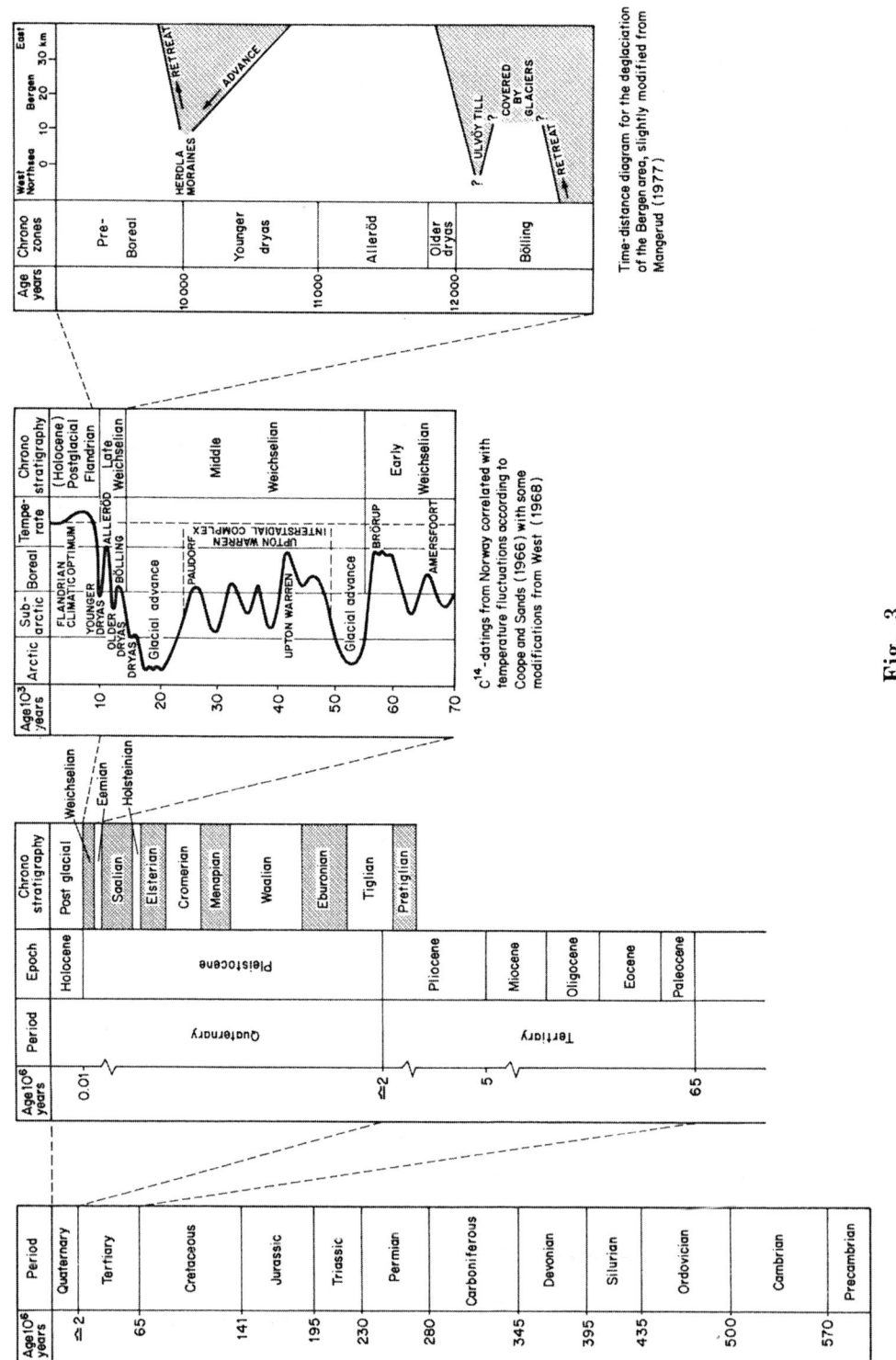

Fig. 3

Description	van Weering	Hovland	Rokoengen	NGI/Geoteam	This study
Laminated marine/ glaciomarine	Unit 1 Holocene	1A Marine	Soft silty Clay	1A Holocene	c. 8000 years BP 1A
Silty/sandy Clay	Unit 2 Late Weichselian	1B Marine / 1C Marine		Zone 1 / 1B Unit 1 / 1C Late Glacial	1B Unit 1 / 1C c. 15000 years BP
Iceberg furrows				—Reflector II—	
Glacial till stiff clay + chalk + flint boulders	Unit 3 no stratification Late Weichselian	boulder clay Quaternary	till-like material	Unit II norm. consolidated overlaying overconsolidated clay	-Int 1— Unit II -Int 2— Weichselian?
Very flat			Level B		
Glacial till Stiff clay boulders	Unit 4 no stratification Late Weichselian?	boulder clay Quaternary	(Base Quaternary) layered sediments Late Pliocene	Reflector IV Unit III overconsolidated clay + boulders	Unit III Weichselian or Saalian?
Layered clays with sand layers marine	not described	Quaternary	layered sediments Late Pliocene	Unit IV clay interbedded	Unit IV layered Early Pleistocene
Sand	not described	Quaternary	basal sand Late Pliocene	with sand	Unit V Early Pleistocene
		Base Quaternary	Level A	Base Quaternary	
Westerly dip strata	not described	Miocene	mid-Pliocene to Jurassic	claystone Miocene	Tertiary?

Fig. 4   Division of soil by different authors

however, the older the sediments, the greater is the disagreement regarding their stratigraphy (Fig. 4). Most authors divide the sediments into four to five major units. The most accepted stratigraphy is discussed in comparison to that of other authors. It should be noted that the unit numbers used for the stratigraphy are those in the literature and then do not necessarily agree with the soil units described later in this paper.

Van Weering's Model

Van Weering (1983) recognized four different units on shallow seismic data shot west of Bergen. These units have also been identified by him in the Skagerrak area.

Unit 1

The uppermost unit, unit 1, is acoustically transparent with either no or few weak internal reflectors. The unit is considered to be a Holocene mud deposit by van Weering, and is described by Rise and Rokoengen (1983) as a soft silty clay.

Unit 2

Unit 2 shows many parallel internal reflectors which can be traced over large distances. Van Weering interpreted this unit as a glaciomarine deposit of late Weichselian age. Rise and Rokoengen (1983) further subdivide this unit into: (a) an upper, thin silty sandy clay, which is well laminated and was deposited around 11 000 years BP; and (b) a lower, thicker laminated clay.

According to van Weering (1983), units 1 and 2 cover the deepest part of the Trench, thinning and eventually pinching out towards the sides. Hovland (1981) describes the two units, units 1 and 2, as a single unit (unit 1) which he divides into sub-units, 1A, 1B and 1C. Hovland's 1A corresponds to unit 1, and 1B and 1C correspond to unit 2 of van Weering (1983). According to Hovland, unit 1 (1A + 1B + 1C) is a marine deposit consisting of soft, silty dark clays with a high water and organic content. The fine laminae in the unit are believed to reflect variations in sedimentation rate and mineral composition. The material was supplied by winnowing from glacial drift deposits in more shallow water and by melting of the ice sheet (Holtedahl and Bjerkli, 1975).

The base of unit 2 is an irregular horizon. The irregularities of this horizon suggest that it is an erosion surface and that the furrows are the result of icebergs ploughing into the sediments as they touched bottom. These large icebergs were present during the deglaciation. This interpretation is agreed upon by all the authors (Belderson and Wilson, 1973; Hovland, 1981; Rise and Rokoengen, 1983; Rokoengen and Rønningsland, 1983; van Weering, 1983; etc.).

Unit 3

Unit 3 is characterized by the absence of internal stratification and by hyperbolic internal reflections (thought to be boulders). The base of this unit is marked by a strong horizontal reflector. Most authors describe this unit as a glacial till consisting of homogeneous stiff grey clay with some chalk and flint fragments (Hovland, 1981).

Unit 4

Unit 4 has a similar seismic character to unit 3. No further information about this unit is available.

Rokoengen and Rønningsland Model

An alternative interpretation of the stratigraphy is provided by Rokoengen and Rønningsland (1983). Their interpretation of the soils will be outlined below and differences from van Weering's interpretation will be mentioned. Rokoengen and Rønningsland (1983) have primarily divided the sediments into three major units.

(1) The upper unit consists of soft silty clay overlying till-like material consisting of stiff silty clay with a variable content of sand, gravel and stones. The base of this unit, which the authors call level B, is a prominent reflector which in some places can be seen to represent a slight angular unconformity between the well-layered sediments below and the till-like deposits above. The authors explain this erosional character by glacial erosion and think that the horizon marks the base of the Quaternary sequence. This is based on biostratigraphical evidence from the Máløy plateau and on seismic character.

The till-like deposit of Rokoengen and Rønningsland corresponds in description to van Weering's unit 3. The overlying soft silty clay unit corresponds to units 1 and 2. As mentioned below under 'Depositional History', the base is not thought to be of Quaternary age by other authors.

(2) Their middle unit is subdivided into:
 (a) An upper sub-unit of sub-horizontal layered sediments. The individual layers can be followed over large areas. This unit is considered as having been deposited primarily in a low-energy environment below wave base and is predominantly clay. The well-developed layering, however, is thought to indicate sandy layers.
 (b) A basal sand layer. The top of the unit is often marked by a very strong reflector of varying intensity, which Rokoengen and

Rønningsland think is formed by combined fluvial and marine erosion, and deposition.

These sub-units are thought to be of Late Pliocene age. The base of this unit, which Rokoengen and Rønningsland called level A, marks a prominent angular unconformity separating older strata with westerly dip from the overlying, more horizontally layered beds.

(3) The lower unit consists of westerly dipping strata, thought to be mid-Pliocene to Jurassic age.

Depostional History

Here we will discuss the depositional history of the near seabed soil units in more detail.

There is some debate within the literature concerning the age and origin of the deepest, very prominent angular unconformity. According to Sellevol and Sundvar (1973) and Floden and Sellevol (1972) this unconformity is the result of differential glacial erosion and marks the base of the Pleistocene. Erosion was greater nearer the coast and thus the unconformity dips gently towards the east. Rokoengen and Rønningsland (1983), however, describe this unconformity (their level A) as the result of fluvial and marine erosion, combined with differential structural subsidence along the coast. These authors consider this unconformity to be of late Tertiary age.

The sediments immediately above this unconformity have been described only by Rokoengen and Rønningsland (1983). The basal sand is thought to be fluvial to marine and may partly have been subjected to marine erosion and reworking during the later marine transgression across this surface. The upper sub-unit is sub-horizontally layered and deposited in a low-energy marine environment below wave base.

The shallower unconformity, which corresponds to Rokoengen and Rønningsland level B and probably to the base of van Weering's unit 3, is thought to be caused by erosion at the base of an ice-sheet moving in

a north to northwesterly direction. Rokoengen and Rønningsland consider that this unconformity marks the base of the Pleistocene sediments. 'During Quaternary times, glacial erosion probably took place several times. Each successive erosion seems more or less to remove the preceding glacial deposits. In the central Trench it is thus possible that the B level represents a hiatus from Late Pliocene to Late Weichselian.' The sediments above this unconformity are therefore glacial deposits.

The interpretation agrees with van Weering's interpretation. He considers both units 3 and 4 as glacial drift deposits separated from each other by a period of erosion. This flat, rather strong horizon at the top of unit 4 contrasts strongly with the hummocky irregular relief of the top of unit 3. The flat reflector was most likely formed by glacial erosion of a grounded ice-sheet. Van Weering suggests that unit 3 sediments were deposited from the ice-sheet when it started to retreat from the western side of the Trench. This took place between 18 000 years BP until 12 600 years BP when the ice had retreated to the present coastline. Since the units above unit 3 are not glacial drift deposits and since no major erosion surface can be seen between the units 3, 2 and 1, it is most likely that unit 3, and possibly unit 4, also were deposited during the Late Weichselian glaciation.

The deposition of the well-layered unit, unit 2, is thought to have started directly after the deglaciation. It is thought by van Weering (1983) to be a glaciomarine sediment, deposited in Late Weichselian and Early Holocene, in the seas which covered the area as the ice retreated. As the sealevel rose, successively larger parts of the North Sea were drowned and became subject to reworking. This produced large amounts of material in suspension which were deposited in the deeper waters of the Norwegian Trench.

The thickness variations within these post-glacial units are mainly attributed to the topographic undulations of the underlying glacial drift (Jansen et al., 1979). Due to the decrease in thickness towards the north-east in the Skagerrak and to the north along the Trench, Jansen *et al.*, favour a southern sediment source area.

According to van Weering, the deposition of unit 1 ceased around 8000 years BP when a tidal current pattern similar to that of the present was established in the North Sea, probably due to the opening of the English Channel (8300 years BP according to Jelgersma, 1979).

It should be noted that the transition from Arctic to Boreal conditions probably occurred before this date. Rise and Rokoengen (1983) have dated this event as 10 000 years BP and suggest a very low sedimentation rate from this date onwards.

According to Holtedahl and Bjerkli (1982), the most important change in the oceanic environment off West Norway took place at the beginning of the Holocene at about 10 000 years BP, but Atlantic water had been reintroduced during the Late Weichselian (i.e. between 13 000 and 10 000 years BP).

DETAILED SOIL STRATIGRAPHY AND GEOTECHNICAL REVIEW

Here the five major soil units, as defined by the variations in seismic character and by geotechnical parameters, are described in detail. Figure 5 illustrates the soil data available and the division into units.

Unit I

- Consists of slightly sandy, very silty clay.
- Occurs throughout the mapped area.
- Thickness varies from 16 to 36 m.
- Undulating strong basal reflector.
- Generally characterized by a multi-layered sequence.
- Variations in geotechnical parameters can be correlated with two or more marked internal reflectors.
- Contains seabed and buried 'pockmarks'.
- With the exception of the area near soil

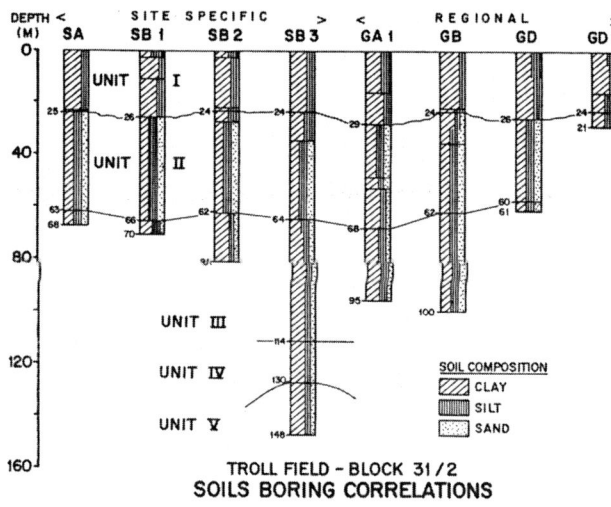

TROLL FIELD – BLOCK 31/2
SOILS BORING CORRELATIONS

Fig. 5

boring GD 1, shows consistent geotechnical characteristics.
- Shear strength (motorvane) increases linearly from mudline with depth (10–40 kN/m²) at the platform site.
- Apparently overconsolidated with OCR from 1 to 8.
- Moisture content decrease with depth from 75 to 55%.
- Plasticity index decrease with depth from 60 to 40%.
- Uunit weight increases with depth from 15 to 16.8 kN/m³.
- Typical glaciomarine sediment.

Unit II

- Consist of sandy, silty clay.
- Occurs throughout the mapped area.
- Thickness varies from 30 to 55 m.
- Flat basal reflector.
- Generally characterized by a threefold division based on two internal reflectors.
- Good correlation between these reflectors and changes of geotechnical parameters, with the exception of soils near soil boring GD 1.
- Shear strength (motorvane): upper sub-unit 40–90 kN/m, middle sub-unit 105–130 kN/m², lower sub-unit ~120 kN/m².
- Above lower internal reflector apparently overconsolidated with OCR from 1.1 to

1.8. Below lower internal reflector normally consolidated.

- Moisture content: upper sub-unit decreasing from 25 to 20%, middle sub-unit 19/20 to 17/18%, lower sub-unit 20%.
- Plasticity index: upper sub-unit around 19–20%, middle sub-unit 21–21%, lower sub-unit 18–19%.
- Unit weight: upper sub-unit increasing from 19.8 to 20.5 kN/m³, middle sub-unit 20.7 kN/m³, lower sub-unit 20.5 kN/m³,
- Rather uniform grain-size distribution, middle sub-unit less sandy (less than 5%).
- Glacial till deposit.

Unit III

- Consists of sandy, very silty clay.
- Occurs throughout the mapped area.
- Thickness varies from 45 to 60 m.
- Base reflector irregular.
- Generally characterized by a threefold division based on two internal reflectors.
- High occurrence of gravel and/or boulders in part of the area.
- Correlation between changing geotechnical parameters and internal reflectors.
- Shear strength (torvane) in upper sub-unit increases from ±300 kN/m² to 500/600 kN/m² at upper internal reflector. Below this reflector, slow increase versus depth.
- Upper sub-unit apparently overconsolidated (max. 1.45), remainder of unit normally consolidated.
- Moisture content increases versus depth from 15 to 20%. Sharp increase around lower internal reflector (96 m depth).
- Plasticity index increases versus depth from 25 to 26% at 99 m depth. Increase of slope of plasticity index versus depth at lower internal reflector.
- Except for lower sub-unit, rather uniform grain-size distribution: 40% clay, 36% silt, 24% sand and gravel.
- Sharp decrease in sand content (25–32%) and increase in silt content (25–39%) versus depth in lower sub-unit.
- Glacial till deposit.

Unit IV

- Consists of slightly sandy, very silty clay.
- Occurs throughout the mapped area.
- Thickness varies from 7 to 55 m.
- Irregular basal reflector.
- Generally characterized by a multi-layered sequence.
- Good correlation with exploration well logs.
- Shear strength (pocket penetrometer) increases versus depth, 510 to 720 kN/m³.
- Apparently normally consolidated.
- Moisture content increases versus depth from 23 to 30%.
- Plasticity index decreases versus depth from 42 to 32%.
- Unit weight of soil decreases versus depth from 20.4 to 17.7 kN/m³.
- Apparently changing grain-size distribution versus depth. Silt content increases from 25 to 50%. Sand and gravel decrease.
- Glaciomarine sediment.

Unit V

- Consists of sandy, silty clay becoming slightly silty fine sand with depth.
- Not present in the southern part of survey area.
- Thickness varies from 0 to a maximum of 37 m.
- Generally characterized by very discontinuous reflectors.
- Basal reflector irregular.
- Limited calibration data available.
- Shear strength around 340–620 kN/m² (undrained triaxial and pocket penetrometer).
- Moisture content decreases versus depth from 15 to 13% (half way).
- Plasticity index decreases versus depth from 23 to 18%.
- Unit weight decreases versus depth from 21.1 to 20.7 kN/m³.
- Increasing sand and gravel content versus depth (35–75%).
- Decreasing silt content versus depth (35–20%).

- Very shallow marine (?), cold environment deposition.

GEOLOGICAL HISTORY

The available data are here developed into a tentative geological depositional history on a unit-by-unit basis.

Unit I

It is suggested that deposition of this unit of marine sediments commenced in Late Weichselian times, immediately after the deglaciation of the area. During deglaciation, the ice front retreated eastward and as it did so the water depths in the area gradually increased. As it melted, the ice apparently did not form large buoyant ice-sheets. Sediments deposited from buoyant ice-sheets are characterized by a fining upwards grain-size sequence, with an abrupt change to predominantly fine material only, when all the ice in the area had melted. Since no characteristic grain-size changes have been noted in unit I, it is thought that the ice-sheet lay directly on unit II as it retreated, i.e. a tidewater front, bordering an expanding sea.

Analyses of dinoflagellates and forams within unit I shows that the unit was deposited in a climate which was becoming warmer, as would be expected after the last glaciation. Sedimentation was relatively constant up to 8000 years BP.

Such a depositional model would suggest that the sedimentary layers of unit I should progressively onlap onto the top of the unit II as the ice retreated eastwards. There is only weak evidence of onlap from the data but the E–W lines which would most clearly show such features are of limited lateral extent. A very rapid ice retreat would obscure onlap features, but this is unlikely since this would produce large amounts of meltwater and thus erosion and there is no evidence of channels in the underlying unit. Even if the underlying soils were frozen, it

is likely that channel formation would occur under such conditions.

The lack of strong reflectors in the lower part of the unit is likely to be the result of higher sedimentation rates during and immediately after the retreat of the ice-sheet. At this time much of the North Sea was still dry land and meltwater flowed into the deeper Norwegian Trench carrying large amounts of sediment. In addition, material was also provided by winnowing of deposits by marine erosion resulting from the ever-increasing sea-level.

It is difficult to date precisely the onset of the glacial retreat due to the large geographical area which the ice covered. The literature suggests it is likely to have started around 18–15 000 years BP on the western side of the Trench and so it must be assumed that unit I sedimentation commenced in the Troll Field after this date. In conflict, however, amino acid geochronology provides a slightly older date for the lower sediments of the unit of 30 000 years BP. However, carbon-14 dating resulted in a date of around 40 000 years BP, but this is most likely due to reworking of the samples.

The two internal reflectors within the unit (1A and 1B) which correlate with geotechnical variations are tentatively attributed to the younger and older Dryas stadials (periods of colder climate). Using this age/depth of burial relationship and others from the literature and previous surveys, a relatively constant rate of sedimentation is predicated throughout unit I to approximately 1.5 m below seabed. A slight increase is apparent in the lower part of the unit, which agrees with the suggested higher sedimentation rate immediately after ice retreat.

The 1.5 m below seabed depth is thought to represent c. 8000 years BP when the English Channel opened and the onset of significant Atlantic influences began. Sedimentation rate was then severely reduced, which may have produced higher compaction of the sediments and thus the higher measured *in situ* CPT values.

It should be noted that van Weering (1977) suggests that the internal layering within the unit was caused by layers of interbedded gravel and pebbles. This is, however, unlikely since such variations should have been detected by the site investigations.

A more plausible alternative explanation of the layering relies upon minor differences in consolidation, which could either be related to times of non-deposition (unlikely in the geological framework) or to differences in mineral content. It is very likely that the layering is the results of a combination of these event.

The boulders within the unit were most likely deposited from drifting icebergs. It is of interest to note that the quantity of boulders does not, as would be expected, decrease from the bottom of the unit upwards. The tentative distribution reveals a maximum at around 7 m depth and a sub-maximum at 18 m depth. These variations in boulder distribution reflect glacial travel paths. Long glacial travel paths abrade material into clay-size particles (Boulton, 1972). Concomitantly, the travel path of ice-sheets during both the Dryas stadials (colder periods) was far less, resulting in more boulder-grade material in the icebergs. The depths of boulder distribution maxima correlate with these stadials. It should be noted, however, that not all the hyperbolae on the seismic records are necessarily caused by boulders but may be due to other strong point reflectors, such as patches of organic-rich material.

Unit I is thought to represent glaciomarine sediments deposited after the last glacial maximum. A low-energy marine environment is likely with progressively increasing water depths and temperatures.

Unit II

The geological history of this unit is more complex than that of unit I. The internal reflectors are difficult to fit into a definitive geological depositional framework.

The very flat character of the base of this unit is likely to be the result of glacial erosion, since it is difficult to explain this character by other processes, such as fluvial–marine or wind erosion. Wave-base erosion in a marine environment is unlikely to create such a flat surface and would imply that the ice-sheet was grounded and eroded the underlying sediments. The amount of sediment removed by this erosion is not known since reliable dates are not available for the underlying soil units. Van Weering *et al*. (1973) suggest the erosion was by the ice-sheet which also was responsible for the deposition of the underlying unit III. However, this study cannot confirm or refute this hypothesis.

Examination of dinoflagellates and forams within this unit suggests a harsh, arctic-like environment. Poor floras are found which are due to unfavourable environmental conditions such as might be expected during periods of glaciation. Periods of some slight climate amelioration are indicated which are thought to represent glacial interstadials.

Unit II is thought to consist of a series of tills deposited by the same ice-sheet as that responsible for the flat erosional base. The study suggests this was most likely to be the last major Weichselian glaciaton whose maximum occurred *c*. 20 000 years BP. The flat base of unit II is thought to be a combination of the effects of compaction and erosion, probably from a dry-based glacier. The areas which show the absence of the reflector at the base can be explained by freezing of the ground below the glacier to the ice-sheet while it was moving.

Although the outer edge of the ice-sheet is thought to have been dry based, Boulton (1972) suggested that Quaternary ice-sheets were characterized by being dry based at their edges but wet based in the internal zone. Powell (1984) states that internal and basal ice states can vary spatially and temporally in glaciers.

As the ice-sheet moved across the area, it subjected the soils to dry-based erosion, followed by wet-based till deposition. It is likely that the lowest sub-unit of unit II is a lodgement till, although its compaction and strength is lower than would be expected. This can be explained by high pore-pressure

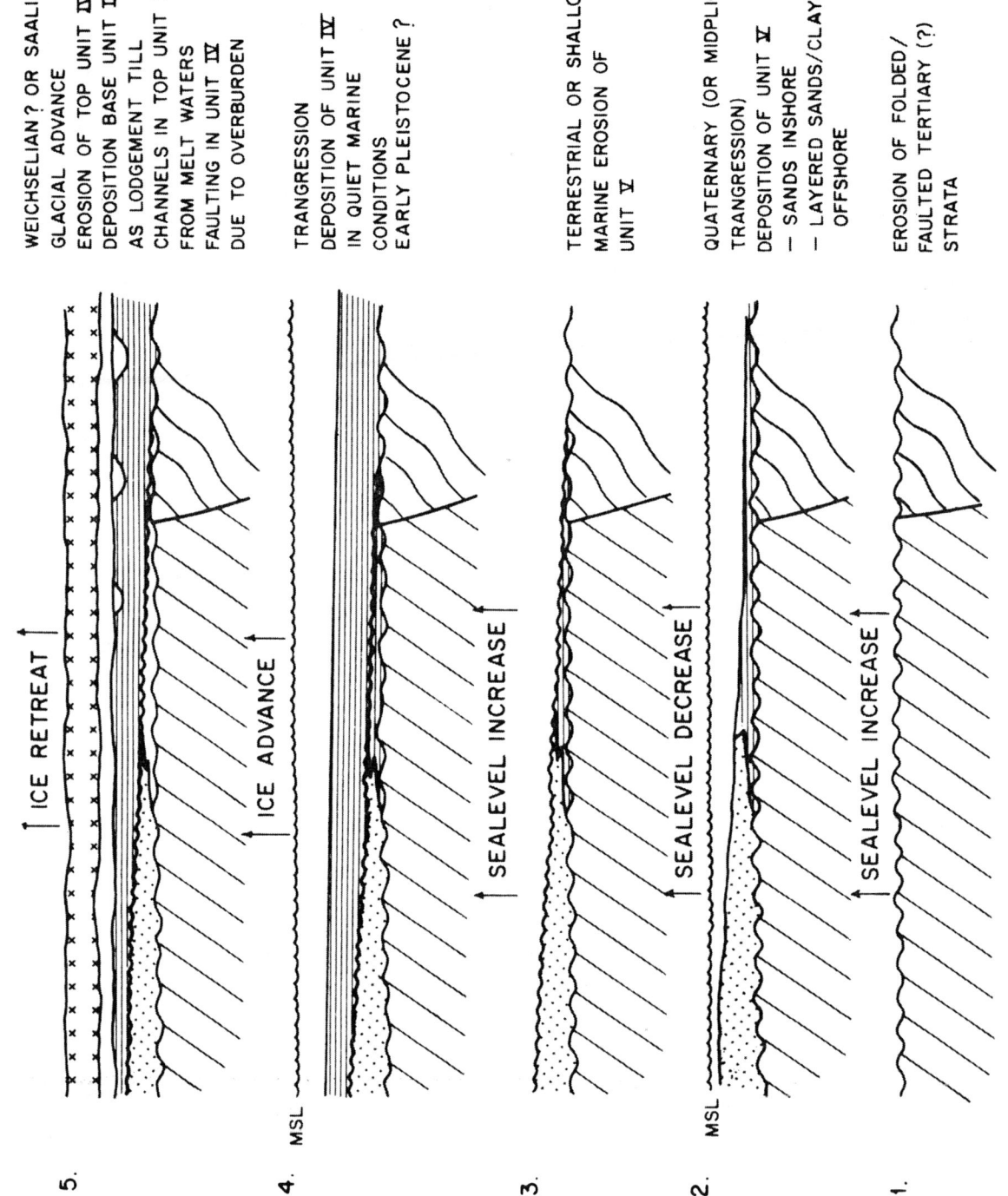

5. WEICHSELIAN? OR SAALIAN?
   GLACIAL ADVANCE
   EROSION OF TOP UNIT IV
   DEPOSITION BASE UNIT III
   AS LODGEMENT TILL
   CHANNELS IN TOP UNIT IV
   FROM MELT WATERS
   FAULTING IN UNIT IV
   DUE TO OVERBURDEN

4. TRANGRESSION
   DEPOSITION OF UNIT IV
   IN QUIET MARINE
   CONDITIONS
   EARLY PLEISTOCENE ?

3. TERRESTRIAL OR SHALLOW
   MARINE EROSION OF
   UNIT V

2. QUATERNARY (OR MIDPLIOCENE
   TRANGRESSION)
   DEPOSITION OF UNIT V
   – SANDS INSHORE
   – LAYERED SANDS/CLAYS
     OFFSHORE

1. EROSION OF FOLDED/
   FAULTED TERTIARY (?)
   STRATA

ICE RETREAT

ICE ADVANCE

SEALEVEL INCREASE

SEALEVEL DECREASE

SEALEVEL INCREASE

MSL

MSL

Fig. 6

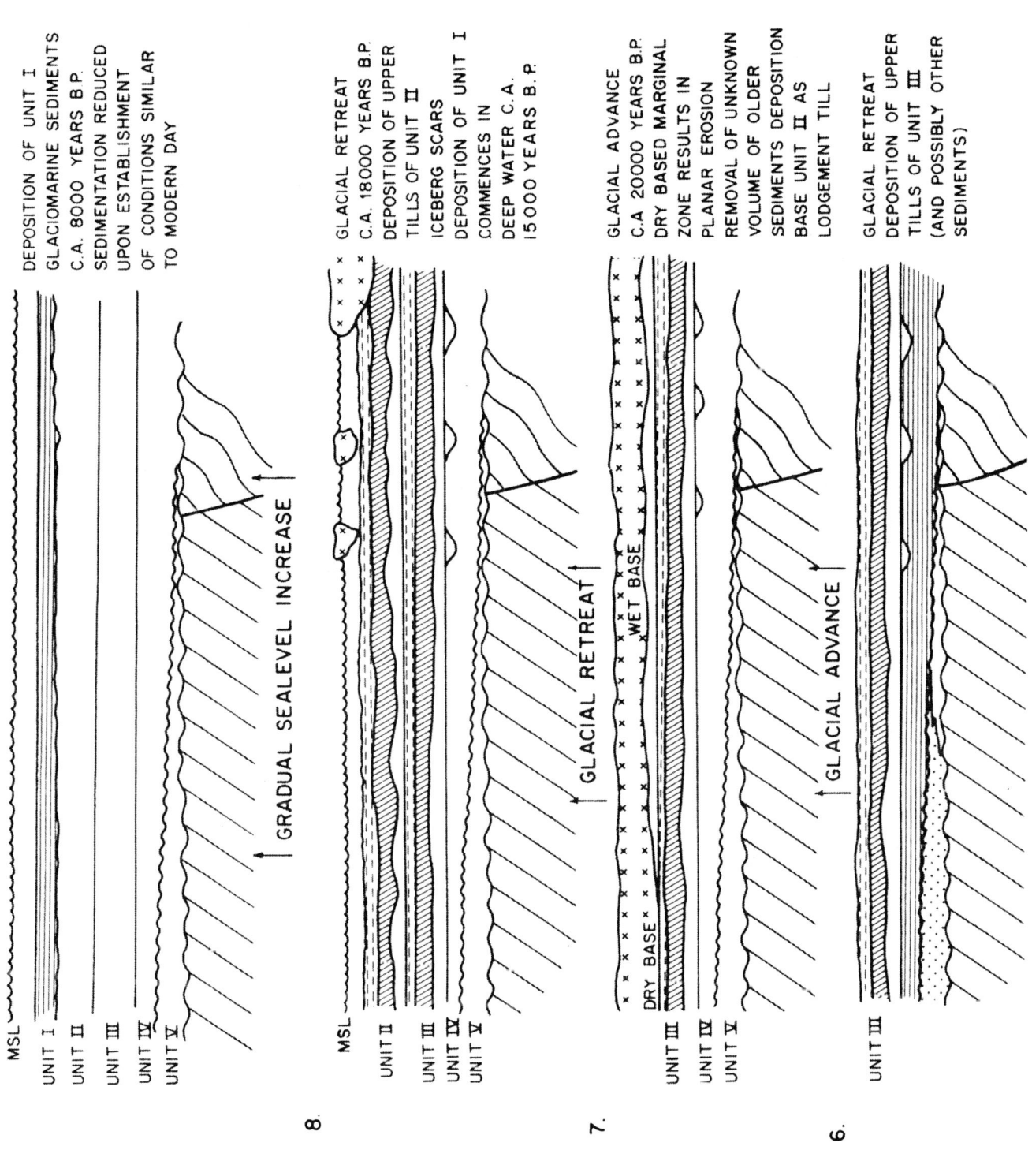

DEPOSITION OF UNIT I
GLACIOMARINE SEDIMENTS
C.A. 8000 YEARS B.P.
SEDIMENTATION REDUCED
UPON ESTABLISHMENT
OF CONDITIONS SIMILAR
TO MODERN DAY

MSL
UNIT I
UNIT II
UNIT III
UNIT IV
UNIT V

GRADUAL SEALEVEL INCREASE

8.

GLACIAL RETREAT
C.A. 18000 YEARS B.P.
DEPOSITION OF UPPER
TILLS OF UNIT II
ICEBERG SCARS
DEPOSITION OF UNIT I
COMMENCES IN
DEEP WATER C.A.
15000 YEARS B.P.

MSL
UNIT II
UNIT III
UNIT IV
UNIT V

GLACIAL RETREAT

WET BASE

DRY BASE

7.

GLACIAL ADVANCE
C.A 20000 YEARS B.P.
DRY BASED MARGINAL
ZONE RESULTS IN
PLANAR EROSION
REMOVAL OF UNKNOWN
VOLUME OF OLDER
SEDIMENTS DEPOSITION
BASE UNIT II AS
LODGEMENT TILL

UNIT III
UNIT IV
UNIT V

GLACIAL ADVANCE

6.

GLACIAL RETREAT
DEPOSITION OF UPPER
TILLS OF UNIT III
(AND POSSIBLY OTHER
SEDIMENTS)

UNIT III

values within the till which would support part of the overburden ice weight. The irregular top relief of this sub-unit can be explained by irregularities at the base of the glacier and limited erosion by meltwaters running down the glacier.

The upper two sub-units may both have been deposited during the subsequent ablation of the ice-sheet. The middle sub-unit is thought to represent a basal melt-out till incorporating englacial sediments which were deposited below the ice-sheet. Limited erosion by meltwaters must have occurred and may be reflected by the irregularities at the top of the sub-unit. The upper sub-unit was then deposited as the ice front retreated and is thought to indicate a dropstone diamicton made up of ice-rafted and suspended sediments. Boulders in the unit have been mapped. These are likely to be glacial erratics which are often found in subaquatic dropstone diamicton and melt-out tills.

The depositional model as outlined above is illustrated in Fig. 6, phases 7 and 8.

It is emphasized again that with the available data exact processes and a definite geological history of the unit are difficult to establish. The history and processes described above are a 'best fit' model based on the data sources available. However, from the results of this study unit II is thought to represent a series of glacial tills deposited as a result of the last major Weichselian glaciation.

Unit III

Although the threefold division of the unit based on seismic character is similar to that of the overlying unit II, due to the complexity of glacial deposits, the depositional history is not necessarily the same.

The unit is thought to be a series of glacial tills which may have been deposited by the same ice-sheet as that responsible for unit II or by one from a previous glaciation. Unfortunately, the limited geological testing would not permit dating of this unit, but an

undifferentiated Middle to Late Pleistocene age is likely.

As with unit II, examination of dinoflagellates and forams within this unit suggest a harsh, arctic-like environment with poor floras due to unfavourable environmental conditions. There are indications of periods of some slight climatic amelioration which are likely to represent minor glacial interstadials.

The base of this unit is not marked by a flat sub-horizontal reflector but is an irregular erosional surface. It forms a well-marked angular unconformity with the underlying sub-horizontal layered unit IV. The base of the unit also shows evidence of channel forms cutting into the underlying unit. There is very little evidence of disturbance of the sediments at the top of the underlying unit, suggesting that the unconformity best characterizes erosion from a wet-base ice-sheet with channels caused by meltwater flowing below the glacier. Unit III then represents tills deposited by this ice-sheet.

The lower sub-unit was probably deposited as a lodgement till. Many such tills are characterized by a finer grained character than other tills, which is true of the lower sub-unit of unit III. Due to the similarity in seismic character between units II and III, it is possible that the depositional history of the other two sub-units was the same as that for unit II, e.g. a melt-out till followed by a dropstone diamicton. However, there is insufficient data for a confirmation of this hypothesis.

The depositional model of this unit is illustrated in Fig. 6, phases, 5 and 6.

From the data available it is thought that unit III represents glacial tills of either the Weichselian or Saalian glaciation.

Unit IV

As there are very few geological test data available on these deeper soils, there remains considerable dispute as to the age and origin of the soils. According to Rokoengen and Rønningsland (1983), their level B, i.e. base unit II, marks the base of the Quaternary and the older units III, IV and V are not glacial sediments but marine deposits of Late Pliocene age. This contrasts with the interpretation of van Weering (1983) and others, who suggests these older units are still part of the Quaternary sequence and are potentially glacial sediments. This is confirmed by the geological test data where Dinoflagellates of Early Pleistocene age and forams likely to be of Early Pleistocene age have been recorded.

The base of unit IV is marked by an irregular, apparently erosional surface. The origin of this erosion is considered by Rokoengen and Rønningsland to be the result of marine erosion and reworking due to the marine transgression. The further transgression then caused a gradual infilling of depressions by the well-layered sediments of unit IV. These sediments are thought to represent a period of relatively quiet marine sedimentation in the Norwegian Trench area, while deltaic sedimentation was restricted to areas close to the Norwegian coast. These authors assume the sea-level rise was due to tectonic movements. However, if the soils are of Quaternary age it may be due to melting ice-sheets.

It is of interest to note that the dinoflagellate and foram analyses indicate an ameliorated climate, warmer than the present day within this unit but becoming colder down the unit.

The soils of unit IV contain a high percentage of clay, indicative of a quiet marine environment. According to palynological data, the maximum depth below sea-level was of the order of 50 m. The transgressive nature of the deposition is confirmed by onlap structures within the unit. Apparently the sea-level rise was not continuous but fluctuated, resulting in minor onlap structures and channel forms.

Minor faulting within the unit is most likely ice loading of the unit by the glaciers responsible for the deposition of units II and III.

Unit IV is therefore thought to represent glaciomarine sediments deposited in prog-

ressively deepening water. These sediments are probably very similar to unit I soils (see Fig. 6, phase 4). They may represent offshore deposits resulting from one of the Early Pleistocene glaciations.

Unit V

As with unit IV, there are few geological test data available on these soils, and considerable dispute remains regarding their age and origin. However, as with unit IV, the geological analyses suggest that these soils are of Early Pleistocene age.

The base of the unit is a very well marked angular unconformity which represents either the base of the Quaternary soils or (according to Rokoengen and Rønningsland, 1983) the level A unconformity of mid-Pliocene age. The sub-seabed depth of Rokoengen and Rønningsland level A correlates very well with the depths as determined in the primary survey area. The unconformity is thought to be formed by fluvial and marine erosion of Tertiary rocks followed by differential structural subsidence along the coast.

The sediments immediately above the unconformity have been identified by Rokoengen and Rønningsland as a basal sand unit. This agrees very well with the seismic character of the unit and the increase in sand content apparent from the field grain-size distribution results from the primary survey. These sandy soils are thought to have been deposited in either fluvial or shallow marine environments. Subsequently, an increase of sea-level, either due to tectonics or to glacial fluctuations, resulted in the deposition of the remainder of the unit, which shows an apparent fining upwards, becoming a sandy silty clay as the water depth increases. This is confirmed by seismic data which show a more continuous layering in the upper part of the unit, indicating a lower-energy environment. Since the depositional environment, as indicated by geological tests, is thought to be very cold, it is suggested that this transgression was, in fact, due to sea-level changes resulting from glacial melting. However, as there is no evidence of till deposits, it is thought that the ice-sheet did not reach this area.

The erosional features noticed within the unit, e.g. channelling and onlap, suggest the marine transgression was not a continuous progress and that minor regressions also occurred. The lateral change of seismic character to the north-west and to the east suggests these areas were deeper water areas with more restricted energy environments.

A rapid regression due to either glaciation or tectonic uplift must have occurred subsequent to the deposition of this unit to account for the unconformable nature of the transition from unit V to unit IV.

Deposition of the unit is illustrated in Fig. 6, phases 1, 2 and 3.

SUMMARY OF GEOLOGICAL HISTORY OF THE SOILS

(1) Deposition of Tertiary (?) strata.
(2) Folding (and faulting?) of these strata.
(3) Fluvial and marine erosion producing deepest unconformity.
(4) Deposition of unit V — fluvial or shallow marine basal sands. Progressively deepening cold water producing layered sandy clays.
(5) Sea-level decrease, possibly glacial-controlled, producing erosion of unit V.
(6) Sea-level increase, possibly glacial-controlled with deposition of well-layered unit IV soils in deeper marine areas. Climate becoming warmer. Early Pleistocene?
(7) Glacial advance (Weichselian or Saalian?). Removal of an unknown quantity of underlying strata. Wet-base ice-sheet producing channel erosion in top unit IV.
(8) Deposition of unit III as a series of glacial tills while the ice-sheet advanced and then subsequently when it melted.
(9) Glacial advance (last major Weichselian glaciation). Dry-based glacier front producing planar erosion. Removal of

unknown quantity of underlying strata.
(10) Deposition of unit II as a series of glacial tills while the ice-sheet advanced then subsequently when it melted. (Glacial maximum c. 20 000 years BP).
(11) Deposition of unit I as glaciomarine sediments while the sea-level increased and water became deeper (commenced c. 15 000 years BP).
(12) Opening of English Channel (c. 8000 years BP) and establishment of present-day conditions. Sedimentation much reduced.

The above depositional history is portrayed in Fig. 6.

POCKMARKS

Pockmarks were first described by King and MacLean (1970) as craterlike depressions on the Scotian Shelf, eastern Canada. From that date onwards pockmarks have been noticed in many places, i.e. the Gulf of Mexico (Sieck, 1973), the Aegean Sea (Newton and Cunningham, 1980), the South China Sea (Platt, 1977) and the North Sea. As their occurrence is apparently associated with hydrocarbon-bearing areas, they have become of interest to oil companies.

Pockmarks are normally formed in soft, silty or clayey sediments. They are erosive features. On the seabed they are round or elliptical, varying in size from 10 to 200 m in diameter and from 1 to 10 m in depth. The extent of a pockmark is well defined by an increase in the dip of the seabed. The dip can reach a maximum value of 36°, which can lead to slumping in the pockmark. Pockmarks can be asymmetrical in a vertical section.

Despite the considerable quantity of literature concerning the occurrence of pockmarks, the results of detailed pockmark investigations have been published by only three organizations: the Bedford Institute of Oceanography (King and MacLean, 1970; Josenhans et al., 1978), the Institute of

Geological Sciences (McQuillin and Fannin, 1979; and various internal reports) and Statoil (Hovland, 1980, 1982). These investigations have been concerned with the Scotian Shelf, the South Fladen area, UK sector, North Sea, and the Norwegian Trench, respectively.

In the 31/2 area one of the most striking features is the lineation of seabed scars associated with the pockmarks. Hovland (1981) in neighbouring areas described orientations of NW-SE (42%), NNW-SSE (30%) and N-S (24%). Similar percentages have been found valid for 31/2 pockmarks. Individual pockmarks comprising these lineations may be 8–60 m in diameter but the total length can commonly attain hundreds of metres.

The largest seabed pockmarks in the Troll Field study area attain diameters of 100 m and depths of 15 m. The most common depth of occurrence is in the 3–5 m region with little evidence of border rims or terraces. (Figure 7 is a boomer record showing these seabed pockmarks in profile.) The average density of seabed pockmarks is 20 per km².

Apart from pockmarks at the seabed, a number of buried pockmarks have also been noted and mapped. These generally occur at the level of reflector IA. The frequency of occurrence of these buried pockmarks is less than those on the seabed. This is partly due to a 100% coverage of layers underneath the seabed using shallow seismic data. Allowing for this, results still differ in frequency since on seismic sections the number of seabed pockmarks is far larger than the number of buried pockmarks (approximately 10 : 1).

It should not be concluded from these figures that the frequency of pockmark formation is greater today than in the past. Geological evidence suggests little sedimentation has occurred in the area over the last 8000 years, so the pockmarks seen today represent many years of formation. It is likely that pockmark formation would be suppressed during periods of rapid sedimentation, and thus they are not found through-

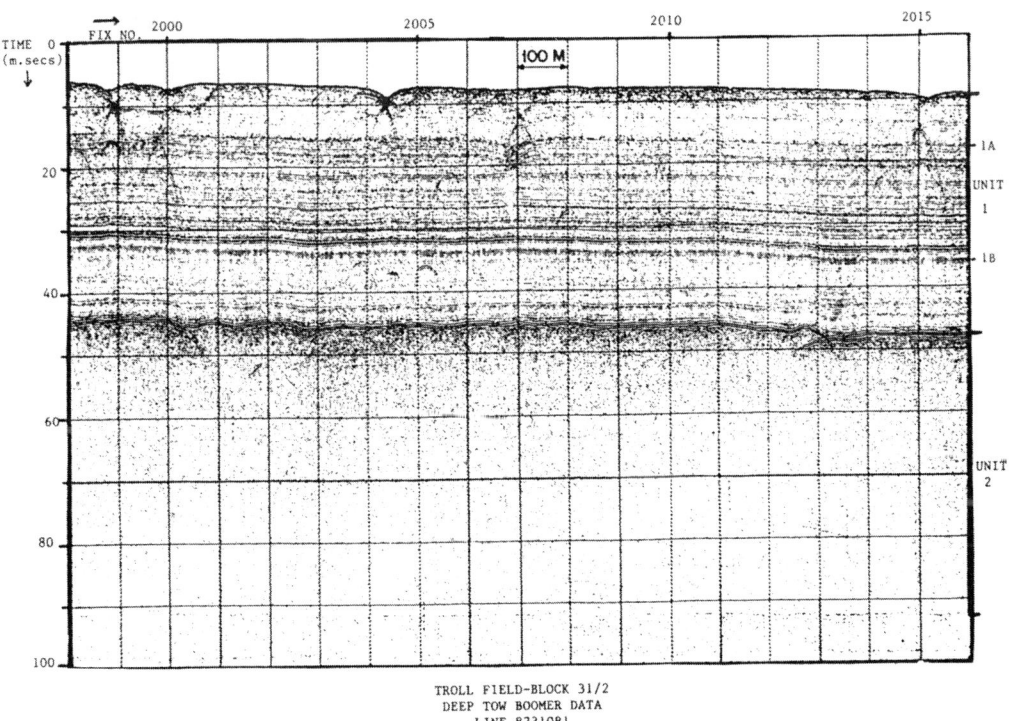

TROLL FIELD-BLOCK 31/2
DEEP TOW BOOMER DATA
LINE 8231081

Fig. 7

out unit I in high concentrations. The levels at which buried pockmarks are found are likely to represent periods of reduced sedimentation.

The erosive nature of some pockmarks is demonstrated by the truncation of acoustic layers on high-resolution seismic reflection profiles. Numerous mechanisms have been suggested to account for the removal of what in many cases is a considerable volume of material. These include suggestions that they are caused by action from above the seabed (meteor impact crater, scour hollows around boulders, iceberg drop-stones, and man-made artifacts such as wrecks, bombs, etc.), by biological activity (e.g. the action of sub-bottom or bottom-drilling creatures), and by mechanisms which are essentially related to the last ice age (e.g. strudel scour, kettlemarks, both of which imply the previous existence of large bodies of ice trapped within the sediment and collapse when the ice melted leaving a relatively easily erodable sediment and subsequently a depression). Of these mechanisms none has been given much credence. The ice age related mechanisms were favoured for some time when most pockmarks reported concerned areas affected by the Pleistocene glaciation. However, when pockmarks were reported in unglaciated areas, such as the South China Sea, it became obvious that alternatives must be considered.

Nowadays it is generally accepted that pockmarks have been formed by the removal of material. The material has been removed by gas or water escaping from the sediment. Although in some specific areas groundwater flow has caused pockmark formation, gas is normally described as being the escaping fluid. Three possible gas sources can be distinguished:

(1) Petrogene gas from deeper units.
(2) Biogene gas from, for example, peat layers.
(3) Volcanic or hydrothermal gas.

If pockmarks were formed by migration of petrogenic gas, this gas would most likely originate from the underlying Troll Field reservoir, the implication being that the gas has migrated through the overlying strata. The simplest case is to assume that the migrating gas tends to follow paths of least resistance upwards, i.e. faults and bedding planes.

A relationship between mapped pockmark lineations and faults would provide useful evidence to support the above suggestion. It is to be expected that the strike of the youngest fault pattern would have the strongest influence on a possible lineation. The youngest major fault pattern has therefore been mapped in the Tertiary sequence at the depth of a prominent reflector called the 'Purple Horizon'. The pattern is shown on Fig. 8.

As is to be seen, the major fault plane lineation is ENE to WSW. This does not correlate with the pockmark lineation (NNW-SSE). This does not necessarily exclude fault-plane migration of petrogene gas. However, a simple migration mechan-

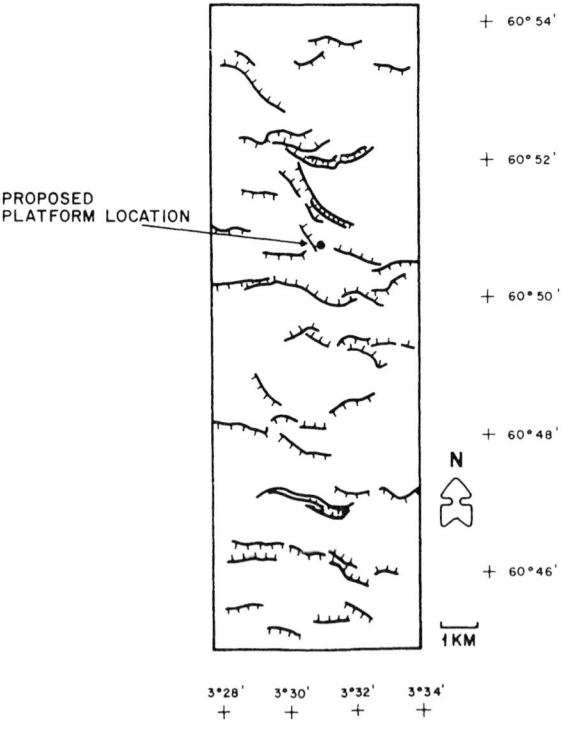

FAULTS IN UPPER TERTIARY

Fig. 8

ism through the Quaternary sequence is not apparent.

The base (worst) case assumed is that pockmarks may create risks for:

(a) short-term settlement of a gravity-based production platform;
(b) spanning of pipelines;
(c) long-term risks (a) and (b) above due to growth of new or enlarged existing pockmarks.

Owing to the uncertainty of numbers, sizes and location of future pockmarks generated within the Troll Field development area, it is appropriate that engineers should express pockmark risk in terms of 'return periods', just as design environmental forces are.

It is desirable for the engineer to have available all the pertinent data and professional judgement of those trained in geophysics and geology in a form most suitable for making a decision. Therefore it is most useful to have information available as cross-plots of pockmark intensity versus age and depth, and pockmark intensity versus average return period. This information can then be used systematically in engineering design procedures.

The methodology to provide an assessment of the risk from pockmarks at a site in these terms has to be considered. Historical geological data must be synthesized. The locations, sizes and activity types may be many and different in kind or may not be even well known.

A method has to be utilized to produce for the engineer the desired relationships between process, geologic history, pockmark age, size and intensity and their average return period for the Troll Field site.

The minimum data needed are the geologist's best estimates of average pockmark numbers with depth on a time-based plot.

The technique has to allow for integrating the individual influences of pockmark sources near and far, and more or less active, into a probability distribution of maximum annual geographic intensity. The average return period follows directly. In determining the distribution of annual intensity, consideration has to be given to the likelihood of occurrence (a) along a neighbouring fault; (b) along a given plane (or planes) of known structural weakness; or, equally likely, (c) anywhere over the area due to an apparent lack of correlation between the geologic structure and the pockmark position formation.

The data required to complete the study are:

(1) intensive geologic dating of unit I soil by pollen, foraminiferal, dinoflagellate and carbon-14 methods;
(2) input of field data required over a 'pockmark study area', providing further data on seabed and sub-seabed pockmarks;
(3) the subsequent integration of data into a report describing average return periods.

Such studies are presently being completed from available data for the Troll Field area.

ACKNOWLEDGEMENTS

The authors wish to express their appreciation to the management of Norske Shell and to the other shareholders of Troll West, namely Statoil, Norsk Hydro Production A/s, Norsk Conoco A/s and Superior Oil Norge A/s for granting permission to submit this paper.

REFERENCES

1. Belderson, R. H. and Wilson, J. B. 1973. Iceberg ploughmarks in the vicinity of the Norwegian Trough. *Norsk. Geol. Tidsskr.* 53, 323–328.
2. Boulton, G. S. 1972. Modern arctic glaciers as depositional models for former ice sheets. *J. Geol. Soc.*
3. Caston, V. N. D. 1977. Quaternary deposits of the Central North Sea. 1: A new isopachyte map of the Quaternary of the North Sea. 2: The Quaternary deposits of the Forties field, northern North Sea. Report No. 77/11, Natural Environment Research Council.

4. Eden, R. A., Holmes, R. and Fannin, N. G. T. 1978. Quaternary deposits of the Central North Sea — depositional environment of offshore Quaternary deposits of the Continental Shelf around Scotland. Report 77/15, Natural Environment Research Council.

5. Fannin, N. G. T. 1979. The use of regional geological surveys in the North Sea and adjacent areas in the recognition of offshore hazards. International Conference Papers.

6. Floden, T. 1972. Notes on the bedrock of eastern Skagerfrak with remarks on the Pleistocene deposits. In *Stockholm Contributions in Geology*, Vol. XXIV, 1971–1973.

7. Floden, T. and Sellevol, M. A. 1972. Two seismic profiles across the Norwegian Channel west of Bergben. *Stockholm Contributions in Geology*, Vol. XXIV, 1971–1973.

8. Holtedahl, H. and Bjerkli, K. 1975. Pleistocene and recent sediments of the Norwegian continental shelf (62°N–71°Z) and the Norwegian channel area. *Norges. Geol. Unders.* 316, 241–252.

9. Holtedahl, H. and Bjerkli, K. 1982. Late Quaternary sediments and stratigraphy of the continental shelf off Trøndelag, W. Norway. *Mar. Geol.* 45, 179–226.

10. Hovland, M. 1980. Detailed seabed mapping for a pipeline across the Norwegian Trench. *Int. Hydrog. Rev.*

11. Hovland, M. 1981a. A classification of pockmark related features in the Norwegian Trench. Institutt for KontinentalSokkelundersøkelser (Continental Shelf Institute), Norway.

12. Hovland, M. 1981b. Characteristics of pockmarks in the Norwegian Trench. *Mar. Geol.* 39, 103–117.

13. Hovland, M. 1982. A coast parallel depression, possibly caused by gas migration, off western Norway. *Mar. Geol.* 50.

14. Hovland, M. Elongated depressions associated with pockmarks in the western slope of the Norwegian Trench. *Mar. Geol.* (preprint).

15. Jansen, J. H. F. and Hensey, A. M. 1981. *Interglacial and Holocene sedimentation in the Northern North Sea — an example of erosion deposits in the Tartan Field.* Spec. Publ. Int. Ass. Sediment. Vol. 5, pp. 323–334.

16. Jelgersma, S. 1979. Sea-level changes in the North Sea Basin. *The Quaternary History of the North Sea* (Eds E. Oele, R. T. E. Schüttenhelm and A. J. Wiggers). Symp. Univ. Uppsala.

17. Josenhans, H. W., King, F. and Fader G. B. 1978. A side-scan sonar mosaic of pockmarks on the Scotian Shelf. *Can. J. Earth Sci.* 15, 831–840.

18. King, L. H. and MacLean, B. 1970. Pockmarks on the Scotian Shelf. *Geol. Soc. Am. Bull.* 81, 3141–3148.

19. Mangerud, J. 1983. The glacial history of Norway. In *Glacial Deposits in North-west Europe* (Ed. J. Ehlers). A. A. Balkema, 1983.

20. McQuillin, R., Fannin, J. and Judd, A. G. (1979). IGS pockmark investigations 1974–1978. Report 98, Natural Environment Research Council.

21. Newton, R. S., and Cunningham, S. 1980. D'Appolonia Consulting Engineers Inc. Mud volcanoes and pockmarks: seafloor engineering hazards or geological curiosities? Offshore Technology Conference.

22. Platt, J. 1977. Significance of pockmarks for engineers. *Offshore Engineer*.

23. Powell, R. D. 1984. Glacimarine processes and inductive lithofacies modelling of ice shelf and tidewater glacier sediments based on Quaternary examples. *Mar. Geol.* 57, 1–52.

24. Rise, L. and Rokoengen, K. 1983. Surficial sediments in the Norwegian sector of the North Sea between 60°30'N and 62°N. *Mar. Geol.* (draft version).

25. Rokoengen, K. and Rønningsland, T. M. 1983. Shallow bedrock geology and Quaternary thickness in the Norwegian sector of the North Sea between 60°30'N and 62°N. *Norsk. Geol.*

26. Sellevol, M. A. and Sundvor, E. 1973. The origin of the Norwegian Channel — discussion based on seismic measurements. *Can. J. Earth Sci.* 11, 224.

27. Sieck, H. C. 1973. Gas-changed sediment cones pose possible hazard to offshore drilling. *Oil Gas J.*, 16 July.

28. Weering, Tj. C. E. van. 1977. Quaternary sedimentation in the Norwegian Channel and the Skagerrak, with emphasis on deposits from the Skagerrak. *Xth INQUIA Congress*, Birmingham, Abstr. 498.

29. Weering, Tj. C. E. van. 1983. Acoustical reflection profiles, sediments and Late Quaternary history of the Norwegian Channel north of Bergen. *Geol. Mijnbouw*.

30. Weering, T. van, Jansen, and Eisma, D. 1973. Acoustic reflection profiles of the Norwegian Channel between Oslo and Bergen. *Neth. J. Sea Res.*

9

Marine Geophysical Site Investigation for Exploration Drilling Rigs: Survey Contracts Appraisal

M. Sarginson and J. Yates, Comap Geosurveys Ltd, UK

ABSTRACT

The very substantial change in the scope of equipment and services offered by the marine geophysical site investigation contractor during the last fifteen years, and the investment costs which this change inevitably must incur, is not in line with the nature of site-specific survey contract awards as practised by the majority of oil companies in the North Sea and many international areas. Comparisons are discussed between site-specific survey contracting in these areas and the situation which prevails in the USA, where there is a requirement to survey the whole of a leased block and present the results to a government-appointed controlling body for approval and comment. In the USA subsequent site-specific analysis can either make use of data acquired in the block survey or, if there is cause for concern, additional investigation of a very restricted area can be commissioned. The deficiencies of site-specific surveys in isolation are discussed in detail and a case is made for the more general adoption of a two-tiered programme of investigation with 'regional' coverage supported, where necessary, by separately commissioned site-specific surveys or drilling investigations designed to further define or substantiate the 'regional' observations.

This should subtantially improve the efficency, value and reliability of marine geophysical site investigations. It will also allow the site survey industry to eventually structure itself into a mix of a relatively small number of large, well-equipped and very efficient data acquisition vessels supported by smaller vessels and specialist project units.

With this sort of structure and the alleviation of the problems accompanying fast turn-around site-specific acquisition and reporting, there is scope for consideration of industry-accepted 'codes of practice' in line with the regulations which control site investigation procedures in civil engineering studies.

The years 1970–1985 have seen a major

expansion in the number of contractors undertaking marine geophysical site investigation surveys, and in the range and complexity of instrumentation and hardware available. The period has also seen a substantial growth in continental shelf exploration, including movement into deeper water and away from the established areas of the USA, the North Sea, the Arabian Gulf, and South-East Asia. The commercial pressures on the contractor to exploit the market and keep pace with the technology have proved challenging, but the financial return on investment is highly unpredictable, though traditionally low. The number of survey companies which have been closed down, endured long periods of severe financial restriction or been subject to takeover and major restructuring, is perhaps too high to be regarded as healthy. If the survey contractor cannot expect an adequate financial return on his investment, then the interests of the equipment supplier/manufacturer and the oil exploration company/drilling contractor and civil engineering contractor must inevitably also suffer when out-of-date equipment becomes the norm and professional practices in acquisition and interpretation are limited by the contractor's own commercial considerations.

This paper describes many of the commercial and technical factors which are considered by the contractor when preparing bids for surveys. The paper also updates and extends some of the points of discussion which closed the geophysical survey section of the 1979 Site Investigation Conference. Some comparisons are drawn between survey specifications and contract negotiation in Northern Europe and other parts of the world. The net result is an argument for a more rational and constructive approach by both the geophysical contractor and by the oil and engineering companies and governmental institutions which direct the scope of work issued.

The paper is structured around the industry of providing geophysical site investigation studies for marine exploration drilling rigs. No accounting is made of similar survey works which can accompany oil production construction and development.

It will be helpful first to state the two aims of the geophysical site investigation survey for exploration rigs:

(1) To define and delineate those features which are a hazard to the safety and integrity of any drilling exploration structure or vessel within a specified area of operation.
(2) To assemble and suitably present as much geological, geotechnical, hydrographic and environmental information as is practical within a certain budget, and agreed scope; and to aid the timely, efficient, safe siting of the rig and its subsequent spudding and early drilling phases.

COMMERCIAL CONSIDERATIONS

In 1970–1975 the standard survey package for a rig-site investigation in the North Sea (Fig. 1) comprised:

- Navigation receiver (Hifix or Mainchain).
- Hydrographic echo-sounder.
- Side-scan sonar.
- Boomer/multi-electrode sparker seismic recording package.
- One or two hydrographic surveyors, a geophysicist and an electronics engineer to operate this equipment.
- A fairly basic vessel of 30–45 m length.

The survey contract was negotiated on a daily charter with the expectation of working only during daylight. Furthermore, the contractor was frequently required to stand by to guide the rig onto its surveyed location on completion of the investigation. This, together with weather limitations on acquisition, daylight working, plus standard day rates recognized by the industry, gave the contractor a respectable income. The 1985 situation shows a dramatic change in hardware and operating procedures. Figure 2 illustrates the development in hardware to meet industry requirements. Improved

1975 NORTH SEA SURVEY VESSEL

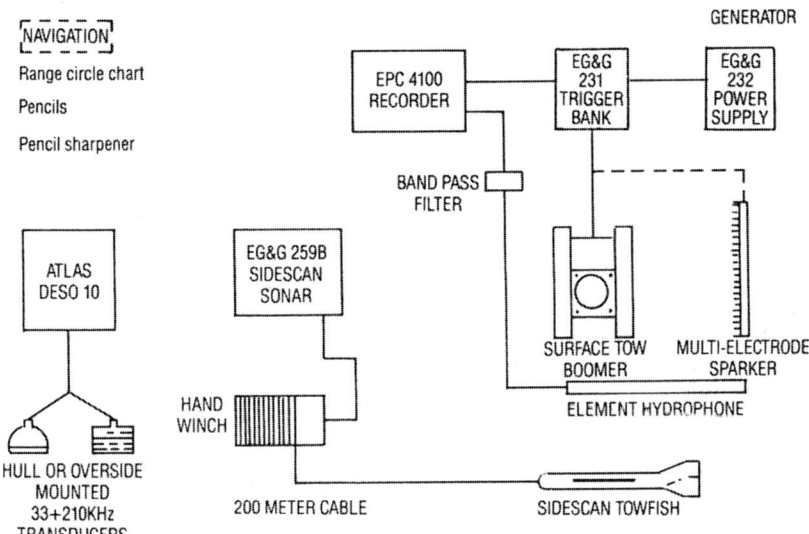

navigation aids and larger vessels mean 24-hour operation with a higher threshold of weather and sea state unsuitability. The survey crew required also shows an increase to 9 or 10. Table 1 shows a guideline comparison of capital costs for the 1975 and the 1985 packages as illustrated, based on 1985 purchase prices in US dollars. The table also indicates the required annual revenue which each of the two investments should recover to be commercially viable. From this, a range of percentage annual utilization figures are quoted and the day rate required to achieve the viable revenue is calculated. Below this the argument is progressed by estimates of other expenditure involved in the survey operation.

The fairly obvious end result of this financial analysis is that the contractor must gamble on high utilization figures if he is to compete at the rates currently prevalent in the established marine exploration areas, or he must seek strategic advantages in new areas of interest. The situation in northwest Europe is of particular interest in this context because there is perhaps a surplus of large, comprehensively equipped survey vessels competing for a share in a very limited market, which is composed primarily

of short-duration site-specific surveys. The fact that the oil industry in this area is geared to the issue of site-specific studies requiring fast turnaround does mean that it is also dependent on this surplus of survey vessels being maintained. As long as workload remains reasonable then the contractors will stand by waiting for numbers to diminish or workload to increase. The end result of a temporary reduction in workload must inevitably be a significant reduction in the number of contractors and a lack of preparedness when workload improves. The ongoing effect of the way the North Sea market is structured is a severe limitation on equipment investment and development.

Bid preparation, in particular the preparation of fixed-price bids, is very demanding on the contractor, requiring extensive amounts of time by experienced people researching logistic and operating costs. The process of decision on award of contract by the oil company should in theory be based upon:

(1) Price quoted.
(2) Availability.
(3) Suitability of vessel and equipment quoted.

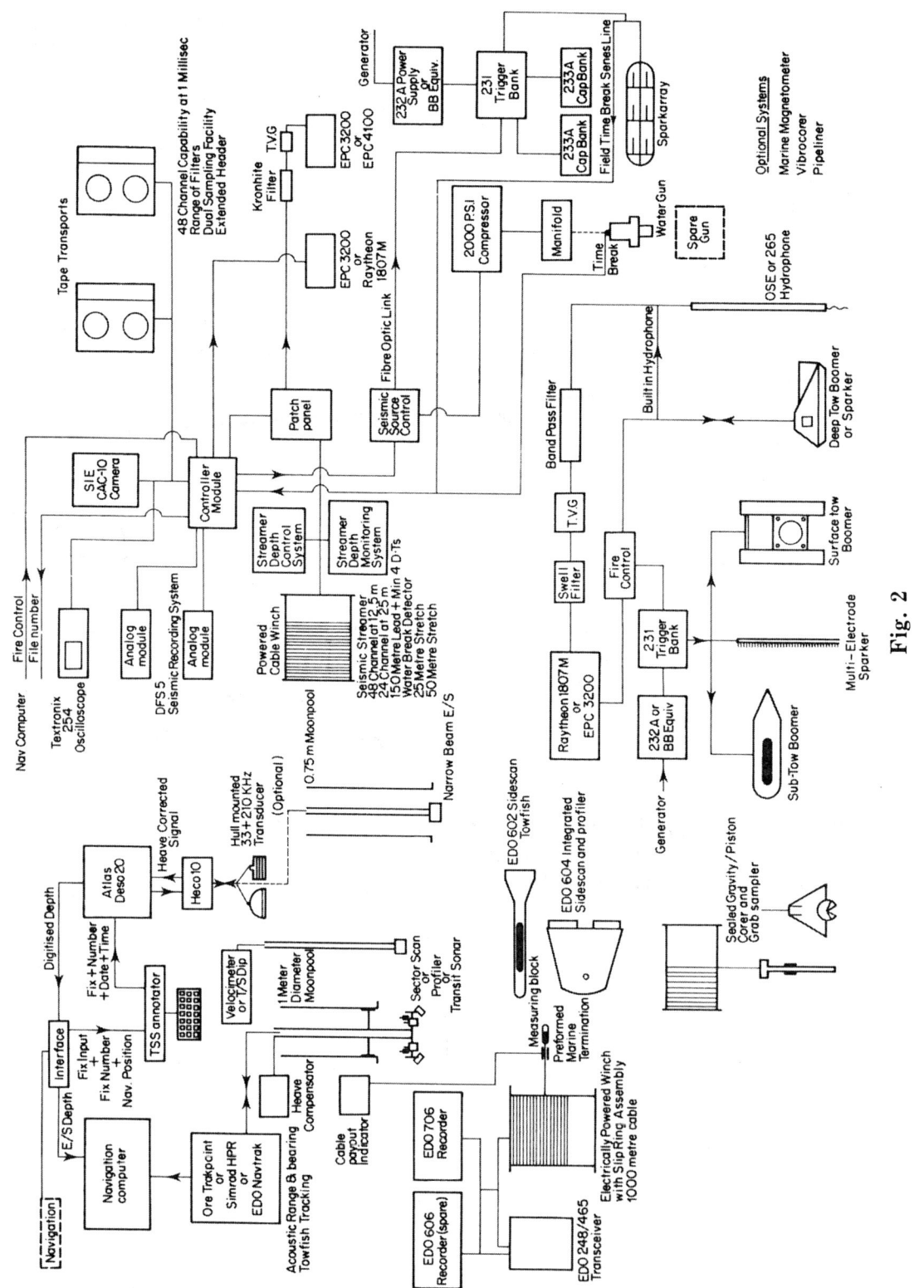

Fig. 2

TABLE 1
1975 Package (using approximate 1985 purchase values and costs)

	New purchase value	Minimum viable annual revenue (note 2)	Minimum day rate revenue required based on percentage annual utilization						
			30%	40%	50%	60%	70%	80%	90%
Survey equipment as illustrated in Fig. 1	115 000	45 430	415	311	249	207	178	156	138
Survey personnel and administrative overheads	—	note 3	712	628	577	534	511	494	475
Total (note 4)	—	—	1127	939	826	714	686	650	613
1985 Package									
Survey equipment as illustrated in Fig. 2	1 100 000	434 500	3968	2976	2380	1984	1700	1488	1323
Survey personnel and administrative overheads	—	note 3	1603	1522	1368	1347	1175	1122	1068
Total (note 4)	—	—	5571	4498	3748	3231	2875	2610	2391

Notes
1. All costs and prices quoted are in US dollars
2. Minimum annual revenue calculation based on: 2.5% insurance; 12% minimum borrowing rate; 5% net service costs: depreciation over 5 years
3. Simple calculation of personnel and administrative costs assumes 13 000 US dollars cost per average survey person, 50% administrative charge, 100% backup on survey personnel at 90% utilization reducing to zero backup at 30% utilization
4. The total calculations are exclusive of vessel charges and navigation hardware and chain rental charges

(4) The quality and experience of the personnel offered to undertake the work.
(4) The previous performances of specific contractors.

When dealing with small site-specific studies the very high ratio of preparation time to acquisition time strongly favours established vessels operating or on standby close to the site; and this is normally reflected in the price quoted. An oil company requesting fixed-price bids, particularly in an active area like the North Sea, must regularly be presented with situations where return bids show a variation between low and high bids of several hundred per cent. In this situation it must be very difficult not to ignore points 3, 4, 5 and possibly also 2, and award to the low bidder. The unhealthy aspect of the business is that the contract award may well go to the survey company which makes an error in its costing calculations and is awarded a contract which can only be made profitable if significant shortcuts are made in the execution of the contract.

SITE-SPECIFIC SURVEYS

At this stage it is worth considering whether the needs of the oil industry, the marine geophysical contractors and other interested bodies are best served by the present generally accepted system of contracting on a site-specific basis. The present system requires the oil company to commit itself to a specific drilling location and use this as the centre of a grid survey with area and grid parameters determined by water depth and the type of drilling rig to be used. The disadvantages of this system are numerous and, we think, highly significant.

(1) The commitment of the oil company to a specific drilling site substantially in advance of the rig becoming available creates problems in any exploration department. Normally target evaluation will continue right up to the arrival of the rig, as evidenced by the occurrence — which all survey contractors have experienced — of changes of location part way through a survey or interpretation. The normal consequence of this is that the oil company will delay the award of the site survey contract as long as it thinks possible and place the burden on the survey contractor, data processing centres and the interpreting geophysical team to complete every phase in as little time as possible. Little, if any, allowance is made for bad weather, delays or equipment failure. The end results are frequently a limitation of choice of survey equipment and equipment to those nearest to the location; limitations on quality control in data acquisition and processing; and the issue of a rushed report which may be sufficient to allay any fears held by the drilling rig operator or the insurance company involved but provides little constructive assistance for the spudding and early drilling phase.

(2) The interpretation of data from a small area in isolation invariably restricts the geophysicist to discussion and illustration of the obvious features of the area, such as seabed topography and the presence of any apparent geological unconformities near to the seabed. Seismic horizons are frequently selected and contoured only to demonstrate the presence or lack of any structural features. The analysis for drilling hazards or constraints is primarily limited to a qualitative assessment of anomalous features

on the seismic sections. It is important to point out that a complete site-specific survey in parts of the North Sea could be contained within the margins of a large infilled glacial channel and, treated in isolation, the interpreting geophysicist may be totally unaware of the unusual nature of the site and the potential problems the drilling contractor may encounter with unusual sediment infill. Similarly, site-specific analysis may identify a relatively strong seismic reflector as a useful but inconsequential lithostratigraphic marker horizon, whereas, if part of a more extensive survey, the horizon may only demonstrate this 'brightness' locally and the interpreter would then have to seek an explanation which may be the presence of cemented material or high-pressure gas.

(3) intensive geophysical investigation of a 3 km by 3 km site may only take 24–36 hours of survey work. However, it is usually accompanied by mobilization and transit to and from the site; calibration and perhaps mobilization of area-specific navigation systems; general geophysical system testing, including such time-consuming duties as seismic-streamer cable balancing and satisfying the client company or its representatives as to the performance of the survey package. All this frequently results in a preparation period which grossly exceeds the actual survey period. Whether the contract has been awarded on a day rate or a fixed sum, these preparation charges — or at least an estimate of their potential cost — are included in charges made to the client oil company.

REGIONAL SURVEYS OR SURVEYS OF COMPLETE LEASE AREAS

The alternative to site-specific surveys is a system of more extensive surveys covering a complete leased area — or at least the whole of the area of potential interest — in one project commissioned a good time before any exploration drilling takes place. The advantages of this approach include:

(1) A more extensive understanding of the geological and geotechnical hazards and constraints. This allows better integration into a regional setting, if known, and allows the interpretation team to set up baselines which can quantify the hazard or constraint potential.

(2) Bigger budgets and the higher ratio of acquisition time to preparation time means a better choice of optimum contractor and survey package.

(3) The larger scale of the project can allow a substantially improved level of preparedness with exploration departments, research units with specific interests or knowledge of the area, and the contractor all contributing to the optimum design and orientation of the survey grid and to the survey data acquisition package.

(4) In many areas the exploration department of the oil company can substantially benefit from good quality processed multichannel data using powerful high-frequency sources. Such data can provide information for: (a) determining drilling hazards; (b) detailing the structure of an area; and (c) assessing hydrocarbon potential.

(5) Government research bodies and, ultimately, the industry in general can benefit by the provision of survey studies which are properly integrated into any known geological framework. The larger survey can be incorporated into existing geological or geotechnical boreholes and study areas.

(6) Because of the much higher ratio of data acquisition time to survey preparation time compared with small site-specific studies, substantially more data can be acquired for the same cost as a small number of separately commissioned site specifics. Moreover, the data from a single extensive investigation should be uniform in quality, presentation and specification in order to ensure direct feature correlation.

Continental shelf legislation in the United States requires that, prior to drilling exploration, a properly conducted geophysical and geological survey on a specified grid pattern must be completed and a report on the potential for geological hazards submitted to the appropriate regional department of the Minerals Management Service for approval. The data from the survey also have to be made available to that department if requested. Approval can be denied for incomplete, inadequate survey work or unsatisfactory interpretation and reporting. The department can recommend additional investigative work if it believes it to be necessary. This control is generally sufficient to persuade most oil companies to commission comprehensive site investigation studies of complete lease blocks shortly after the award of the lease to avoid potential difficulties in any planned drilling schedule later on. The Minerals Management Service use the data acquired and the reports submitted to prepare useful regional studies and guidance notes and charts on problems to be expected in blocks released in subsequent lease sales. If the investigation reveals possible hazard features at or near a planned drilling site which the oil company wishes to use, then the company will invariably commission a site-specific survey which may involve either additional geophysics or a geotechnical drilling investigation.

CONCLUSIONS

This two-tiered approach to site investigation has substantial merits when considering the primary roles of the investigation in minimizing risk to any exploration drilling project and providing useful data to optimize site selection, rig-type selection, and associated preplanning. The approach is also beneficial to the site investigation industry because it improves the integration of geophysical and geotechnical investigation and allows the geophysical site investigation industry to structure itself. A move away from a large number of small sites — frequently with tight deadlines — to a smaller number of extensive surveys supported, where necessary, by specialist site-specific investigations will create a market for a limited number of large, well-equipped, and efficient geophysical data acquisition vessels and packages, with a second level of lower-cost specialist investigation packages and companies. The present situation dictates that virtually every vessel must be equipped with a suite of very comprehensive and expensive equipment; yet the majority of time is spent standing by, preparing, or in transit to short-duration contracts.

Looking beyond this efficiency problem and the possible alleviation of the tight deadline policy, there is a need for the contractor's geophysical team and specialist consultancy groups to create a more responsible and respected role for marine geophysics in site investigation, and exploration work in general. Some degree of standardization in the broader aspects of data acquisition, interpretation and the content and presentation of reports should eventually be consolidated into a 'code of practice' similar in concept to that which acts as a control for most civil engineering site investigation projects. The more responsible role requires a more positive and quantified approach to site hazard and constraint assessment than is normal in present practice. The latter would be significantly improved by the adoption of more extensive surveying. It would also benefit from better dialogue between geophysical contractors and exploration drilling engineers and rig supervisors. A third area of benefit would be in regional control by an individual or institution with a brief to define hazard potential by correlating survey reports with observations recorded during rig siting and drilling.

With a more responsible and respected role, there may be much more enthusiasm for dialogue from the oil company exploration and drilling departments concerned. instead of a pro-forma issue of a request to conduct an investigation survey on a particular site, there could be direct liaison with the contractor's geophysical team or a consultant directly involved in acquisition and interpretation. This should greatly improve the value of the report supplied. The supply or discussion of proprietary seismic or nearby drilling data to which the survey can be tied, and open discussion on the potential problems of the area, particularly with reference to the type of drilling rig and drilling schedule, can greatly improve the methods of acquisition and report presentation.

10

In Situ *Testing and Sampling Offshore in Water Depths Exceeding 300 m*

A. F. Richards and H. Zuidberg, Fugro BV, Leidschendam, The Netherlands

ABSTRACT

In this paper 'deep water' is defined as being at depths greater than 300 m. A few past events in deep-water testing and sampling which are relevant to the development of new equipment are reviewed; and some deep-water geotechnical problems that require the development of new equipment for their investigation are discussed. The paper also contains a summary of new testing and sampling equipment either at present in use or under development for deep-water site investigations. It looks at future developments by industry in water depths of less than 1000 m and by government and academe in water depths exceeding 4000 m. Most new testing and sampling equipment discussed has been developed or deployed since about 1982.

INTRODUCTION

There are five principal objectives in writing this invited paper on *in situ* testing and sampling in deep water:

(1) to discuss the connotation of the words 'deep water';
(2) to review briefly a few past events in deep-water testing and sampling that relate to the development of new equipment;
(3) to present briefly some current deep-water geotechnical problems;
(4) to cite new deep-water geotechnical equipment very recently developed and presently under development by industry, government and academe for site investigations; and
(5) to look to the future regarding possible industrial developments in water depths of less than about 1000 m and likely development by government and academe in water depths greater than about 4000 m.

Rock sampling equipment used offshore, and shipboard or shore laboratory techniques — such as the use of X-radiography for

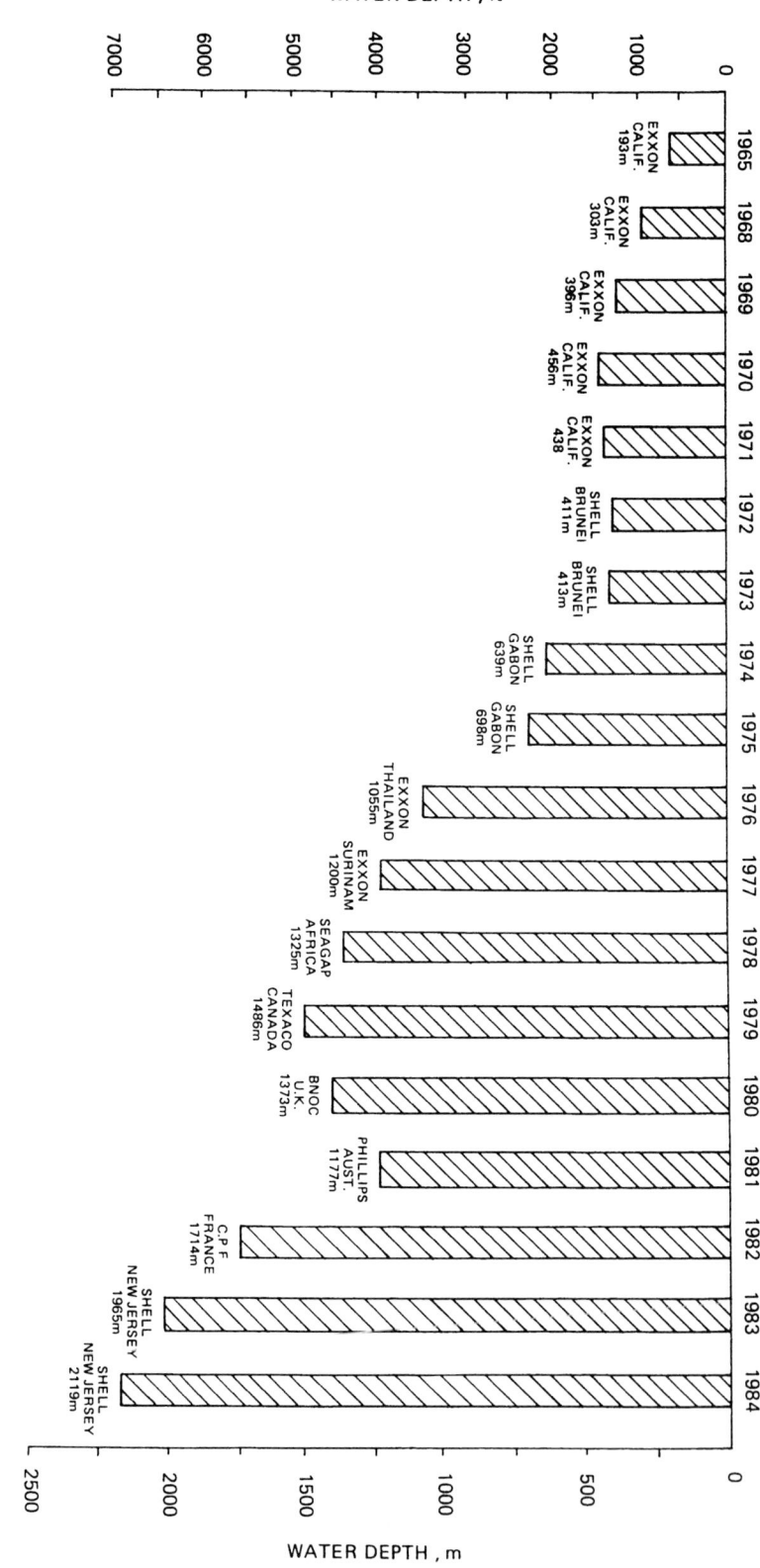

Fig. 1 Water-depth records by year for drilled oil wells. (Modified from a Shell Oil Company illustration)

the softer soils — are not discussed in this paper.

The words 'deep water' and 'deep sea' are almost meaningless because of widely differing usage in oceanography and geotechnical ocean engineering. In the offshore oil and gas industry the words 'deep water' usually refer to the water depth at which exploration and exploitation are occurring within a particular geographic region at a given time. Figure 1 shows that the maximum deep-water depth has been generally increasing during the past twenty years or so. In 1965 the drilling record was almost 200 m. In early 1985 the record is more than ten times deeper. This depth trend is likely to increase in the future with water depths of 3000 m being predicted by the year 2000. 'Deep water' in the oil and gas industry in 1985 generally refers to a maximum water depth of about 300 to 1000 m, depending on the region. People in the nascent polymetallic ('manganese') nodule and polysulphide mining industry use the term 'deep water' to refer to operational water depths of about 3000 to 5000 m.

Academic oceanographers and ocean engineers have very different requirements in the 'deep sea' compared to industry. They usually refer to the deep sea floor as being a province separate from the continental mar-

gin and one characterized by water depths greater than about 4000 m (Fig. 2). The specific deep sea floor bathymetric provinces adjacent to the continents usually are an abyssal plain next to a passive margin and a trench next to an active margin. The industrial use of the name of the provinces shown in Fig. 2 is strongly recommended because each has a universally understood water-depth range associated with it.

It is recommended that the general expression 'deep water' be abandoned whenever possible in favour of citing a specific water depth or depth range. Deep water in this paper signifies water depths in excess of 300 m. This is about the minimum limit of the transition range of depths for using most conventional industrial offshore geotechnical equipment (Table 1).

The selection of 300 m precludes a description or discussion of many new and fascinating instrumentation designed to operate at lesser depths, such as the Dutch and Brazilian geotechnical diving bells (Graaf and Smits, 1983; Hunt, 1984), new drilling, sampling and testing systems (e.g. Okumura and Matsumato, 1981; Ardus *et al.*, 1982; de Haas, 1983; Beard and Johnson, 1984; Pheasant, 1984), mobile geotechnically instrumented crawlers (Benoit, 1983), most pressuremeters, and the dilatometers. On the other hand, innovative equipment presently operated in shallow water will be briefly included if it serves to illustrate what can or should be done in water depths greater than 300 m.

Emphasis has been placed on new developments in water depths greater than 300 m in this paper. But it is very important to keep firmly in mind that there are a substantial number of older tools and techniques that are used offshore in water depths greater than 300 m by the geotechnical consulting companies. This use will be discussed later.

A number of useful reviews of offshore sampling and *in situ* testing equipment have been made during the past few years (Andresen, 1981; The Sub-Committee on Soil Sampling, 1981; Vollset and Gunleik-

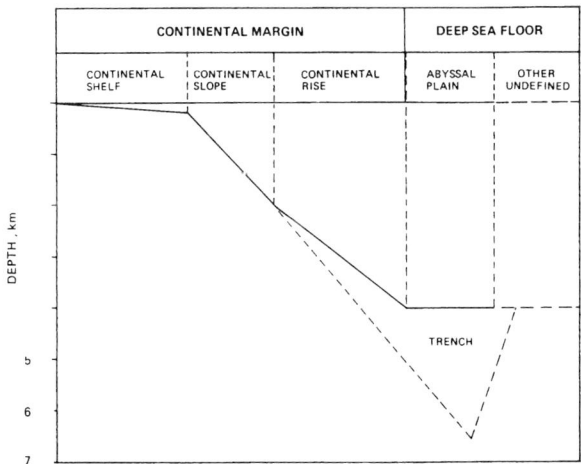

Fig. 2 Bathymetric provinces of the continental margin and adjacent deep sea floor. The ater depths given are the usual averages

TABLE 1
Water depth related to industrial geotechnical drilling, sampling and testing use

Water depth (m)	Geotechnical use
<30	Approximate maximum water depth to use conventional land drilling rigs and associated sampling and *in situ* testing equipment efficiently
30 to 300	Approximate range of water depths to use almost all conventional offshore geotechnical drilling rigs and associated sampling and *in situ* testing equipment very efficiently
300 to 1000	Approximate transitional range of water depths to use some conventional offshore geotechnical drilling rigs and associated sampling and *in situ* testing equipment relatively inefficiently, as well as the range to use other conventional equipment efficiently
>1000	Approximate water depth requiring specialized drilling rigs used for oil and gas exploitation rather than geotechnical drilling rigs; downhole samplers only used by the Ocean Drilling Project; no downhole geotechnical *in situ* testing equipment available (1984); only research-type, seabed *in situ* testing and sampling equipment available

srud, 1982; Marsland and Windle, 1982; Schultheiss, 1982; Briaud and Meyer, 1983; Toolan, 1983; Aas *et al.*, 1984; Wroth, 1984; Lunne *et al.*, in press). It should be noted that a number of the machines described in several of these and other references were never built; some land equipment planned for offshore use and some designed and built for offshore use were never deployed offshore; and some equipment that was used offshore has been retired. Richards and Zuidberg have summarized generic methods of offshore sampling and testing (in press, b) and methods of offshore *in situ* strength determination (in press, a). These two papers contain many references to recent reviews of *in situ* testing methods not cited in this paper. Lee (in press) has done the same for sampling and laboratory testing. In addition, Zuidberg *et al.* (in press) review the outfitting of a geotechnical vessel for advanced drilling, sampling and *in situ* testing in deep water.

SOME SAMPLING AND TESTING PRECEDENTS

Arrhenius (1952) initiated the first geotechnical work in ocean-basin water depths in 1947–1948 on the Swedish *Albatross* Expedition around the world. He used the new Kullenberg (1947, 1955) piston corer and a new deep-sea winch to obtain long cores. The Swedish Fall Cone (Hansbo, 1957) was used in the ship's laboratory to obtain from the cores 'relative shear strength' values, which were later converted into conventional strength units (Moore and Richards, 1962).

It may be useful to remember that many of the basic principles of *in situ* testing and sampling are not new. Schlosser (1978) summarizes a number of *in situ* tests on land and their development dates in the United States and Europe. Flodin and Broms (1981) and Andresen (1981) expand on details. For instance, the drive-wheel or roller method of continuously driving test rods into soil on land dates back to 1950 when the Swedish Geotechnical Institute's sounding machine was developed (Kallstenius, 1961; Dahlberg, 1974). This was the beginning of the modern offshore continuous wheel-drive seabed systems built in Sweden and the Netherlands in the 1980s.

There are three basic types of continuous drive systems designed for seabed machines. The first is the hydraulic push-and-puller apparatus built by Borros AB in Sweden for use on land. It consists of four hydraulic cylinders operating two by two in pairs connected with one automatic test rod clamp under each pair. Penetration or retraction of the test rod is said to be continuous. In 1984, this machine was success-

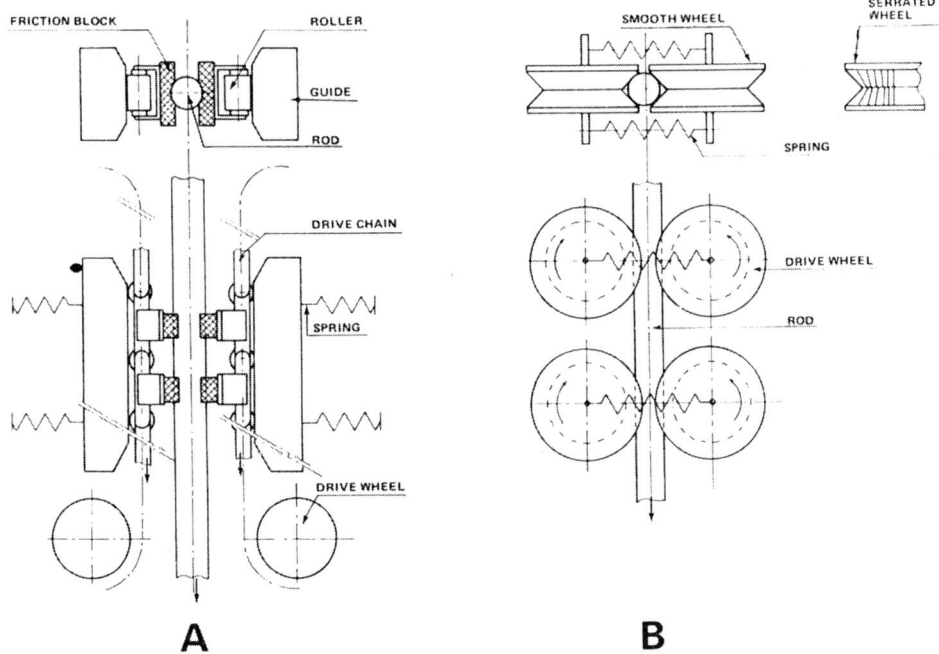

Fig. 3 Two basic types of wheel-drives to obtain continuous penetration of test rods in a seabed penetration apparatus. A, friction generated by elastic deformation of the material pressed against the rods. B, friction generated by slight plastic deformation of the test rods driven by smooth or serrated drive wheels

fully adapted by McClelland for offshore operation. The second type of drive-wheel mechanism drives the test rods by the friction of continuously moving pads pressed against the test rods (Fig. 3A). The pads are composed of an elastic material such as neoprene or synthetic rubber. A 'Catapuller' drive system of this type was devised by N. D. Remy & Associates for the Lehigh University Geotechnical *In situ* Probe System (GIPSY) that was designed and constructed in the 1970s and early 1980s. Laboratory tests showed that the type of neoprene used in the pressure pads, which in this machine pressed against a hollow square test rod, was critical. Material that was too soft was deformed plastically or stripped off the pads. This type of drive does not mark the test rod. The third type depends upon developing friction by a very small plastic deformation caused by the wheel pushing against the test rod (Fig. 3B), which tends to mark the rod with flat faces or teethmarks and generally requires more frequent

rod replacement if substantial drive forces are to be developed. The drive-wheel can be smooth or serrated. This type of drive system is used by Borros for land penetration equipment, for the offshore Van den Berg ROSON, and for Fugro's offshore DORA and Sea Lion that will be subsequently described. There are now so many different types of deployment methods used for *in situ* testing offshore in water depths greater than 300 m that it was deemed useful to summarize most of them (Table 2).

People who are not intimately connected with the industrial offshore geotechnical profession frequently lose sight of the fact that sampling and *in situ* testing equipment is usually developed to meet the specific needs of clients, particularly in the oil and gas industry. It is not generally appreciated that the oil and gas industry will use almost any new piece of equipment once, but if it does not provide useful data or is difficult to use at sea it may rarely be used again. This provides a useful filter for the retention of

TABLE 2

Examples of deployment methods used for in situ *testing*

System and subsystems	Example	References
Downhole		
Carrier tool, probe, sensor(s)	Various	Richards and Zuidberg (in press, b); Zuidberg *et al.* (in press); this paper
Seabed		
Penetration apparatus		
Cable type		
multiple	Seacalf	Zuidberg (1975)
	Stingray	Ferguson *et al.* (1977)
single electromechanical	Tower	Richards *et al.* (1972)
single mechanical; acoustic telemetry	DORA	This paper
single mechanical; seabed recording	Sea Lion	This paper
no cable, seabed recording	ISV	Babb and Silva (1983)
	ISHTE	Bennett *et al.* (1983a)
Test rod drive type		
rigid test rod		
screw rod	Tower	Richards *et al.* (1972)
chain drive	XSP	Beard and Johnson (1984)[a]
discontinuous jack	Seacalf	Zuidberg (1975)
	Stingray	Ferguson *et al.* (1977)
continuous jack (push and puller)	CPM	Amundsen *et al.* (in press)
drive wheels or 'roller jack'	ROSON	Van den Berg (1984); this paper
flexible test rod	PAM	Le Tirant *et al.* (1981)
Bottom crawler	RUM	Andersen *et al.* (1972)
Free-fall (fixed penetration depth)	PUPPI	Schultheiss *et al.* (in press)
Gravity emplaced or 'impact'	–	Richards *et al.* (1975)
	–	Chari *et al.* (1979)
	GISP	Pringle and Lopez (1983)
Penetrator (variable penetration depth)		
Free-fall (no wire)		Beard (1977; in press); Freeman *et al.* (1984); Talbert (1984a)
free-fall (wire for communications)	–	Colp *et al.* (1975); Ingram (1982)
gun-launched ISP	ISP	James and Calloway (1983)[a]
Submersible		
Manned	–	Inderbitzen and Simpson (1972); Richards (1972); Lambert (1982); Geise and Kolk (1983)
Unmanned	–	Geise and Kolk (1983)[a]

[a]Not designed for water depths greater than 300 m, but probably capable of operation at such water depths with modifications

TABLE 3
Sampling and in situ *testing equipment most commonly used by the offshore geotechnical industry*

		Examples[a] *Fugro*	*McClelland*
I.	Sampling equipment		
	A. Downhole		
	1. Hammer or percussion sampler	x	x
	2. Push sampler	Wipsampler	Stingray
	3. Piston sampler	Hydraulic piston sampler	Dolphin
	B. Seabed		
	1. Drop or gravity corer	x	x
	2. Piston corer	x	x
	3. Vibrocorer and vibrohammer corer	x	x
II.	*In situ* testing equipment		
	A. Downhole		
	1. Cone penetrometer	Wison	Stingray
	2. Piezocone penetrometer	x	Dolphin/Stingray
	3. Vane	*In situ* vane	Remote vane
	B. Seabed		
	1. Cone penetrometer	Seacalf	Stingray
	2. Piezocone penetrometer	Seacalf	Stingray
III.	Seabed testing systems	Seacalf	Stingray
IV.	Seabed drillstring clamp	Seaclam	Stingray

[a]Water and penetration depth compilations of various offshore equipment have been published by Richards (1982), Vollset and Gunleiksrud (1982) and Briaud and Meyer (1983)

x, unnamed equipment in common use

useful, easy to use, economical and successful equipment and the rejection of equipment which is not useful, difficult to use, uneconomical (prone to failures) and unsuccessful. The equipment listed in Table 3 has survived the test of time and has gained the acceptance by the oil and gas industry for common use offshore. It will be noted that this list is purposely restrictive and does not include the pressuremeter, which is used only 2–3% of the available time by industry in water depths greater than about 30 m (Richards and Zuidberg, in press, a), or the seabed vane. But push-in downhole pressuremeters capable of operation in water depths greater than 300 m do exist (Hender-

son *et al.*, 1980; Fyffe *et al.*, 1982; and others), as do self-boring machines such as the PAM, which will be mentioned later. Many shallow-water self-boring pressuremeters, of course, are in use by industry (e.g. Hughes *et al.*, 1984).

In Table 3 and elsewhere in this paper, the word 'downhole' refers to sampling or *in situ* testing equipment that is lowered by wireline or umbilical through the drillpipe to the bottom of the drillstring. The word 'seabed' refers to any equipment lowered through the water column to rest on or in the seabed. Examples of the latter include seabed jacking or penetration machines, conventional piston corers and so forth.

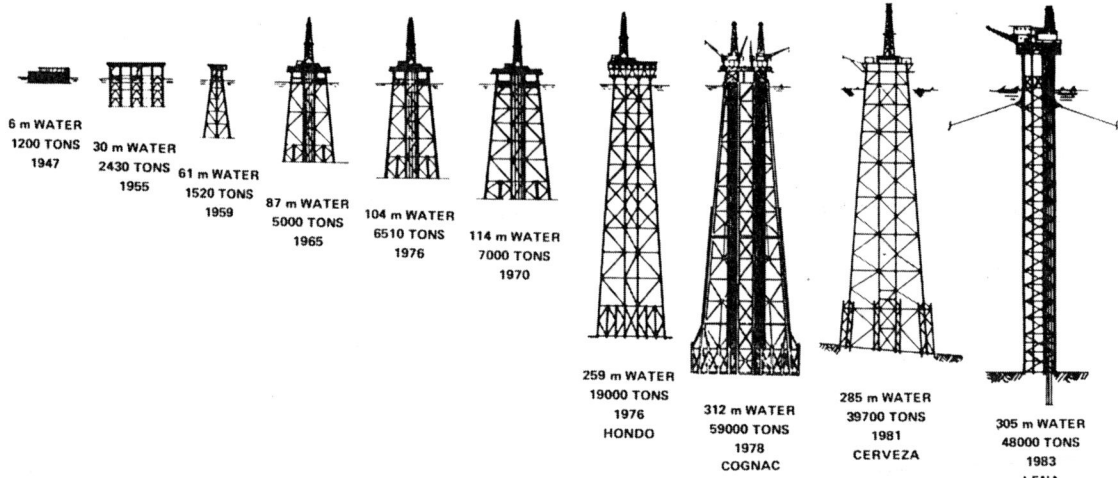

Fig. 4 Significant offshore platforms considered to be 'deep water' in the year listed that have been installed in the Gulf of Mexico and offshore of California. (Modified from a Shell Oil Company illustration)

Details are given by Richards and Zuidberg (in press, b).

The reason for the restricted number of samplers, probes and sensors used in the offshore industry is related to the nature of the foundations to be investigated. The development of significant jacket structures over almost forty years in the Gulf of Mexico and offshore of California is shown in Fig. 4. These structures require piles. The geotechnical site investigations to determine vertical and lateral pile capacities are made by drilling and downhole sampling, which has been augmented by cone penet-

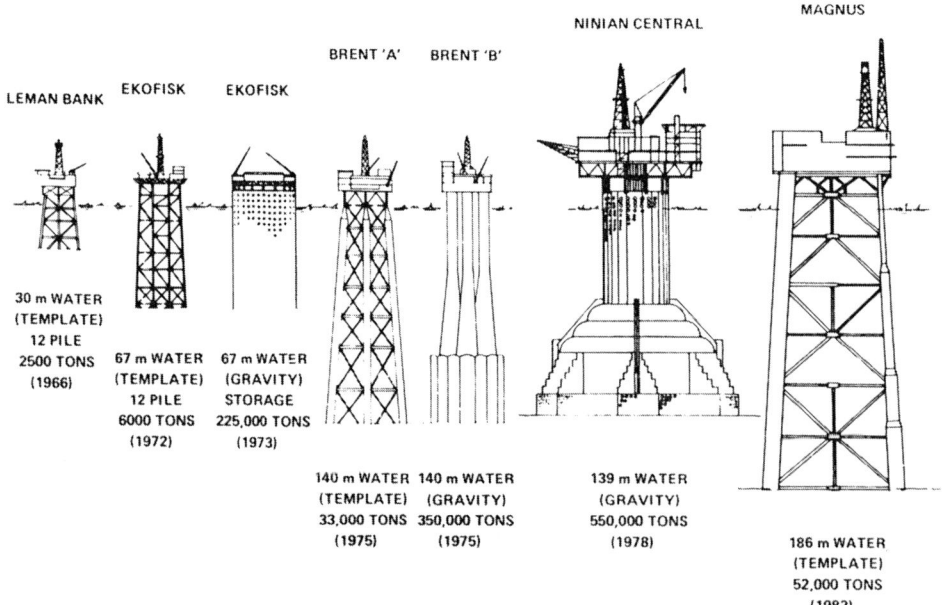

Fig. 5 Significant offshore platforms considered to be 'deep water' in the year listed that have been installed in the North Sea. (Modified from a Shell Oil Company illustration)

rometer or piezocone tests in recent years. Comparable significant platforms considered to be 'deep water' at the time of emplacement in the North Sea have included jacket structures and also gravity-base structures (Fig. 5). Geotechnical site investigations for these structures have required the use of a greater range of tools, including both downhole and seabed equipment (Table 3). Examples of site investigations in water depths greater than 300 m have been given by Vyas *et al.* (1983), Moeyes and Hackley (1983) and Amundsen *et al.* (in press), which show the application of testing and sampling equipment. Lunne *et al.* (in press) discuss how a number of offshore *in situ* tools are used to obtain design parameters.

In summary, the improvement of existing tools and deployment systems and the development of new tools and deployment systems in water depths greater than 300 m for the oil and gas industry is likely to follow closely the developments for operations in lesser water depths in this highly competitive field. On the other hand, academic and governmental engineers and scientists, who are not as severely constrained to keep costs as low as possible while maintaining a full production capacity, are likely to continue to develop a different variety of tools and techniques for other applications.

CURRENT GEOTECHNICAL PROBLEMS

Geotechnical instrumentation is a means to an end and not an end itself. Progress is made when an effective new tool is designed and successfully used to solve an existing problem. There is no shortage of significant industrial or governmental–academic geotechnical problems offshore in water depths extending from 300 to over 6000 m.

Industrial oil and gas related geotechnical problems tend to be closely associated with investigations of site-specific areas for the foundations of steel jacket structures, gravity-base structures, hybrid (jacket and gravity-base) structures (Figs 4 and 5) and

TABLE 4
Marine soil distribution by generalized types[a]

Continental slope
Calcareous oozes
Ocean basins
Water depth <4–5 km: calcareous oozes
Water depth >4–5 km: pelagic clays

[a]Modified from Richards and Zuidberg (in press, a)

special types of structures, such as the tension leg platform and the guyed-wire towers. Pipelines from these structures or from subsea production structures to shore or elsewhere have their own particular foundation problems. In water depths greater than about 300 m, problems for both structures and pipelines include soft soils, which may be comprised of the remnants of calcareous and siliceous organisms (Table 4), and soils that contain dissolved gas and perhaps gas hydrates or clathrates, which have recently been found within a few metres below the seabed in the northwestern Gulf of Mexico (Brooks *et al.*, 1984) and elsewhere on the continental margin (Kvenvolden and Barnard, 1982). In addition, some soils may be so rich in organic matter that their sensitivity increases and they become susceptible to fluidization (Keller, 1982), and quick clays having normal porewater salinities of about 3.5% may be found on the continental shelf and on the deep sea floor (Richards and Zuidberg, in press, a). On the other hand, Chaney and Fang (in press), who have examined the geotechnical behaviour of a number of shallow- and deep-water marine soils, conclude that all marine soils appear to behave in accordance with the principles of conventional geotechnical theory.

Governmental and academic geotechnical investigators at present are concerned with at least four specific sets of problems and one general area of activity. Since Carson (1977) demonstrated that surficial seabed soils could be consolidated as the result of horizontal compressive forces at plate boundaries, there has been a spate of papers dealing with the geotechnical problems of the active or convergent continental margins or trenches (Von Huene, 1984). These problems include consolidation and deformation (Carson et al., 1982), venting of pore waters from subducted soils with attendant methane-derived carbonate cementation (Carson et al., 1984; Kulm et al., 1984; Suess and Massoth, 1984) and relating the state of overconsolidation to the rate of convergence (Shepard and Bryant, 1983).

The second set of problems has to do with the recent and unexpected discovery of extensive and, presumably, major faulting of soils that does not appear to be related to basement topography in two abyssal plains of the North Atlantic Ocean (Duin et al., 1984; Schüttenhelm et al., in press). It is likely that a geotechnical process has caused the normal faulting, although it is not yet known which one. The proven existence of normal faulting implies the existence of forces capable of generating excess pore pressures and consequent hydraulic flow of pore fluids, which in turn is very relevant to the high-level radioactive waste disposal investigations (NEA, 1984). This would tend to corroborate a hypothesis of excess pore pressures in Indian Ocean soils (Abbott et al., 1981) and the proposed paradigm that all cohesive marine soils deeper than a few metres from the seabed may exhibit excess pore pressures in situ (Richards, 1984)

The third set of problems concerns the venting of hot, mineral-rich pore fluids at divergent plate boundaries, such as the hydrothermal vents in the middle of the Red Sea (Guney et al., 1984), in the East Pacific Rise in the Pacific Ocean and elsewhere (Rona et al., 1984; Rona, 1984). This type of venting may have produced the mineralization causing the highly unusual brittle soils having very high water contents and also high shear strengths in the Pacific Ocean nodule province (Richards and Parks, 1977).

The fourth specific set of problems is concerned with the in situ measurement of shear waves to determine the shear modulus and other parameters reflecting dynamic

forces of concern to the geotechnical engineer. A later section will discuss new instrumentation to investigate shear waves.

The general area of activity concerns the formation of a (Marine) Geotechnical Consortium to investigate in detail selected International Project of Ocean Drilling (IPOD) cores up to 200 m long from the deep sea floor in different oceans. To date, the Walvis Ridge in the South Atlantic Ocean has been studied (Geotechnical Consortium, 1984), the investigations of the North-West Pacific core have been completed (The Marine Geotechnical Consortium, in press), and the cores collected from the Mississippi Fan in the Gulf of Mexico are presently being analysed. These studies in particular have yielded much information about soils that are apparently 'over consolidated' near the seabed and 'underconsolidated' at greater depths. A recent discussion (Richards, 1984) speculated that this relationship was real. Unpublished information, on the other hand, suggests that for the deep sea floor (Chaney, 1985, personal communication) and for nearshore soils (Sills, 1985, personal communication) the apparent over and under consolidation states may be artifacts of the test method in a laboratory consolidometer or oedometer. In addition, other scientists and engineers have studied the geotechnical properties of various Deep Sea Drilling Project (DSDP) and IPOD cores. Bennett et al. (1984) summarize strength measurements on only the DSDP cores.

These four sets of problems and one general area of activity have already generated new types of in situ testing equipment for site investigations, which will be described subsequently. When earth scientists better understand that marine geotechnology is a means of quantifying many seabed geological, geophysical and geochemical processes it is likely that additional equipment will be constructed for these investigations on the continental slope and rise and on the deep sea floor, which will complement the ongoing geotechnical engineering studies.

NEW DEVELOPMENTS

The following list of new developments of *in situ* testing and sampling equipment in water depths greater than 300 m is almost certainly not exhaustive. Nevertheless, the selected tools presented should at least convey the range of developmental activity. In the discussion that follows, *in situ* testing equipment will come first, sampling equipment second and certain geophysical tests for geotechnical purposes third. Summary tables are presented for each major category. Within each of the three categories, downhole equipment will be considered first and then seabed equipment. Conventional downhole logging methods are not included because they have been adequately covered by Threadgold (1980).

Before proceeding, one new shipboard development is first discussed: vertical heave compensation of the drillstring when drilling is difficult in very soft soils because there is not enough soil–drillstring reaction for even the normally efficient crown-block, or any other drillstring heave compensator to respond to. To cure this problem, Fugro engineers developed a 'hard-tie' rigging system that successfully ties together the seabed clamp or jack line-tensioner system and the drillstring heave compensator. The Fugro hard-tie system was first used from the MV *Bucentaur* in water depths of over 300 m in the Troll Field in 1983 (Moeyes and Hackley, 1983). When drilling the drillstring vertical movement was found to vary from about 10 cm when the ship was heaving about 50 cm, and 30 cm when the *Bucentaur* was heaving about 3 m. Seabed measurements were made using a television camera mounted on the seabed clamp or jack. The hard-tie system performed satisfactorily with a load range on the wire of up to 4 tonnes.

Downhole *In Situ* Testing Equipment

Two new improvements in downhole geotechnical testing have appeared in the last several years. The McClelland Sword-

fish system has been augmented by the Dolphin, which is a mud-driven system to operate push and piston samplers, cone and piezocone penetrometers, and the Remote Vane. The Dolphin has been used in 750 m of water and is designed for 900 m (Amundsen *et al.*, in press; Bayne, 1985, personal communication).

The Oyo PS Suspension Logger will be discussed later in the section on offshore shear-wave measurements.

Seabed *In Situ* Testing Equipment

Seabed Systems Used

The variety of new seabed testing equipment used — or capable of being used — in water depths greater than 300 m is large (Table 5). Equipment listed in this table is generally in addition to the list of older equipment having a depth capability in excess of 300 m previously published (Richards, 1982; Vollset and Gunleiksrud, 1982; Briaud and Meyer, 1983). Short discussions of each piece of equipment listed will supplement the tabulated information.

Pressiomètre Autoforeur Marin, or PAM, is the name of the offshore self-boring presuremeter developed by the French Institute of Petroleum near Paris. First described by Le Tirant *et al.* (1981), this machine has been further discussed in many publications. A paper on the PAM is included in this volume (Faÿ *et al.*, 1985). Le Tirant (1984, personal communication) considers the PAM to be primarily a research tool and not a production piece of equipment.

The A. P. van den Berg Ingenieursburo BV Rotating Seabed Static Cone Penetrometer, or ROSON, consists of one to twelve ROSON 25 kN driving modules, each of which has two grooved drive wheels that continuously penetrate or retract a test rod in the manner shown in Fig. 3B. The power source is electrical rather than hydraulic (van den Berg, 1984). The ROSON was the first seabed apparatus to use drive wheels offshore for continuous movement of the

TABLE 5
Summary of new developments: seabed in situ *testing equipment*[a]

Name		Water depth (m) Used	Designed	Penetration depth (m) Used	Designed	References
Seabed penetration systems						
PAM	Pressuremeter	625	1000	51	?	Brucy *et al.* (1984)
ROSON	CPT	350	1000	?	?	Van den Berg (1984)
CPM	CPT, PCPT[b]	610	900	>40	50	Amundsen *et al.* (in press)
NOAA	CPT, resist./cond.	1450	?	1.3	1.3	Lambert (1982)
IMAGES	CPT, vane, etc.	5200	6000	?	0.3	Tsurusaki *et al.* (1984)
ISV	Vane	5845	6000	1.5	1.5	Silva (in press)
PUPPI	Piezometer	5407	6000	3	3	Schultheiss *et al.* (in press)
ISHTE	Piezometer	1450[c]	6000	?	1.0	Bennett, *et al.* (in press, a)
GISP	Piezometers	19	450	10.5	10.5	Prindle and Lopez (1983)
UP[d]	CPT, PCPT	25	600	38	50	Bruzzi (1983); this paper
USGS	Piezometers	*c.* 1000	2000	12?	12	This paper
GIPSY	CPT, etc.	–	1000	–	6	Richards and Chaney (1981); this paper[e]
DORA	PCPT, etc.	–	6000	–	50	This paper
Sea Lion	PCPT	–	6000	–	5	This paper
Dogfish	Vane	–	6000	–	2–5	This paper
Seabed probes and sensors						
DSML	Nuclear density	220	1000	35	50?	Tjelta *et al.* (in press)
I-sV	Vane	334	1000	22	30	This paper
Penetrators						
(US Navy)	Shear strength	4908	6000	12?	?	Beard (in press)
(Sandia)	Shear strength	5400	6000?	>30	<50?	Talbert (1984a)
(BRE/JRC)	Shear strength	5400	6000?	>30	<50?	Freeman *et al.* (1984a)

[a]The water depths and penetration depths have been obtained fron manufacturers or operators, and may be optimistic in some instances because 'used' does not necessarily mean used successfully
[b]PCPT, piezocone penetrometer
[c]Sucessfully tested in a hyperbaric chamber at a pressure of 55 MPa (Bennett *et al.* in press, a)
[d]UP, underwater penetrometer
[e]The GIPSY is unlikely ever to be used because of a terminated project

test rods. The ROSON has worked well in water depths of less than 300 m, but not very well at greater depths.

McClelland recently adapted the Borros (Sweden) land 20-ton Static Sounding Push-and-Puller Apparatus for use offshore. This equipment continuously penetrates and retracts test rods using four hydraulic cylinders operating two and two in pairs connected to a single automatic rod clamp under each pair. The equipment, called a Continuous Penetration Machine, or CPM, by McClelland was successfully operated in the North Sea in 1984 from the Stingray (Amundsen *et al.*, in press).

A specially developed miniaturized geotechnical system was devised by National Oceanic and Atmospheric Administration (NOAA personnel for use on the Woods Hole Oceanographic Institution's submersible *Alvin* (Lambert, 1982). The system included a coring machine (described later), a cone penetrometer, a resistivity–conductivity probe, the *In Situ* Heat Transfer Experiment, or ISHTE (Percival *et al.*, 1984), a miniature piezometer and a pendulum-type inclinometer. During the dives, the piezometer was used only once. The probes performed successfully. All of the equipment is understood to have been transferred from NOAA to the Naval Ocean Research and Development Activity in Mississippi a few years ago.

Tsurusaki *et al.* (1984) report on the development and use at sea of an *In Situ* Measuring Apparatus of Geotechnical Elements of Sea-floor (IMAGES). The system, consisting of a plate-bearing test unit, vane, cone penetrometer, blade dragging test unit and fixed piston corer (discussed later), is designed to obtain seabed information relative to polymetallic nodule mining. In sea trials northwest of Tahiti, only the bearing capacity unit functioned as designed. Modifications and subsequent tests are planned by the National Research Institute for Pollution and Resources and the Government Industrial Research Institute in Tokyo.

An *In Situ* Vane (ISV) has been built at the University of Rhode Island for the ISHTE. Silva *et al.* (in press) decided to make the ISV system autonomous so that it could be used separately from the ISHTE equipment. Operations are microprocesor controlled. Data are stored on magnetic tape on the seabed unit. The equipment operated successfully in 1984 (Silva, in press).

The Pop-Up Pore Pressure Instrument, or PUPPI, is a highly innovative instrument designed and constructed at the Institute of Oceanographic Sciences (IOS) in England to measure differential pore pressures *in situ* (Schultheiss *et al.*, in press). Figure 6 gives general information about this equipment. It will be used for measuring ambient soil pore pressures and pore water hydraulic gradients, particularly in the abyssal plains of the North Atlantic as part of the IOS high-level radioactive waste disposal programme. The probe containing the pore pressure port is 3 m long and 5 cm in diameter. Pore pressure resolution is considered to be 50 Pa over a 200 kPa range. The instrument system also contains an accelerometer. The PUPPI emplaces itself after free-fall from a hip to the seabed. Release of the PUPPI

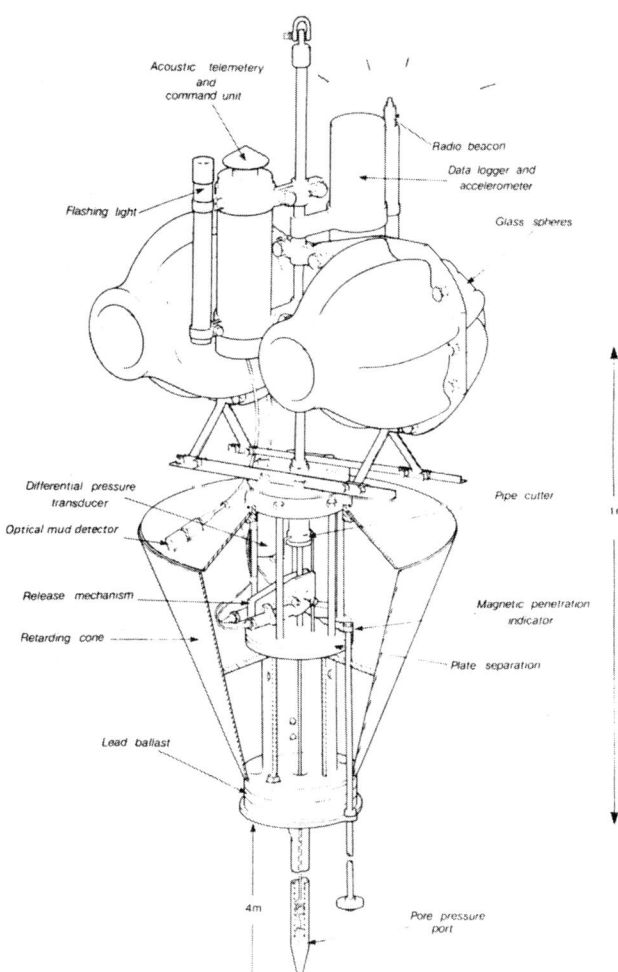

Fig. 6 Pop-Up Pore Pressure Instrument (PUPPI), Institute of Oceanographic Sciences, for a water depth of 6000 m (Schultheiss *et al.*, in press)

from its weights for 'pop-up' recovery is by the transmission of an acoustic signal from the ship. While results of tests at sea are still being evaluated, Schultheiss *et al.* (in press; 1984, written communication) reports that there were mainly no excess pore pressures in areas of thick soil cover and near faults in the Great Meteor East area, Madeira Abyssal Plain, and that in areas of thin soil cover the measured pore pressures were slightly negative. These pore pressure measurements are believed to be the first to be made in ocean-basin water depths.

The ISHTE piezometer design and construction was started at NOAA. Later, all geotechnical operations were transferred to the Naval Ocean Research and Development Activity, as previously mentioned. A differential semiconductor transducer was used on the *Alvin* dives (Table 5) to measure differential pore pressure. The sensor malfunctioned and then failed in the laboratory at a pressure of 68.9 MPa. This type of transducer was subsequently abandoned and a Validyne variable reluctance differential pressure transducer was selected that had a precision of 0.34 to 0.69 kPa over the same 69 MPa range (Bennett *et al.*, in press, a). Laboratory test results indicated 'better than expected performance'. There have been zero-shift problems, but these are not considered detrimental to the measurement of ambient pore pressures (Bennett *et al.*, in press, a). The present unit is shown in Fig. 7. The probe part of the unit, only 8 mm in diameter, is made of a titanium tube. Recently, Bennett *et al.* (in press, b) have shown how to derive shear strength and permeabilities from pore-pressure data.

Both the PUPPI and the ISHTE piezometers have a single pore pressure port. The earlier NOAA piezometer system, designed for water depths of less than 200 m, had multiple ports (Bennett *et al.*, 1982). Another new multiple-port piezometer system is the Sandia National Laboratories' Geotechnically Instrumented Sea-floor Probe, or GISP (Reece *et al.*, 1979). Abso-

lute pressure, quartz-crystal, oscillating-beam pressure transducers, having an operating range of 0 to 6.2 MPa, sense pore pressures at three locations 3.3, 6.4 and 9.5 m below the mudline (Prindle and Lopez, 1983). A hydrostatic pressure transducer is mounted just above the sea floor on the Data Gathering Subsystem (DAGS). This unit acoustically telemeters data to a Command and Recording Subsystem (CARS) on the surface. The DAGS, emplaced using heavy weights, is designed to remain on the sea floor for an extended period of time.

Late in 1984, Ismes (Bergamo, Italy) personnel successfully operated their new Underwater Penetrometer offshore in shallow water. Bruzzi (1983) has described an earlier version of this machine. The new equipment automatically levelled on a 25° slope and no problems were experienced. The current cable and handling system is limited to a water depth of 350 m, but the unit is designed to operate down to 600 m (Bruzzi, 1985, personal communication).

A differential piezometer probe having six levels of pore pressure ports, usually arranged with a pair on opposite sides, was tested at sea in 1983 by the US Geological Survey in Woods Hole, Massachusetts (Booth, 1985, personal communication). The probe was lowered on a mechanical cable and used acoustic telemetry to communicate data to the ship. On the sea trial the acoustic link performed well until near the seabed when the signals faded out. The problem is believed to have been solved, according to Booth, but the machine has not yet been tested again. Eventually the plan is to have the equipment free-fall to the seabed.

This concludes the descriptions of new seabed equipment that has actually operated offshore in water depths greater than 300 m. Three units will now be discussed briefly that have yet to be tested at sea. Each has some interesting design features.

Seabed Systems under Development

In the early 1970s, a multiprobe seabed sys-

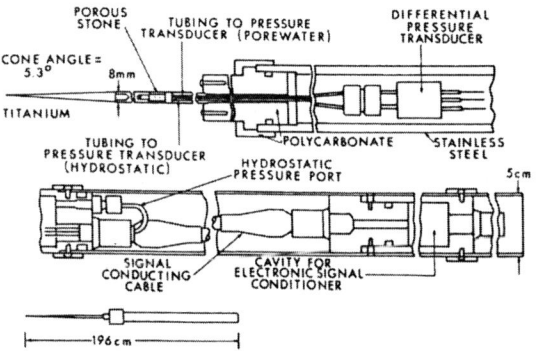

Fig. 7   ISHTE piezometer developed by NOAA and NORDA for a water depth of 6000 m. (Courtesy of NORDA)

TABLE 6

Geotechnical In situ *Probe System (GIPSY) specifications*[a]

1. Operational water depth[b]	1000 m
2. Penetration below sea floor	6 m
3. Penetration method, continuous	Catapuller (for probes)
4. Probes is system	Nuclear densitometer, cone penetrometer, piezometer, vane
5. Corer	Large diameter fixed piston with chain drive
6. Control and data presentation	Cable-transmitted, multichannel digital telemetry for real-time control and the display of data in SI units, using a microprocessor, from probe sensors
7. Power	Lead–acid storage batteries mounted on the chassis

[a]Modified from Richards and Chaney (1981)
[b]Extendable to 6000 m by replacing pressure-protected cans for the electronics

tem called the Geotechnical *In situ* Probe System or GIPSY was conceptuallized at Lehigh University. The GIPSY was designed and constructed with one probe module in the late 1970s and very early 1980s. Table 6 lists specifications. The four probes were each designed to operated with the 'Catapuller' continuous drive system (Fig. 3A) and hollow, square test rods. Each of the four probes and the fixed piston corer modules were designed to be simply plugged in electrically and mechanically to the main frame containing the battery pack and electronics. Data acquisition and control was to be by digital telemetry over an electromechanical cable. A novel feature was the utilization of a microprocessor to provide positive feedback to the ship that any command sent to the machine on the seabed was actually carried out, rather than only an electrical circuit being activated. The entire machine was essentially completed, including one probe and the entire GIPSY data acquisition and control system, when the project was abandoned at the time the Marine Geotechnical Laboratory was disbanded in 1982.

A consortium consisting of Fugro, Laboratorium voor Grondmechanica (Delft Soil Mechanics Laboratory), Marine Structures Consultants and Ismes (Italy) has completed a detailed feasibility study for a

Deep Ocean Research Apparatus, or DORA. An artist's impression of the DORA is shown in Fig. 8. Table 7 lists specifications. The DORA is designed to use a

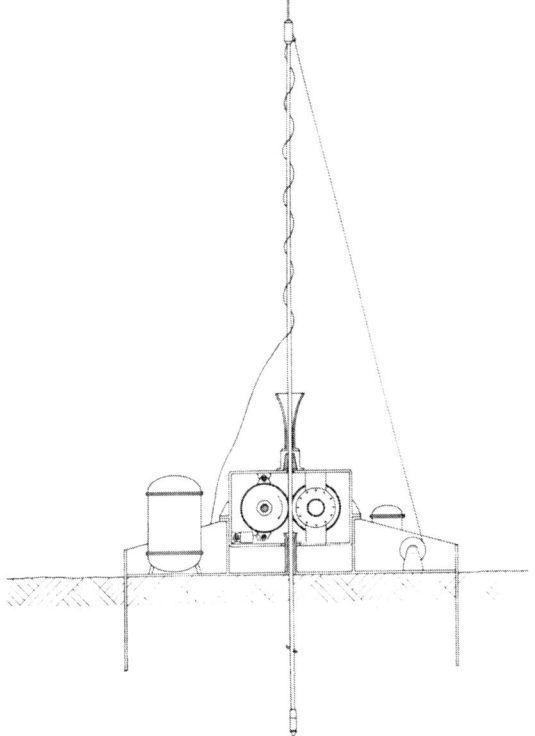

Fig. 8 An artist's impression of the Deep Ocean Research Apparatus (DORA) designed to push test rods 50 m into the seabed in a water depth of 6000 m

TABLE 7
DORA specifications

I. General	
Maximum water depth	6000 m
Penetration depth	50 m
Foundation loading, maximum	50 kN
Penetration method (speed)	Drive wheels, continuous (2 cm/s)
Data acquisition and control	Microprocessor, acoustic telemetry to vessel
Power supply	Lead-acid storage batteries
Power pack	Hydraulic, pressure compensated
II. Piezocone	
Cone-resistance range	7.5 kN
Maximum shear strength of soil	140 kPa
Pore pressure range	1 MPa
Diameter of cone	46 mm
Cone base area	15 cm^2

TABLE 8
Sea Lion specifications

I. General	
Maximum water depth	6000 m
Penetration depth	5 m
Maximum weight	1000 kg in air
	1500 kg in air with ballast
Dimensions: base plate	2.4 × 2.4 m square
height	6.8 m without skirt
Additional bearing plates	4 plates (2.7 × 0.3 m)
Skirt	0.5 m (optional use)
Penetration method (speed)	Drive wheels, continuous (about 4 cm/s)
Operation time	About 10 penetration–retractions per lowering
Control after touchdown on seabed	Automatic (programmable logic controller), penetration–retraction sequence initiated after a trigger touches the sea floor
Data storage	Multichannel, solid-state recorder on Sea Lion
Pinger	Detects sea floor, advises go–no go on inclination and when piezocone not in full up position using different repetition rates
Power supply	Lead–acid storage batteries
Power pack	Hydraulic, pressure compensated
II. Piezocone	
Cone resistance range	0–2.3 kN
Maximum shear strength	50 kPa
Pore pressure range	0–600 kPa
Diameter of cone	44 mm
Cone base area	15 cm^2

double-wheel drive system (Fig. 3B) and 50 m of test rods in a water depth of 600 m. Data acquisition and control is planned to be by two-way acoustic telemetry through the water column. Control of the 50-m long up-right string of 36 mm diameter test rods is planned to have buoyancy at the top of the test rods and a length of buoyant cable to a clump weight, which should be emplaced within about 100 m of the DORA.

Fugro engineers have designed and contructed a Sea Lion, which is scheduled to be used in a water depth of 6000 m in the Banda Sea during May 1985. Table 8 gives specifications, which include a double wheel drive system (Fig. 3B) and a test rod five metres long. A specially designed pinger will detect the seabed, advise the attitude of the Sea Lion on the seabed to a microprocessor and, if the machine is within acceptable tilt limits, communicate acoustically when the penetration–retraction cycle is automically activated. Piezocone and other data are stored in a solid-state recorder (Schaap and Liefting, 1984). The piezometer transducer will be of the total pressure type.

The Dogfish vane has been designed and built at the Building Research Establishment for use in the summer of 1985 on the *Marion Dufresne* cruise. This vane is designed to operate in a manner similar to the Sea Lion, including a special multipurpose pinger, microprocessor control of operations and data storage on the unit (Burdett, 1985, personal communication). The vane is designed to penetrate automatically and take torque measurements, after being lowered by a wire to the seabed, about every 0.3 m to a penetration depth of two metres. Ultimately, the Dogfish will be mounted in a penetrator to extend five metres out of the back end after penetration has ceased for the purpose of evaluating hole closure.

Seabed Probes

The Delft Soil Mechanics Laboratory used offshore for the first time in 1984 their nuclear backscatter density probe (Fig. 9) from the Fugro Seacalf (Tjelta *et al.*, in press). At

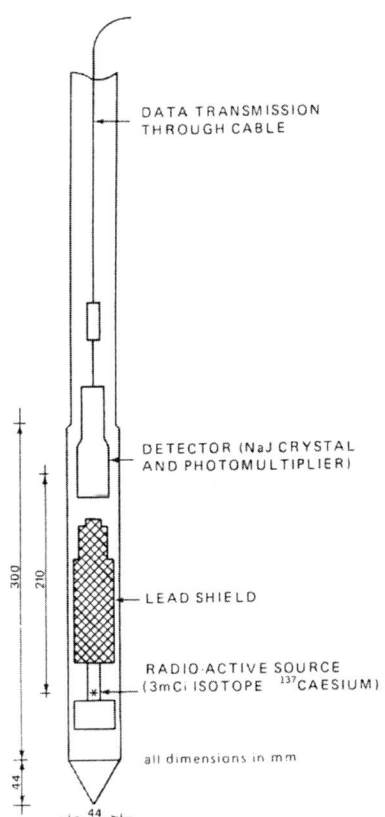

Fig. 9 Delft Soil Mechanics Laboratory nuclear backscatter density probe for seabed penetration apparatus deployment (Tjelta *et al.*, in press)

the second ESOPT conference, two similar density-CPT probes were described (Ledoux *et al.*, 1982; Nieuwenhuis and Smits, 1982). The density part of the later conceptualization has been adapted for use offshore. The radioactive source is 3 mCi of Cs-137. A combination cone penetrometer or piezocone and nuclear backscatter densitometer has yet to be used offshore.

Fugro successfully used their *In situ* Vane in 1983 from the Seacalf. This is believed to be the first time a vane has been pushed deeply into the soil using seabed penetration apparatus without using the more customary downhole technique.

It has been reported (Jamiolkowski, 1985, personal communication) that the Japanese Public Water Research Institute has developed a vibrating cone with a vibrator mounted within the tip. This cone has been used for assessing the liquefaction potential of soils onshore. It is not known if the cone can be used offshore.

Seabed Penetrators

At present, three organizations or groups of organizations are using penetrators in ocean-basin water depths. A penetrator, which is not a penetrometer — although a penetrometer may be attached to a penetrator — is a free-fall from a ship or fully dynamic instrumentated device. The modern Doppler penetrator is an expendable projectile containing accelerometers and a sound source that sends an acoustic signal to a hydrophone and a receiver for data processing aboard a ship. The US Navy Civil Engineering Laboratory (Beard, 1977; in press) originated the Doppler penetrator and these devices are now manufactured and sold by Sonatech (Cyr, 1984). True (1975) and Beard (in press) have presented information on how shear-strength data are determined from an integration of velocity–time information to obtain depth–time data, which is integrated to obtain soil shear strength.

At about the same time as the Navy penetrator was being developed, Sandia National Laboratories personnel were experimenting with shallow-water penetrators that trailed a thin wire back to the ship (Colp *et al.*, 1975). These devices evolved into the gun-launched penetrator (James *et al.*, 1981; Prindle and Lopez, 1983). The penetrators were dropped in a free-fall mode in ocean-basin water depths in the early 1980s for the high-level radioactive waste disposal investigations (NEA, 1984, pp. 110–119). An innovative aspect of the current Sandia penetrators is the Explosive Acoustic Telemetry System (EATS), which is a microprocessor-controlled method to encode data from the time intervals between explosive detonations (Talbert, 1984b, and 1984, personal communication).

In the early 1980s, the Building Research Establishment (BRE) in England joined

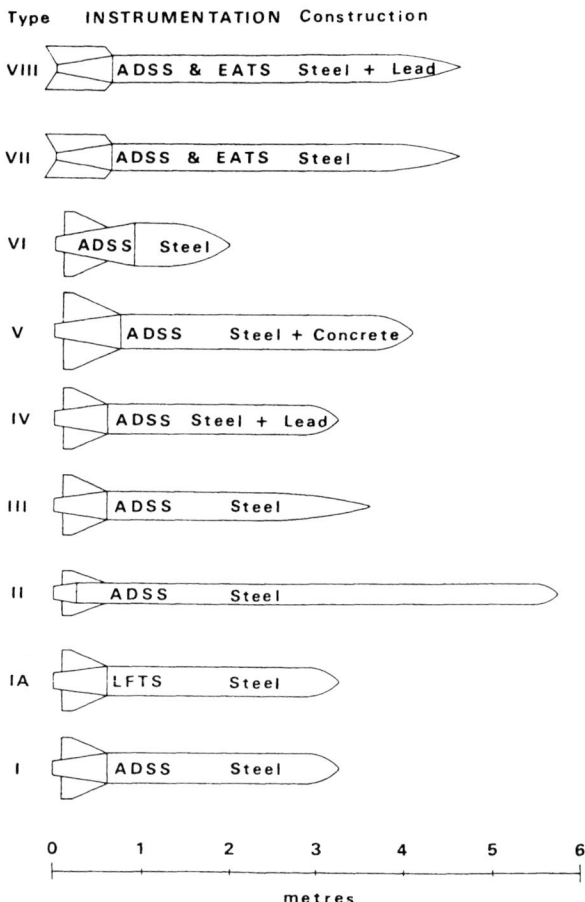

Fig. 10 Penetrator designs for DOMP II trials performed in the Nares Abyssal Plain, March 1984 (Freeman *et al.*, in preparation, a). See text for abbreviations

with the Commission of European Communities' Joint Research Centre (JRC) in Italy to construct and test at sea a series of different types of penetrators as part of the European effort in the high-level waste disposal programme (Freeman *et al.*, 1984b). Deep Ocean Model Penetrator (DOMP) trials were made in 1983 and 1984. The penetrators used in the 1984 series are shown in Fig. 10. In this figure 'type' refers to the shape; 'instrumentation' refers to the Sonatech Acoustic Doppler Shift System (ADSS), the Sandia Explosive Acoustic Telemetry System (EATS) and the Institute of Oceanographic Sciences Low Frequency Transponder System (LFTS) — which

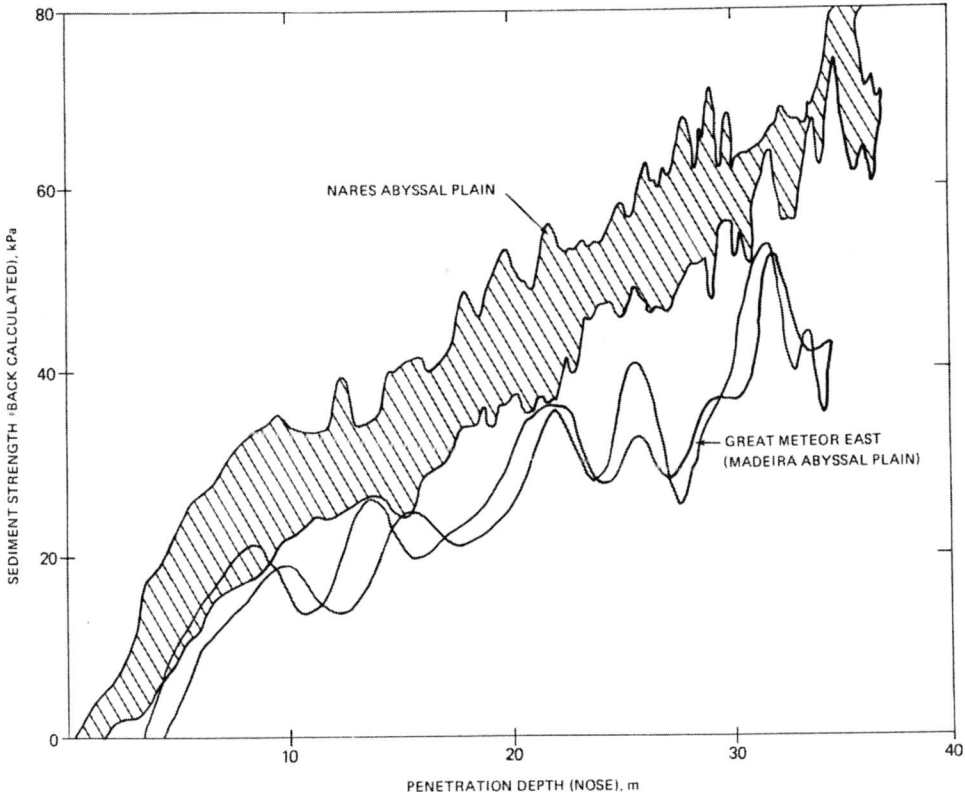

Fig. 11 Soil strength (s_u) profiles back-figured from penetrator data (Freeman *et al.*, in preparation, a, b) using the BRE/Fugro time-stepping computer programme 'PREPEN' (Freeman *et al.*, in preparation, b)

operates on a frequency of 3.5 kHz rather than the 12 kHz of the Sonatech equipment; and 'construction' refers to the material used, which was concrete, lead and steel. Grouped results (Fig. 11) from two tests of type I penetrators (Fig. 10) in the Great Meteor East area and six tests of type I, III. IV, VI, VII and VIII penetrators in the Nares Abyssal Plain during the DOMP trials in the Atlantic show an average linear increase of strength with increasing depth of penetration in the abyssal plain soils, which are highly anisotropic because they are composed of alternating layers of turbidites and pelagic soils. The implications of this discovery can soon be expected to be discussed by Freeman and his colleagues. Future plans call for further geotechnical and geochemical instrumentation, including measurement of shear strength and pore ressure (Freeman *et al.*, 1984a). Hunting

Engineering in England have completed a feasibility study for instrumenting penetrators that includes a cone penetrometer, a differential pressure piezometer and a bulk-density transducer (Rodgers, 1984, oral presentation at the JRC).

Downhole Samplers

Four new downhole samplers have been developed and successfully used at sea in water depths greater than 300 m in the past four years (Table 9). The hydraulic piston corer (HPC), designed at the Deep Sea Drilling Project of the Scripps Institution of Oceanography, obtains a core 9.5 m long and was first operated at sea in 1979–1980 (Storms *et al.*, 1983). The successor to the HPC is the Variable Length HPC (VLHPC), which adjusts the core length to the resisting force of corer penetration

TABLE 9
Summary of new developments: Samplers

Name/ (Organization)	Sampler Type	Downhole	Seabed	Water depth (m)		Maximum core length (m)		References
				Used	Design	Obtained	Design	
VLHPC	Piston	x		>3500?	6000?	9.5?	9.5	Storms *et al*. (1983)
(Fugro)	Piston	x		340	>700	3	3	Toolan (1983)
(McClelland)	Piston	x		750	900	2	2	Bayne (1985, personal communication
APS	Pressurized soil	x		340	>700	65 cm^3	sample	Zuidberg *et al.* (1984)
(NOAA)	Multiple gravity		x	1450	?	0.7	1.3	Lambert (1982)
IMAGES	Fixed piston		x	5200	6000	'0'	0.3	Tsurusaki *et al.* (1984)
STACOR	Fixed piston		x	1225	6000	18	30	Montarges *et al.* (1983); Faÿ *et al.* (in press)
(LDGO)	Hydrostatic		x	1000	?	2	3	McCoy and Selwyn (1984)
HCL	Hydrostatic		x	5845	6000	1.4	1.6	Percival *et al.* (1984); Silva (in press)
(Long core facility)	Fixed piston		x	–	6500	–	50	Driscoll (1981)
Advanced piston corer	Fixed piston		x	–	6000	–	30	This paper
GIPSY	Fixed piston		x	–	1000	–	6	This paper

(Storms *et al.*, 1983). The VLHPC (Fig. 12) has been in operation since 1981.

The Fugro hydraulic piston sampler is an adaptation of the Wipsampler (wireline push sampler) to use a fixed piston (Toolan, 1983). It has been in routine operation since 1982. It is likely that the improvement in soft soil sample quality using this sampler may be as great, compared to the push sampler, as the change from using the percussion sampler to the push sampler reported by Young *et al.* (1983).

McClelland also have a new piston sampler for downhole use with their Swordfish or Dolphin, which has been routinely used since 1984.

Texas A & M University personnel pioneered the design and use of the wireline geotechnical sampler to obtain samples of gas-containing soils from boreholes at their ambient pressure (Denk *et al.*, 1981; Johns *et al.*, 1982) in shallow water. In 1983 Fugro successfully designed and used a new and simpler sampler, the Ambient Pressure Sampler (APS), in the Troll Field offshore of Norway (Zuidberg *et al.*, 1984). Consultation with geochemical engineers led to the conclusion that a sample volume of 30 to 50 cm^3 was adequate for soil gas analysis. This led to a conservative design of the APS to sample 65 cm^3 of soil with as little disturbance as possible using a single ball-valve adapted for actuation by the Wipsampler (Fig. 13). A feature of the APS is that it collects a non-pressurized sample 40-cm long adjacent to the pressurized part of the sampler contained in the ball-valve.

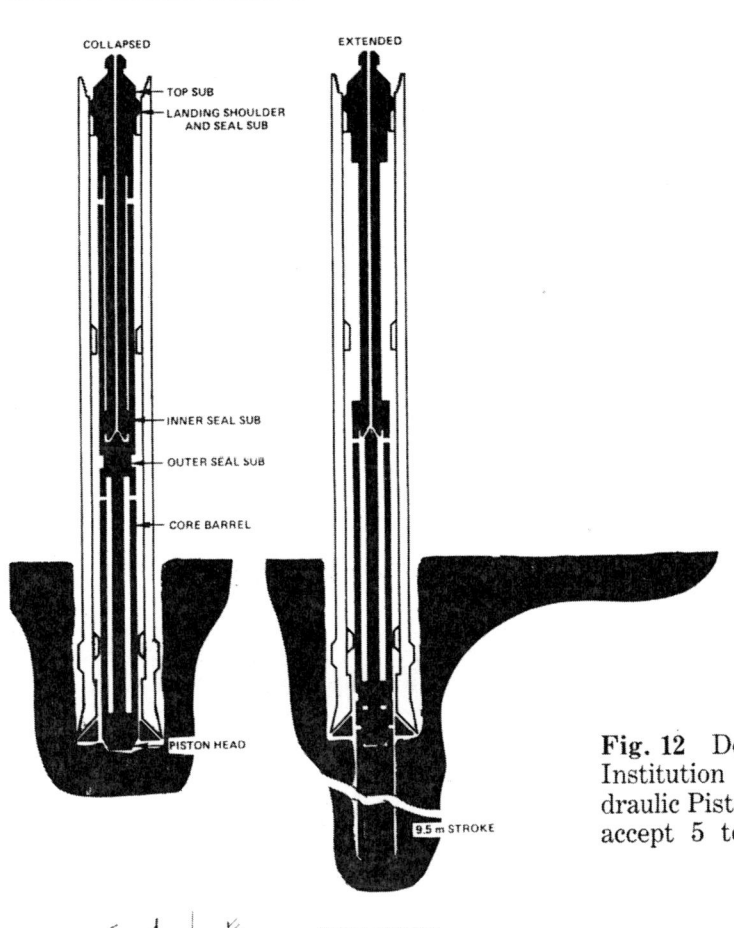

Fig. 12 Deep Sea Drilling Project, Scripps Institution of Oceanography, Very Long Hydraulic Piston Corer, which can be configured to accept 5 to 9.5 m of core. (Courtesy of the DSDP)

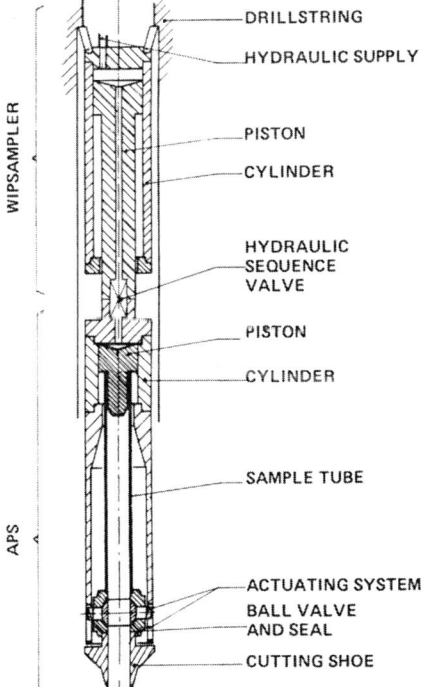

Fig. 13 Fugro Ambient Pressure Sampler (Zuidberg *et al.*, 1984)

Seabed Samplers

Eight new seabed corers have been selected to illustrate developments in water depths greater than 300 m (Table 9). In 1970, the Lockhead Ocean Laboratory was already successfully using, in a water depth of 1235 m, a six-unit Hydraulically operated piston coring system from the submersible *Deep Quest* (Inderbitzen and Simpson, 1972). More recently, Lambert (1982) described an eight-unit electrically operated gravity (no piston) coring system to take cores 1.3 m long from the submersible *Alvin*. The core tubes had an inside diameter of about 32 mm. Lambert (1982) reported that operation was only partly successful in as much as only a 70 cm maximum length of core could be recovered.

The *In Situ* Measuring Apparatus of Geotechnical Elements of Sea-floor (IMAGES) fixed-piston corer apparently did not sample because of a malfunctioning drive motor (Tsurusaki *et al.*, 1984).

Montarges *et al.* (1983) and Fäy *et al.* (in press) have described the design and successful use at sea of the STACOR, which is believed to be the first use of the fixed-piston principle to take long cores in water depths greater than about 1000 m. The STACOR — developed by the French Institute of Petroleum, Elf Aquitaine and the French Petroleum Company (Total) — is based on the design of the SACLANT sphincter corer with recoilless piston (Kermabon and Cortis, 1969). The advantage of the STACOR is that it obtains a high-quality 11 cm diameter core up to 25 or 30 m long; however, a disadvantage is that the equipment is a very heavy (5 to 10 tonnes), requiring at least a 10 to 20 tonne winch and handling frame and a substantial coring cable for the STACOR pullout from the seabed. Side-by-side comparisons of geotechnical data obtained from cores raised by Kullenberg and STACOR corers at the same location show a substantially higher quality for the STACOR data, as would be expected using a truly fixed piston. Plans exist to use the STACOR in the summer of 1985 to obtain 30 m long cores from the French Antarctic vessel *Marion Dufresne* in the Madeira Abyssal Plain as part of the high-level radioactive waste disposal programme.

A number of people have tried rather unsuccessfully to harness the great hydrostatic pressure in deep water to operate equipment. McCoy and Selwyn (1984) report on the successful operation of a Lamont–Doherty Geological Observatory of Columbia University gravity corer driven by a hydrostatic motor in 1980. These authors, however, have not demonstrated the potential of this corer to outperform conventional gravity and piston corers.

Percival *et al.* (1984) have described a hydrostatic actuated corer designed to be used in the ISHTE experiment. The actuator operates in a manner similar to a hydraulic cylinder. High-pressure seawater replaces hydraulic oil as the working fluid. Six corers are planned for use on the ISHTE platform. Silva (in press) sites the successful use of these corers in a 1984 trial.

The innovative development and subsequent use of the Woods Hole Oceanographic Institution's Giant Piston Corer (Hollister *et al.*, 1973) has led to the establishment of a Long Core Facility at the University of Rhode Island to design and construct a Long Core Facility Corer (Driscoll, 1981). Present plans call for a corer weight of 13 to 15 tonnes and a free activating piston controlled by valves to approximately fix the piston relative to the sea floor (Driscoll, 1985, personal communication); the cable will be Kevlar, 54 mm in diameter. This machine has not been constructed.

C. Karnes announced at a meeting of the Engineering Studies Task Group, Nuclear Energy Authority Seabed Working Group, in late 1984 that Sandia National Laboratories was developing its own 30 m 'Advanced Piston Corer' for the 1985 cruise of the *Marion Dufresne*. The piston is planned to be controlled by a servovalve that permits water to flow into a partially air-evacuated container in response to an overpressure developed by the movement of the core into the core barrel (Karnes, 1984, oral communication). No other details are available.

The proposed Lehigh University fixed-piston corer module for the Geotechnical *In Situ* Probe System (GIPSY) was designed by N. D. Remy Associates but not constructed. The method of core-tube penetration and retraction was by a chain drive, similar to the Remy design for the XSP cone penetrometer equipment (Beard and Lee, 1982). It is unlikely that the GIPSY corer will be built in the future for the reasons previously given.

Offshore Shear Wave Measurements

Geotechnical engineers tend to think only of *in situ* testing and sampling equipment for offshore site investigations. Wave, earthquake, ice and other environmental forces on offshore structures are cyclic and, in consequence, knowledge of the shear modulus is important in assessing soil–structure interaction under cyclic loading conditions. In addition, Taylor Smith (1983) has summarized how seismo-acoustic information can be used to predict soil engineering properties and to estimate consolidation behaviour. For these reasons, examples of four different recent developments (Table 10) of shear-wave measurements in water depths greater than 300 m — or applicable to these water depths — are considered relevant to this review.

Ohya *et al.* (1984) of the Oyo Corporation in Japan have successfully used a downhole logging tool that can propagate and receive both P and S waves. This tool provides improved resolution more appropriate for geotechnical purposes than many of the conventional logging tools used for oil and gas exploration. The vibration source is an electromagnetic solenoid hammer contained within the probe. Results are said to compare favourably with downhole and crosshole methods (Ohya *et al.*, 1984). The tool has been used offshore, but details are not given.

The US Defense Advanced Research Projects Agency has developed a marine seismic system and deployed it in a dedicated borehole in collaboration with a number of other organizations (Foss and Wallerstedt, 1984). The seismometers will be used to listen for natural seismic events. A somewhat comparable method that has

TABLE 10
Summary of new developments: offshore shear-wave measurements

Name	Method	Downhole	Seabed	Water depth (m)	References
Suspension PS logging system	Transmits and receives P and S waves from instrument	x		500 (maximum)	Ohya *et al.* (1984)
Marine seismic system	Seismometers placed in a dedicated borehole	x		5620	Foss and Wallerstedt (1984)
–	'Pop-up' bottom seismograph records S waves produced by dropping heavy weight on seabed		x	5260	Whitmarch and Lilwall (1984)
Seismic cone penetrometer	S waves produced by external surface sound source, received by seismometers built into a cone penetrometer		x	Not used offshore, but such use seems reasonable	Campanella and Robertson (1984)

been used in the shallow water of the Beaufort Sea (Justice *et al.*, 1984) may also be applicable to the water depths of concern in this paper.

Ocean-bottom seismometers and seismographs have been used for some time (Prothero, 1984). Whitmarsh and Lilwall (1984) of the Institute of Oceanographic Sciences (IOS) in England describe the successful use of the IOS pop-up bottom seismograph to record S waves generated when a heavy weight is dropped to the seabed within 1000 m of the seismograph.

Campanella and Robertson (1984) measure the dynamic modulus of soil *in situ* on land using a University of British Columbia cone penetrometer. Small seismometers are installed triaxially in the same probe. Shear waves are generated at the surface and are received by the seismic cone penetrometer every one metre after penetration has been stopped for the measurement. Excellent agreement is claimed with conventional crosshole techniques (Robertson *et al.*, 1985). While the technique has not been used offshore, there is no apparent reason why it could not be used.

THE FUTURE

After reviewing deep-water problems and equipment one might well wonder what it means and what the future portends. It has been established that science fiction writers generally do a much better job at divining the future than scientists and engineers. Nevertheless, some trends appear for industry, working at depths less than 1000 to 2000 m, and for many university and governmental investigators, working in water depths between about 4000 and 6000 m.

In water depths greater than 1000 m, the development of the GLORIA, SeaMARC II, Sea Beam and the Bathymetric Swath Survey Systems (see Richards and Hartevelt, 1984) have very great potential or delineate clearly areas for geotechnical site investigations. Prior and Doyle (1984) present an

excellent example of how to combine many tools for site-specific surveys in water depths of 2000 m.

Studies by Kirkpatrick and Khan (1984), as well as others, on the stress relief caused by sampling appear to show effects that may be greater on some soils than previously believed. This is additional evidence to bolster the conclusion by Richards and Zuidberg (in press, a) that *in situ* strength testing offshore involves fewer significant problems than sample collection and subsequent laboratory testing. Stress relief, soil degassing and other deleterious phenomena are likely to increase with increasing water depths. This is ample reason why *in situ* testing may gradually replace many sampling applications in the years to come. For some problems sampling under ambient pressure followed by geotechnical and geochemical testing under the same pressure, without having to resort to personnel working in hyperbaric chambers, is likely to increase.

Site investigations made by the geotechnical consulting industry show clearly two different methods of testing and sampling: (1) by work downhole and (2) by work on the seabed. For the first, greater electromechanical sophistication can be expected utilizing the mighty microprocessor and digital telemetry. For the second, we can expect more automated machines operating from a single electromechanical cable or perhaps utilizing only a mechanical cable and communicating data and control signals by acoustic telemetry. This may be through the test rods, as has been done by Jefferies and Funegard (1983) in shallow water, or through the entire water column.

It is likely that the piezocone penetrometer will soon replace the cone penetrometer offshore because the former provides much more useful information at little additional cost. Figure 14 shows the results from a Fugro piezocone used from the Seacalf in the North Sea in which there is a moderate amount of variability in the soil. Both Fugro and McClelland have used their piezocones in water depths greater than 300 m. It is

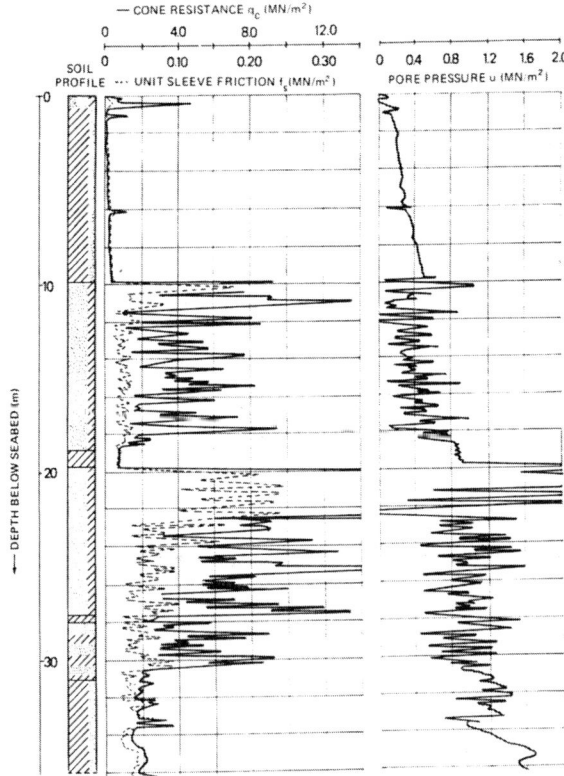

Fig. 14 Results from a Fugro piezocone penetrometer used from the Seacalf in the North Sea in 1984. The water depth was 217 m

also likely that the pressuremeter may gain more widespread use offshore in deep water than the dilatometer because the former yields more information. A Fugro full-displacement pressuremeter, operating behind a cone or piezocone penetrometer, has been designed and built for initial tests on land. It is the intention to adapt such a tool for offshore work in shallow and deep water. The costs of working in water depths greater than 300 m are already so great that the cost differential between the pressuremeter and the dilatometer is likely to be negligible.

There are a variety of probes and sensors that have been used offshore in shallow or deep water as well as some that probably could be used offshore (Table 11). In this table a combination acoustic cone incorporating both transmitting and listening

functions was being built a few years ago for the National Bureau of Standards by Hogentogler & Company. A number of tools listed by Campanella and Robertson (1982) have not been included in Table 11 because of a lack of suitability perceived by the author for one reason or another. Nevertheless, enough examples remain to provide a fruitful field for the solution of a large variety of industrial and governmental–academic problems. It is highly probable that the trend of using multiple sensors in one probe system will continue.

Site investigations made by government and university researchers at present lack downhole geotechnical testing tools to complement the HPC sampler for deep-sea drilling under the new Ocean Drilling Programme. A 1980 report or White Paper, 'Report by the *Ad Hoc* Working Group on Long-Range Plans, Sedimentary Petrology and Physical Properties Panel', Joint Oceanographic Institutions Deep Earth Sampling (JOIDES), recommended the technical development of the following downhole equipment:

(1) sand-core catcher for the Hydraulic Piston Corer;
(2) *in situ* vane shear meter;
(3) *in situ* pore-pressure measuring device;
(4) *in situ* hydrofracturing device; and
(5) *in situ* stress–strain meter.

Interestingly, in just five years all equipment in this list are in use offshore — except perhaps the first — in water depths of a few hundred meters or so. There is no technical reason why any of these tools could not be adapted to work downhole in 6000 m at relatively modest cost.

Seabed site investigations by the governmental and academic groups are likely to be orientated towards studies on trenches, the abyssal plain and hydrothermal vents, all of which involve measurements of *in situ* pore pressures to determine hydraulic gradients for the study of vertical and horizontgal pore-water fluid movement and to elucidate the consolidation and eventual lithification of deep sea floor soils.

TABLE 11

Geotechnical probes, sensors and relevant tests which are used or could be used offshore

Probes and sensors	Parameters measured or (derived)[b]	Used offshore	Water depth >300 m	Selected example references[c]
Acoustic[d]				
Passive comp. wave	Acoustic emissions	x	?	Tringale and Mitchell (1982); Schaap and Hoogendoorn (1984)
Active comp. wave	Acoustic emissions			This paper
Shear wave ('seismic')	Shear modulus	x	x?	Reece et al. (1979); Campanella and Robertson (1984)
Densitometer, nuclear	Bulk density			
Backscatter		x		Keller (1965);
		x	x	Rose and Roney (1971)
		x	x	Tjelta et al. (in press)
Transmission	Bulk density	x	x	Hirst et al. (1975)
Dielectric cone	Capacitance			Singh and Chomg (1984)
Dilatometer	Stress–strain	x		Burgess et al. (1983)
Electrical resistivity	Resistivity/conductivity	x		Rietsema and Viergever (1979)
		x	x	Hulbert et al. (1982); Bennet et al. (1983b)
Hydraulic fracture test	Fracture pressure	x	x	Toolan (1983)
Penetrator, instrumented	Shear strength	x	x	Beard (in press); Freeman et al. (1984)
Penetrometer, 'static'				
Cone/friction sleeve	Tip/sleeve resistance	x	x	de Ruiter (1982)
Piezocone	'Dynamic' pore pressure	x	x	de Ruiter (1982)
Penetrometer, 'dynamic'[e]	Tip/sleeve resistance	x		Chari et al. (1979)
Permeameter	Permeability			Capelle (1984)
Piezometer	Pore pressure			
Differential pressure		x	x	Richards et al (1975)
Total pressure		x	x	Presland and Babb (1979) Bennet et al. (in press, a)
Pressuremeter	Stress–strain			
Push-in		x	x	Reid et al. (1982)
Self-boring		x	x	Faÿ et al. (in press)
		x		Hughes et al. (1984)
Full displacement				This paper
Temperature cone	Temperature	x		Schaap and Hoogendoorn (1984)
Thermal conductivity cone	Thermal conductivity			Schaap and Hoogendoorn (1984)
Vane	Shear strength			
Downhole		x	x	Kraft et al. (1976)
Seabed		x	x	Demars and Taylor (1971)
		x	x	Richards et al. (1972)
		x		Lewis et al. (1970)
		x	x	Silva (in press)
		x	x	This paper

[a]Many sensors can be used in combination. Downhole and seabed use is not differentiated
[b]Only the principal parameter is listed
[c]Generally the most recent general description is given, or references relating to work by different organizations. The general reference cited may not relate to offshore use
[d]'Passive' refers to listening only and 'active' refers to transmitting and listening in the same probe
[e]The free-fall type from the seabed, not the type used on land

At these great depths, the trend is likely to include more acoustic telemetry through the water column to avoid the expense and complexity of electromechanical cables; also the use of fully autonomous free-fall (from the ship) and 'pop-up' (to the ship) equipment having self-contained recorders or using acoustic telemetry is likely to increase.

A new generation of manned and unmanned submersibles for water depths of 6000 m is near at hand. The PLA2, developed by a number of French organizations, is a fully pre-programmed autonomous ROV designed to descend to 6000 m and automatically harvest a small load of polymetallic nodules. The prototype has been built and is expected to be ready for testing at sea very soon. The INFREMER (ex. CNEXO) organization soon plans to operate its manned submersible SM79, which is designed for 6000 m, a depth at which the autonomous *Epaulard* ROV already operates. The larger submersibles could easily carry geotechnical instrumentation, some of which is in an early stage of development.* This activity in Europe is expected to be an interesting addition to the *Atlantis II* and *Alvin* submersible operations in the United States.

SUMMARY AND CONCLUSIONS

1. There are two moderately distinct testing and sampling activities in water depths exceeding about 300 m: (1) industrial, in water depths usually less than about 1000 m, and (2) governmental–academic, in water depths generally greater than about 4000 m. Some equipment is shared between the two activities and some is distinctly different.
2. Most systems, probes and sensors have been gradually developed over long time periods before being used in water

depths exceeding 300 m. A few, however, have been rapidly developed specifically for these water depths.
3. Industrial-type problems for site investigations in water depths greater than 300 m tend to be different from site investigations conducted by governmental–academic organizations in response to their particular problems.
4. Many more seabed penetration systems, probes and sensors have been improved or developed during the past three to five years than comparable downhole equipment. The reason for this probably is that downhole geotechnical investigations are not usually available to governmental–academic groups because of the cost. Three notable exceptions, the DSDP, IPOD and presently the ODP, are in general under the control of scientists who are only now beginning to understand what geotechnical engineering can contribute to science.
5. The seabed testing systems utilize a variety of different deployment methods, which are very briefly described in Table 2. Particularly rapid development is occurring in the continuous drive-wheel devices and in the penetrators. The latter have the potential of revolutionizing ocean basin soil investigations in the geoscience and geotechnical fields. This is because the penetrator is a means of emplacing, relatively cheaply, a seabed laboratory deeply into the seabed to receive data at the surface by acoustic telemetry either in real time or until it is recalled from a memory system. The receiver may be aboard a ship, or on a buoy for transmission ashore via a satellite.
6. Seabed samplers currently of great interest are the various devices to obtain high-quality cores about 30 m long. Three different organizations have active development plans. At least two groups will test new or improved equipment at abyssal plain depths in 1985, if present plans come to fruition.
7. One downhole and one seabed develop-

*From presentations at the Society for Underwater Technology's meeting in December 1984 on objectives and technology for ocean instrumentation deeper than 200 m.

ment are described to measure seismic events or shear waves in ocean-basin water depths. In addition, one downhole and one seabed-type development exists actually or potentially to measure shear waves *in situ*. These developments are the likely precursors of routine shear-wave measurements leading to an improved understanding of the range of shear moduli to be found in soils of the deep sea floor.

8. Many probes and sensors exists for *in situ* testing in water depths exceeding 300 m. Many others, as well as certain tests — such as the hydraulic fracture test — are applicable for use offshore in these water depths. Table 11 summarizes both. It is concluded that both industrial and governmental–academic organizations will be combining more and more sensors into a single probe system for use either downhole or from seabed equipment.

9. The future is examined in light of the various developments reported. It is concluded that because of stress relief, soil degassing and so forth, as samples are brought from water depths exceeding 300 m to the surface, *in situ* testing may replace many requirements for sampling in the future. Industrial testing is likely to be characterized by increased use of the piezocone penetrometer downhole and from continuous-drive seabed devices. Governmental–academic testing is likely to increase with the development of downhole and seabed equipment for use in water depths exceeding 4000 m. The measurements *in situ* of true ambient pore pressures to calculate the existence of any hydraulic gradients in ocean basin soils is likely to receive emphasis during at least the next few years.

ACKNOWLEDGEMENTS

Mr T. Freeman, Building Research Establishment, is thanked for inviting me to prepare this paper, for kindly making available Figs 10 and 11 in advance of formal publication and for valued discussions on various topics. Drs P. J. Schultheiss, Institute of Oceanographic Sciences, and R. H. Bennett, Naval Ocean Research and Development Activity, made available unpublished manuscripts and permitted publication of Figs 6 and 7, respectively. Mr R. L. Geer kindly made available Figs 1, 3 and 4 from the Shell Oil Company. Mr D. Cover, Scripps Institution of Oceanography, Deep Sea Drilling Project, kindly made available Fig. 12. Mr A. W. W. Tieges, Delft Soil Mechanics Laboratory, kindly permitted the use of Fig. 9. Mr J. M. Bayne, McClelland; Dr J. Booth, US Geological Survey; Dott. D. Bruzzi, Ismes; Dr J. Burdett, Building Research Establishment; Drs A. Driscoll and A. Silva, University of Rhode Island; Dr I. Noorany, San Diego State University; Dr G. Singh, Leeds University; Mr P. Le Tirant, French Institute of Petroleum; and others, as well as colleagues at Fugro, graciously provided much useful information. Mrs E. A. Richards assisted with the references and illustrations. Mr D. W. Eijmaal modified several illustrations and drafted two. Messrs J. M. Geise, H. J. Kolk, J. de Ruiter, J. van der Wal and H. M. Zuidberg reviewed the manuscript and made helpful comments. The opinions in this paper are those of the author and do not necessarily represent those of Fugro BV or any other organization or person.

REFERENCES

1. Aas., J., Lacasse, S., Lunne, T. and Madshus, C. 1984. *In situ* testing: new developments. *Nordiska Greteknikermøtet.* Vol. 2: *Linköping*, pp. 705–716 (Reproduced in Norwegian Geotechnical Institute Publication Nr. 153, 1984).
2. Abbott, D., Menke, W., Hobart, M. and Anderson, R. 1981. Evidence for excess pore pressures in southwest Indian Ocean sediments. *J. Geophys. Res.* **86** (B3), 1813–1827.

3. Amundsen, T., Lunne, T., Christopherson, H. P., Bayne, J. M. and Barnwell, C. 1985. Advanced deep-water soil investigation at the Troll East Field. Paper 11, this volume.

4. Anderson, V. C., Clinton, J. R., Gibson, D. K. and Kirsten, O. H. 1972. Instrumenting RUM for *in situ* subsea soil surveys. *Underwater Soil Sampling, Testing, and Construction Control*. Special Technical Publication 501, American Society for Testing and Materials, Philadelphia, pp. 216–231.

5. Andresen, A. 1981. Exploration, sampling and *in situ* testing of soft clay. In *Soft Clay Engineering* (Eds E. W. Brand and R. P. Brenner). Elsevier, Amsterdam, pp. 241–308.

6. Ardus, D. A., Skinner, A., Owens, R. and Pheasant, J. 1982. Improved coring techniques and offshore laboratory procedures in sampling and shallow drilling. *Oceanology International Exhibition and Conference Papers*, Vol. 2, Paper 5.8.

7. Arrhenius, G. 1952. Sediment cores from the east Pacific. In *Reports of the Swedish Deep-Sea Expedition 1947—1948*, Vol. 5, pp. 1–227.

8. Babb, J. D. and Silva, A. J. 1983. An *in situ* vane system for measuring deep sea sediment shear strength. In *Oceans 83 Proceedings*, Vol. 1, pp. 598–607.

9. Beard, R. M. 1977. Expendable doppler penetrometer: a performance evaluation. Report No. TR-855, US Navy Civil Engineering Laboratory.

10. Beard, R. M. In press. Expendable doppler penetrometer for deep ocean sediment strength measurements. In *Strength Testing of Marine Sediments: Laboratory and In-Situ Measurements* (Eds R. C. Chaney and K. R. Demars). Special Technical Publication 883, American Society for Testing and Materials, Philadelphia.

11. Beard, R. M. and Johnson, B. A. 1984. XSP cone penetrometer: a performance evaluation. Report No. TR-911, US Navy Civil Engineering Laboratory.

11a. Beard, R. M. and Lee, H. J. 1982. A 40-foot static cone penetrometer. *Offshore Technology Conference Proceedings*, Vol. 2, Paper 4300, pp. 809–818.

12. Bennett, R. H., Burns, J. T., Clarke, T. L., Faris, J. R., Forde, E. B. and Richards, A. F. 1982. Piezometer probes for assessing effective stress and stability in submarine sediments. In *Marine Slides and Other Mass Movements* (Eds S. Saxov and J. K. Nieuwenhuis). Plenum Press, London, pp. 129–161.

13. Bennett, R. H., Burns, J. T., Lipkin, J. and Percival, C. N. 1983a. Piezometer probe technology for geotechnical investigations in coastal and deep-ocean environments. In *Twelfth Transducer Workshop Proceedings* (Eds L. Bates and K. D. Cox). Secretariat, Range Commanders Council, Telemetry Group, White Sands Missile Range, New Mexico, pp. 377–404.

14. Bennett, R. H., Lambert, D. N., Hulbert, M. H., Burns, J. I., Sawyer, W. B. and Freeland, G. L. 1983b. Electrical resistivity/conductivity in seabed sediments. In *CRC Handbook of Geophysical Exploration at Sea* (Ed. R. A. Geyer). CRC Press, Boca Raton, p. 333–375.

15. Bennett, R. H., Nastav, F. L. and Bryant, W. R. 1984. Strength measurements. *Sedimentology, Physical Properties, and Geochemistry in the Initial Reports of the Deep Sea Drilling Project*. Vols 1–44: *An Overview* (Ed. G.R.). Report MGG-1, US Department of Commerce, Boulder, Colorado, pp. 129–146.

16. Bennett, R. H., Burns, J. T. and Nastav, F. L. In press, a. Deep-ocean piezometer probe technology. *IEEE J. Oceanic Engng*.

17. Bennett, R. H., Huon, L., Valent, P. J., Lipkin, J. and Esrig, M. I. In press, b. *In situ* undrained shear strengths and permeabilities derived from piezometer measurements. In *Strength Testing of Marine Sediments: Laboratory and In-Situ Measurements* (Eds R. C. Chaney and K. R. Demars). Special Technical Publication 883, American Society for Testing and Materials, Philadelphia.

18. Benoit, C. 1983. Einsatz eines ferngelenkten Unterwasser-Boden-Untersuchungsgerätes (First experiences with an underwater soil investigation crawler). *Meerestechnik* **14**(1), 3–7.

19. Berg, A. P. van den, 1984. Developments in static cone penetrometering of the seabed. *Underwater Systems Design* **6**(5), 28–31.

20. Briaud, J.-L. and Meyer, B. 1983. *In situ*

tests and their application in offshore design. *Geotechnical Practice in Offshore Engineering* (Ed. S. G. Wright). American Society of Civil Engineers, New York, pp. 244–266.

21. Brooks, J. M., Kennicutt, M. C., II, Fay, R. R. and McDonald, T. J. 1984. Thermogenic gas hydrates in the Gulf of Mexico. *Science* 225 (4660), 409–411.

22. Brucy, F., Faÿ, J. B. and Le Tirant, P. 1984. Three years' experience with the offshore self-boring pressuremeter 'PAM' *Offshore Technology Conference Proceedings*, Vol. 1, Paper 4677, pp. 265–274.

23. Bruzzi, D. 1983. Underwater static penetrometer. *International Symposium Soil and Rock Investigations by In-Situ Testing*, Vol. 2, pp. 223–226.

24. Burgess, N. C., Hughes, J. M. O., Innes, R. and Gleadowe, J. 1983. Site investigation and *in-situ* testing techniques in Arctic seabed sediments. *Offshore Technology Conference Proceedings*, Vol. 3, Paper 4583, pp. 27–34.

25. Campanella, R. G. and Robertson, P. K. 1982. State of the art in *in-situ* testing of soils: developments since 1978. Soil Mechanics Series No. 56, The University of British Columbia, Vancouver.

26. Campanella, R. G. and Robertson, P. K. 1984. A seismic cone penetrometer to measure engineering properties of soil. Paper presented at the Society of Exploration Geophysicists' 54th Annual Meeting, Atlanta.

27. Capelle, J.-F. 1984. A new instrument for the *in situ* measurement of the permeability of clays: the self boring non-clogging permeater. *Field Measurements in Geomechanics*, Vol. 1 (Ed. K. Kovari). A.A. Balkema, Rotterdam, pp. 49–57.

28. Carson, B. 1977. Tectonically induced deformation of deep-sea sediments off Washington and northern Oregon: mechanical consolidation. *Mar. Geol.* 24, 289–307.

29. Carson, B., von Huene, R. and Arthur, M. 1982. Small scale deformation structures and physical properties related to convergence in Japan Trench slope sediments. *Tectonics* 1 (3), 277–302.

30. Carson, B., Ritger, S. D. and Suess, E. 1984. Precipitation of carbonate crust associated with subduction-induced pore-water expulsion: Washington–Oregon conti-

nental slope. Abstract, *EOS Trans Am. Geophys. Union* 65(45), 1089–1090.

31. Chaney, R. C. and Fang, H.-Y. In press. Static and dynamic properties of marine sediments. *First Shanghai Symposium on Marine Geotechnology and Nearshore/Offshore Structures* (Eds R. C. Chaney and H.-Y. Fang). Special Technical Publication, American Society for Testing and Materials, Philadelphia.

32. Chari, T. R., Abdel-Gawad, S. M. and Chaudhuri, S. N. 1979. Geotechnical survey of the seafloor with a free fall penetrometer. POAC 79, *Fifth International Conference on Port and Ocean Engineering Under Arctic Conditions Proceedings*, Vol. 2. The Norwegian Institute of Technology, Trondheim, pp. 833–843.

33. Colp, J. L., Caudle, W. N. and Schuster, C. L. 1975. Penetrometer system for measuring *in-situ* properties of marine sediments. In *Oceans 75 Conference Records*, Institute of Electrical and Electronics Engineers, and Marine Technology Society, pp. 405–411.

34. Cyr, R. J. 1984. Sea-bed surveys by acoustic penetrometer. *Oceanology International*, Paper 1.7.

35. Dahlberg, R. 1974. Penetration testing in Sweden. In *Proceedings of the European Symposium on Penetration Testing*, Vol. 1, pp. 115–131.

36. Demars, K. R. and Taylor, R. J. 1971. Naval seafloor soil sampling and in-place test equipment: a performance evaluation. Report No. 730, US Navy Civil Engineering Laboratory.

37. Denk, E. W., Dunlap, W. A., Bryant, W. R., Milberger, L. J. and Whelan, T. J., III, 1981. A pressurized core barrel for sampling gas-charged marine sediments. In *Offshore Technology Conference Proceedings*, Vol. 4, pp. 43–52.

38. Driscoll, A. H. 1981. The long coring facility, new techniques in deep ocean coring. In *Oceans 81 Conference Records*, Vol. 1. Institute of Electrical and Electronics Engineers, and Marine Technology Society, pp. 404–410.

39. Duin, E. J. Th., Mesdag, C. S. and Kok, P. T. J. 1984. Faulting in Madeira Abyssal Plain sediments. *Mar. Geol.* 56, 299–308.

40. Faÿ, J. B., Montarges, R., Le Tirant, P. and Brucy, F. 1985. Use of the PAM self-

boring pressuremeter and the STACOR large-size fixed-piston corer for deep seabed surveying. Paper 12, this volume.

41. Ferguson, G. H., McClelland, B. and Bell, W. D. 1977. Seafloor cone penetrometer for deep penetration measurements of ocean sediment strength. *Offshore Technology Conference Proceedings*, Vol. 1, Paper 2787, pp. 471–478.

42. Flodin, N. and Broms, B. 1981. History of civil engineering in soft clay. *Soft Clay Engineering* (Eds E. W. Brand and R. P. Brenner). Elsevier, Amsterdam, pp. 27–156.

43. Foss, G. N. and Wallerstedt, R. L. 1984. Operational aspects of borehole deployment of a marine seismic system in deep water. *Offshore Technology Conference Proceedings*, Vol. 2, Paper 4757, pp. 397–406.

44. Freeman, T. J. and Burdett, J. R. F. In preparation, a. Deep ocean penetrator experiments. Final report to the Commission of the European Communities, Contract 392-83-7 WAS, UK.

45. Freeman, T. J., Carlyle, S. G., Francis, T. J. G. and Murray, C. N. 1984a. The use of large-scale penetrators for the measurement of deep-ocean sediment properties. *Oceanology International*, Paper 1.8.

46. Freeman, T. J., Murray, C. N., Francis, T. J. G., McPhail, S. D. and Schultheiss, P. J. 1984b. Modelling radioactive waste disposal by penetrator experiments in the abyssal Atlantic Ocean. *Nature* 310(5973), 130–133.

47. Freeman, T. J., Murray, C. N. and Talbert, D. M. In preparation, b. Penetrator experiments in the Nares Abyssal Plain of the Atlantic Ocean.

48. Fyffe, S., Reis, W. M. and St. John, H. D. 1982. The use of the push-in pressuremeter in offshore site investigation. *Oceanology International Exhibition and Conference*, Vol. 1, Paper 4.5.

49. Geise, J. M. and Kolk, H. J. 1983. The use of submersibles for geotechnical investigations. *The Design and Operation of Underwater Vehicles, Subtech '83*. Paper 7.3, Society of Underwater Technology, London.

50. Geotechnical Consortium, 1984. Geotechnical properties of sediments from Walvis Ridge. *Deep Sea Drilling Project, Leg 75, Hole 532A, Initial Report of the Deep Sea Drilling Project*, Vol. 75 (Eds W. W. Hay, J.-C. Sibuet *et al.*). US Government Printing Office, Washington, pp. 1109–1127.

51. Graaf, H. C. van de and Smits, A. P. 1983. Offshore site investigations by rotary drilling from a diving bell. *Ground Engng* 16(1), 18–19.

52. Guney, M., Nawab, Z. and Marhoun, M. A. 1984. Atlantis-II-Deep's metal reserves and their evaluation. *Offshore Technology Conference Proceedings*, Vol. 3, Paper 4780, pp. 33–44.

53. Haas, W. J. M. de, 1983. A seabed sampler for the dredging industry. *World Dredging Congress 1983*, Paper E3, pp. 237–249.

54. Hansbo, S. 1957. A new approach to the determination of the shear strength of clay by the fall-cone test. *R. Swed. Geotech. Inst. Proc.* 14, 1–47.

55. Henderson, G., Smith, P. D. K. and St. John, H. D. 1980. The development of the push-in pressuremeter for offshore site investigation. *Offshore Site Investigation* (Ed. D. A. Ardus). Graham & Trotman, London, pp. 159–167.

56. Hirst, T. J., Perlow Jr., M. and Richards, A. F. 1975. Improved *in situ* gamma-ray transmission densitometer for marine sediments. *Ocean Engng* 3(1), 17–27.

56a. Hollister, C. D., Silva, A. J. and Driscoll, A. 1973. A giant piston-corer. *Ocean Engng* 2, 159–168.

57. Hughes, J. M. O., Jefferies, M. G. and Morris, D. L. 1984. Self-bored pressuremeter testing in the Arctic offshore. *Offshore Technology Conference Proceedings*, Vol. 1, Paper 4676, pp. 255–264.

58. Hulbert, M. H., Bennett, R. H. and Lambert, D. N. 1982. Seabed geotechnical parameters from electrical conductivity measurements. *Geo. Mar. Lett.* 2, 219–222.

59. Hunt, R. E. 1984. *Geotechnical Engineering Investigation Manual*. McGraw-Hill, New York, pp. 103–104.

60. Inderbitzen, A. L. and Simpson, F. 1972. A study of the strength characteristics of mining sediments utilizing a submersible. *Underwater Soil Sampling, Testing, and Construction Control*. Special Technical Publication 501, American Society for Testing and Materials, Philadelphia, pp. 204–215.

61. Ingram, C. 1982. Expendable penetrome-

ter for seafloor classification. *Geo. Mar. Lett.* **2**, 239–241.

62. James, L. T. and Calloway, T. M. 1983. Initial field tests of ISP-1 and ISP-2 — systems that extend the technology of instrumented seabed penetrators. *Oceans 83 Proceedings*, Vol. 1. Institute of Electrical and Electronics Engineers, and Marine Technology Society, pp. 608–612.

63. James, L. T., Edrington, T. S., Reis, G. E. and Suazo, J. E. 1981. Development of a gun-launched, instrumented seabed penetrometer system. *Oceans 81 Conference Record*, Vol. 2. Institute of Electrical and Electronics Engineers, and Marine Technology Society, pp. 656–661.

64. Jefferies, M. G. and Funegard, E. 1983. Cone penetration testing in the Beaufort Sea. *Geotechnical Practice in Offshore Engineering* (Ed. S. G. Wright). American Society of Civil Engineers, New York, pp. 220–243.

65. Johns, M. W., Taylor, E. and Bryant, W. R. 1982. Geotechnical sampling and testing of gas-charged marine sediments at *in situ* pressure. *Geo. Mar. Lett.* **2**, 231–236.

66. Justice, J. H., Hinds, R. and Stirbys, A. F. 1984. The use of vertical seismic profiling in geotechnical site investigation. *Offshore Technology Conference Proceedings*, Vol. 2, Paper 4756, pp. 391–396.

67. Kallstenius, T. 1961. Development of two modern continuous sounding methods. In *Fifth International Conference on Soil Mechanics and Foundation Engineering Proceedings*, Vol. 1, pp. 475–480.

68. Keller, G. H. 1982. Organic matter and the geotechnical properties of submarine sediments., *Geo. Mar. Lett.* **2**, 191–198.

69. Keller, G. H. 1965. Nuclear density probe for in place measurement in deep-sea sediments. *Transactions of the Conference and Exhibit*, Marine Technology Society and American Society of Limnology and Oceanography, Vol. 1, pp. 363–372.

70. Kermabon, A. and Cortis, V. 1969. A new Sphincter corer with a recoilless piston. *Mar. Geol.* **7**, 147–159.

71. Kirkpatrick, W. M. and Khan, A. J. 1984. The reaction of clays to sampling stress relief. *Géotechnique* **34**(1), 29–42.

72. Kraft, L. M., Jr., Ahmad, N. and Focht, J. A. 1976. Application of remote vane results to offshore geotechnical problems. *Offshore Technology Conference Proceedings*, Vol. 3, Paper 2626, pp. 75–96.

73. Kullenberg, B. 1947. The piston core sampler. Svenska Hydrografisk — Biologiska Kommissionens Skrifter, Tredje Serien, Hydrografi, Band 1, Häfte 2.

74. Kullenberg, B. 1955. Deep-sea coring. Reports of the Swedish Deep-Sea Expedition, 4(2), Fasc. 1, pp. 35–96.

75. Kulm, L. D., Thornburg, T. M. and Carson, B. 1984. Distribution of lithologies and fluid vents from the Oregon underthrust margin: implications for active subduction of the Juan de Fuca plate. Abstract, *EOS Trans Am. Geophys. Union* **65**(45), 1090.

76. Kvenvolden, K. A. and Barnard, L. A. 1982. Hydrates of natural gas in continental margins. *Studies in Continental Margin Geology* (Eds J. S. Watkins and C. L. Drake), AAPG Memoir No. 34, American Association of Petroleum Geologists, Tulsa, pp. 631–640.

77. Lambert, D. N. 1982. Submersible mounted *in situ* geotechnical instrumentation. *Geo. Mar. Lett* **2**, 209–214.

78. Ledoux, J. L., Menard, J. and Soulard, P. 1982. The penetro-gammadensimeter. *Penetration Testing*, Vol. 2 (Eds A. Verruijt, F. L. Beringen and E. H. de Leeuw). A.A. Balkema, Rotterdam, pp. 679–681.

79. Lee, H. J. In press. State of art: laboratory determination of the strength of marine soils. *Strength Testing of Marine Sediments: Laboratory and In-Situ Measurements* (Eds R. C. Chaney and K. R. Demars). Special Technical Publication 883, American Society for Testing and Materials, Philadelphia.

80. Le Tirant, P., Faÿ, J. B., Brucy, F. and Jezequel, J.-F. 1981. A self-boring pressuremeter for deep sea soils investigations. *Offshore Technology Conference Proceedings*, Vol. 2, Paper 4019, pp. 115–126.

81. Lewis, L., Nacci, V. and Gallagar, J. 1970. *In situ* investigations of ocean sediments. *Civil Engineering in the Oceans II Proceedings*. American Society of Civil Engineers, New York, pp. 641–654.

82. Lunne, T., Lacasse, S., Aas, G. and Madshus, C. In press. Design parameters for offshore sands; use of *in situ* tests. Paper 17, this volume.

83. The Marine Geotechnical Consortium. In press. Geotechnical properties of north-

west Pacific pelagic clays: Deep Sea Drilling Project Leg 86, Hole 576A. *Leg 86 Initial Reports of the Deep Sea Drilling Project*, Vol. 86 (Ed. G. R. Heath). US Printing Office, Washington.

84. Marsland, A. and Windle, D. 1982. Developments in offshore site investigation. *Oceanology International Exhibition and Conference*, Vol. 1, Paper 2.7.

4a. McCoy, F. W. and Selwyn, S. 1984. The hydrostatic corer. *Mar. Geol.* **54**, 33–41.

85. Moeyes, G. and Hackley, M. 1983. Soil investigations in the Troll area. *Offshore Northern Seas Advanced Project Conference*, Stavangar, Paper T6.

86. Montarges, R., Le Tirant, P., Wannesson, J., Valéry, P. and Bethon, J. L. 1983. Large-size stationary-piston corer. In *Deep Offshore Technology 2nd International Conference and Exhibition*, pp. 63–74.

87. Moore, D. G. and Richards, A. F. 1962. Conversion of 'relative shear strength' measurements by Arrhenius on east Pacific cores to conventional units of shear strength. *Géotechnique* **12**, 55–59.

88. NEA, 1984. *Seabed Disposal of High-Level Radioactive Waste*. Nuclear Energy Organisation for Economic Co-operation and Development, Paris.

89. Nieuwenhuis, J. K. and Smits, F. P. 1982. The development of a nuclear density probe in a cone penetrometer. Penetration Testing, Vol. 2 (Eds A. Verruijt, F. L. Beringen and E. H. de Leeuw). A.A. Balkema, Rotterdam, pp. 745–749.

90. Ohya, S., Ogura, K. and Imai, T. 1984. The suspension PS velocity logging system. *Offshore Technology Conference Proceedings*, Vol. 1, Paper 4680, pp. 291–298.

91. Okumura, T. and Matsumoto, K. 1981. Marine auto sampler and sample quality. *Soil Mechanics and Foundation Engineering Tenth International Conference Proceedings*, Stockholm, Vol. 2, pp. 537–540.

92. Percival, C. M., McVey, D. F., Olson, L. O. and Silva, A. J. 1984. *In situ* heat transfer experiment (ISHTE). *Mar. Geotech.* **5**(3–4), 361–377.

93. Pheasant, J. 1984. A microprocessor controlled seabed rockdrill/vibrocorer. *Underwater Technol.* **10**(1), 10–14.

94. Preslan, W. L. and Babb, L. 1979. Piezometer measurement for deep penetration marine applications. *Offshore Technology Conference Proceedings*, Vol. 2, Paper 3461, pp. 901–908.

95. Prindle, R. W. and Lopez, A. A. 1983. Pore pressures in marine sediments — 1981 test of the geotechnically instrumented seafloor probe (GISP). *Offshore Technology Conference Proceedings*, Vol. 1, Paper 4463, pp. 173–180.

96. Prior, D. B. and Doyle, E. H. 1984. Geological hazard surveying for exploratory drilling in water depths of 2000 meters. *Offshore Technology Conference Proceedings*, Vol. 2, Paper 4747, pp. 311–318.

97. Prothero, W. A., Jr. 1984. Ocean bottom seismometer technology. *EOS Transactions, Am. Geophys. Union* **65**(13), 113–116.

98. Reece, E. W., Ryerson, D. E., Kestly, J. D. and McNeill, R. L. 1979. The development of *in situ* marine seismic and geotechnical instrumentation systems. POAC 79. In *Fifth International Conference on Port and Ocean Engineering Under Artic Conditions Proceedings*, Vol. 1. The Norwegian Institute of Technology, Trondheim, pp. 331–344.

99. Reid, W. M., St. John, H. D., Fyffe, S. and Rigden, W. J. 1982. The push-in pressuremeter. In *Symposium on the Pressuremeter and Its Marine Applications*, *37*. Editions Technip, Paris, pp. 247–261.

100. Richards, A. F. 1972. Instrumentation of two submersibles for *in situ* geotechnical measurements in cohesive sea floor soils. *2nd International Ocean Development Conference Preprints*, Vol. 2, pp. 1329–1346.

101. Richards, A. F. 1982. Review of marine *in situ* geotechnical testing equipment. *Oceanology International Exhibition and Conference Papers*, Vol. 1, Paper 2.1, 10 pp.

102. Richards, A. F. 1984. Modelling and the Consolidation of Marine Soils. In *Seabed Mechanics* (Ed. B. Denness). Graham & Trotman, London, pp. 3–8.

103. Richards, A. F. and Chaney, R. C. 1981. Present and future geotechnical research needs in deep ocean mining. *Mar. Min.* **2**(4), 315–337.

104. Richards, A.F. and Hartevelt, J. J. A. 1981. Marine engineering geology; scope and new developments. In *27th International Geological Congress Proceedings*,

Vol. 17. NVU Press, Utrecht, The Netherlands, pp. 261–275.

105. Richards, A. F. and Parks, J. M. 1977. Geotechnical predictor equations for east central North Pacific nodule mining area sediments. *Offshore Technology Conference Proceedings*, Vol. 1, Paper 2773, pp. 377–386.

106. Richards, A. F. and Zuidberg, H. M. In press, a. State of art: *in situ* determination of the strength of marine soils. In *Strength Testing of Marine Sediments: Laboratory and In Situ Measurements* (Eds R. C.Chaney and K. R. Demars). Special Technical Publication 883. American Society for Testing and Materials, Philadelphia.

107. Richards, A. F. and Zuidberg, H. M. In press, b. Sampling and *in situ* geotechnical investigations offshore. *First Shanghai Symposium on Marine Geotechnology and Nearshore/Offshore Structures* (Eds R. C. Chaney and H.-Y. Fang). Special Technical Publication, American Society for Testing and Materials, Philadelphia.

108. Richards, A. F., McDonald, V. J., Olson, R. E. and Keller, G. H. 1972. In-place measurement of deep-sea soil shear strength. *Symposium on Underwater Soil Sampling, Testing and Construction Control.* Special Technical Publication No. 501, American Society for Testing and Materials, Philadelphia, pp. 55–68.

109. Richards, A. F., Øien, K., Keller, G. H. and Lai, J. 1975. Differential piezometer probe for an *in situ* measurement of sea-floor pore-pressure. *Géotechnique* 25, 229–238.

110. Rietsema, R. A. and Viergever, M. A. 1979. *In situ* measurement of permeability. In *Design Parameters in Geotechnical Engineering*, Vol. 2. Seventh European Conference on Soil Mechanics and Foundations Engineering. British Geotechnical Society, London, pp. 261–264.

111. Robertson, P. K., Campanella, R. G. and Gillespie, D. 1985. Seismic CPT to measure *in-situ* shear wave velocity. Paper presented at the American Society of Civil Engineers, April 1985. Denver.

112. Rona, P. A. 1984. Hydrothermal mineralization at seafloor spreading centers. *Earth-Sci. Rev.* 20, 1–104.

113. Rona, P. A., Bostrom, K., Lucien, L. and Smith, K. L. (Eds) 1984. *Hydrothermal Processes at Seafloor Spreading Centers.* Plenum Press, New York.

114. Rose, V. C. and Roney, J. R. 1971. A nuclear guage for in-place measurement of sediment density. *Offshore Technology Conference Proceedings*, Vol. 1, Paper 1329, pp. 43–52.

115. Ruiter, J. de 1982. The static cone penetration test state-of-the-art-report. *Penetration Testing* (Eds A. Verriujt, F. L. Beringen and E. H. de Leeuw). *Proceedings Second European Conference on Penetration Testing*, Vol. 2. A.A. Balkema, Rotterdam, pp. 389–405.

116. Schaap, L. H. J. and Hoogendoorn, H. G. 1984. A versatile measuring system for electric cone penetration testing. *Field Measurements in Geomechanics*, Vol. 1 (Ed. K. Kovari). A.A. Balkema, Rotterdam, pp. 313–324.

117. Schaap, L. H. J. and Liefting, J. H. 1984. Geotechnical applications of solid state recorders. *Field Measurements in Geomechanics*, Vol. 1 (Ed. K. Kovari). A.A. Balkema, Rotterdam, pp. 325–333.

118. Schlosser, F. 1978. Recent advance in *in situ* testing. *Bulletin de Liaison des Laboratoires des Ponts et Chaussées*, Special Issue VI E, 15–43.

119. Schultheiss, P. J. 1982. Geotechnical properties of deep sea sediments: a critical review of measurement techniques. Report No. 134, Institute of Oceanographic Sciences, Wormley.

120. Schultheiss, P. J., McPhail, S. D., Packwood, A. R. and Hart, B. In press. An instrument to measure differential pore pressures in deep ocean sediments: Pop-Up-Pore-Pressure Instrument (PUPPI). Report No. DoE/RW, Institute of Oceanographic Sciences, Wormley.

121. Schüttenhelm, R. T. E., Kuijpers A and Duin, E. J. Th. 1985. The geology of some Atlantic abyssal plains and the engineering implications. Paper 3, this volume.

122. Shephard, L. E. and Bryant, W. R. 1983. Geotechnical properties of lower trench inner-slope sediments. *Tectonophysics* 99, 279–312.

123. Silva, A. 1985. The Comparison of *in situ* and ship-board vane measurements on a deep-sea clay. Paper 14, this volume.

124. Silva, A. J., Babb. J. D., Lipkin, J., Pietryka, P. and Butler, D. In press. *In situ* vane system for seafloor strength investigations. *IEEE J. Oceanic Engng.*

125. Singh, J. and Chomg, M. K. 1984.

Development of a site investigation tool based on quasi static cone penetrometer. Internal Report, University of Leeds, Department of Civil Engineering.

126. Storms, M. A., Nugent, W. and Cameron, D. H. 1983. Hydraulic piston coring — a new era in ocean research. *Offshore Technology Conference Proceedings*, Vol. 3, Paper 4622, pp. 369–378.

127. The Sub-Committee on Soil Sampling, 1981. *International Manual for the Sampling of Soft Cohesive Soils*. Tokai University Press, Tokyo.

128. Suess, E. and Massoth, G. 1984. Evidence for venting of pore waters from subducted sediments of the Oregon continental margin. Abstract, *EOS Trans., Am. Geophys. Union* 65(45), 1089.

129. Talbert, D. M. 1984a. Report of the Fourth Interim Meeting of the Seabed Working Group Engineering Studies Task Group. Report SAND83-2333, Sandia National Laboratories.

130. Talbert, D. M. 1984b. Shallow-water free-fall penetrator test (FPT-1). Appendix IV, *Report of the Fourth Interim Meeting of the Seabed Working Group Engineering Studies Task Group* (Ed. D. M. Talbert). Report SAND83-2333, Sandia National Laboratories, pp. 42–45.

131. Taylor Smith, D. 1983. Seismo-acoustic wave velocities and sediment engineering properties. In *Acoustics and the Sea-Bed* (Ed. N. G. Pace). Bath University Press, Bath, pp. 9–17.

132. Threadgold, P. 1980. Borehole logging for offshore site investigation for engineering purposes. In *Offshore Site Investigation* (Ed. D. A. Ardus). Graham & Trotman, London, pp. 133–142.

133. Tjelta, T. I., Tieges, A. W. W., Smits, F. P., Geise, J. M. and Lunne, T. In press. *In situ* density measurements by nuclear backscatter for an offshore soil investigation. *Offshore Technology Conference Proceedings*, Paper 4917.

134. Toolan, F. E. 1983. Recent improvements in soil investigation techniques. In *Design In Offshore Structures*. Thomas Telford, London, pp. 29–36.

135. Triangale, P. T. and Mitchell, J. K. 1982. An acoustic cone penetrometer for site investigations. In *Penetration Testing*, Vol. 2 (Eds A. Verruijt, F. L. Beringen,

and E. H. de Leeuw). A.A. Balkema, Rotterdam, pp. 909–194.

136. True, D. G. 1975. Penetration of projectiles into seafloor soils. Report No. R-822, US Navy Civil Engineering Laboratory.

137. Tsurusaki, K., Itoh, F. and Yamazaki, T. 1984. Development of *in situ* measuring apparatus of geotechnical elements of sea floor (IMAGES). *Offshore Technology Conference Proceedings*, Vol. 1, Paper 4681, pp. 299–307.

138. Vollset, M. and Gunleiksrud, T. 1982. Offshore geotechnical investigations with the use of *in situ* measurements. Report No. P-381/1/82, Institutt for Kontinentalsokkelundersøkelser, Trondheim.

139. Von Huene, R. 1984. Tectonic process along the front of modern convergent margins — research of the past decade. *A. Rev. Earth Planet. Sci.* 12, 359–381.

140. Vyas, Y. K., Angemeer, J., Murff, J. D., Neuberger, C. A. McNeilan, T. and Klejbuk, L. W. 1983. Deepwater geotechnical site investigations: Santa Ynez Unit, offshore California. *Offshore Technology Conference Proceedings*, Vol. 1, Paper 4467, pp. 217–226.

141. Whitmarch, R. B. and Lilwall, R. C. 1982. A new method for the determination of *in situ* shear-wave velocity in deep-sea sediments. *Oceanology International Exhibition and Conference Papers*, Vol. 1, Paper 4.2, 21 pp.

142. Wroth, C. P. 1984. The interpretation of *in situ* soil tests. *Géotechnique* 34(4), 449–489.

143. Young, A. G., Quiros, G. W. and Ehlers, C. J. 1983. Effects of offshore sampling and testing on undrained soil shear strength. *Offshore Technology Conference Proceedings*, Vol. 1, Paper 4465, pp. 193–204.

144. Zuidberg, H. M. 1975. Seacalf: a submersible cone-penetrometer rig. *Mar. Geotechnol.* 1(1), 15–32.

145. Zuidberg, H. M., Schrier, W. H. and Pieters, W. H. 1984. Ambient pressure sampler system for deep ocean soil investigations. *Offshore Technology Conference Proceedings*, Vol. 1, Paper 4679, pp. 283–290.

146. Zuidberg, H. M., Richards, A. F. and Tsuzuki, M. In press. Vessel outfit for advanced geotechnical site investigations in deep water. *International Symposium on Ocean Space Utilization*, Tokyo.

11

Advanced Deep-water Soil Investigation at the Troll East Field

T. Amundsen, Norsk Hydro A/S, Norway, *T. Lunne*, Norwegian
Geotechnical Institute, Norway, *H. P. Christophersen*, Saga Petroleum,
J. M. Bayne and C. L. Barnwell, McClelland Ltd, UK

INTRODUCTION

Norsk Hydro, Saga Petroleum and Statoil
are operators for the Troll East Field, one of
the largest offshore gas fields in the world.
The field is situated in the Norwegian
Trench in water depths of 300 to 330 m. Figure 1 shows the location of the Troll East
Field. This field is planned to come into production in the mid-1990s. The development
of Troll East Field in deep water with soft
sea-floor sediments presents a new
engineering challenge in the North Sea.
Because of this, present engineering feasibility studies encompass a wide range of
platform concepts, including both gravity-
base and piled structures. All concepts
require a comprehensive knowledge of the
soil conditions to evaluate their feasibility.
Norsk Hydro have been responsible for the
geotechnical aspects of site investigation
and, in 1983, a preliminary soil survey was
carried out which served as a basis for
detailed planning of a major 1984 soil investigation described in this paper. The Nor-

wegian Geotechnical Institute (NGI) was
awarded the main geotechnical contract for
this investigation, with McClelland Ltd sub-
contracted to perform sampling and *in situ*
testing. Anton von der Lippe's soil survey
ship *Bucentaur* was chartered for drilling.
Racal Survey provided positioning services.
Stressprobe Ltd performed pressuremeter
testing. This paper describes the planning of
a technically advanced soil investigation
which in several aspects represents the
state of the art. Offshore soil investigation
equipment and procedures are then
described and results of *in situ* and laboratory tests are presented. Finally, operational
performance during the fieldwork is discussed.

PLANNING

In order to evaluate the soil conditions of a
large field like Troll East, which has an area
of approximately 600 square kilometers, site
investigation was planned in four stages.

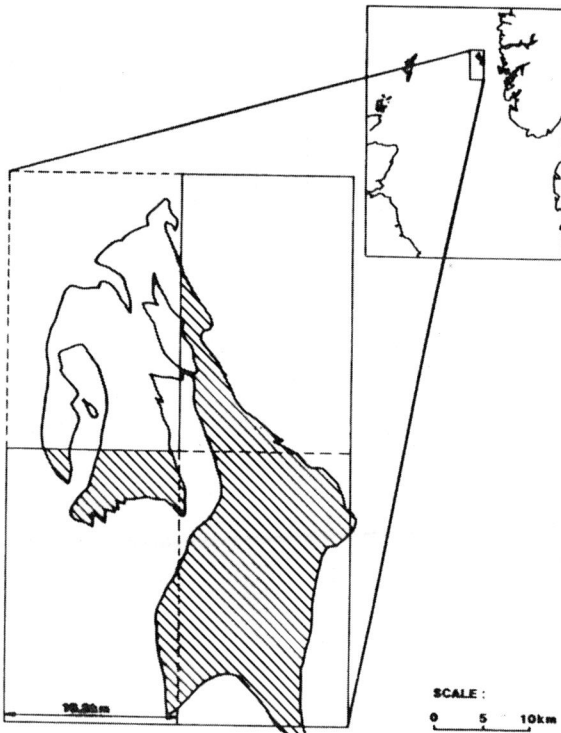

SCALE :

0   5   10km

Fig. 1   Location map, Troll East Field

Firstly, in 1983, a high-resolution geophysical survey was undertaken followed, secondly, by a preliminary soil investigation. Information from this preliminary soil investigation was used to interpret the geophysical profiles. Figure 2 shows an example of how the results of the early investigation were interpreted to develop preliminary soil stratification in the upper 100 to 120 m in Block 31/3 (Østmo and Kleiven, 1984). Thirdly, based on this information, a detailed regional soil investigation was planned during the autumn of 1983. This was designed to provide a relatively detailed knowledge of soil stratification across the field as well as preliminary soil design parameters at likely structure locations. The regional soil investigation was performed during the summer of 1984. Fourthly, detailed local soil investigations will be performed at the final platform locations after these locations have been selected.

The field programme for the regional inves-

tigation was developed through an iterative process between geophysical and geological interpretation, the structural requirements for the various platform concepts and the overall geotechnical planning strategy. Number and density of sample points were carefully considered. Parametric variations, for example, expected soil conditions, practical time constraints and other factors influenced the final plan. Figure 3 shows the final investigation programme. The main location for a platform is around Areas 1 and 2. Other possible alternatives are Areas 3 and 4. Test Areas 5 and 6 were chosen to supplement the regional data base. The soil investigation was designed to be flexible to allow for continuous updating throughout the field work, based on the accumulated soil data at any given point in time.

The investigation was divided into three contracts: drillship charter, positioning services, and main geotechnical contractor services. The main geotechnical contract included soil sampling and *in situ* testing, offshore and onshore laboratory testing, and interpretation and reporting of soil parameters.

QUALITY ASSURANCE AND CONTROL

With increasing water depth and softer soil conditions, soil design parameters become increasingly significant in platform design. In the optimization of foundation geometries and dimensions of a gravity-base structure, the accuracy of the soil design parameters has direct and significant impact on the structural cost. High quality in all technical aspects related to soil sampling, laboratory testing and *in situ* testing was therefore selected as the main criterion in planning a detailed soil investigation.

The following three areas were chosen as particularly significant for the achievement of the established quality requirements:

(1) the provision of a quality-control system throughout the project;

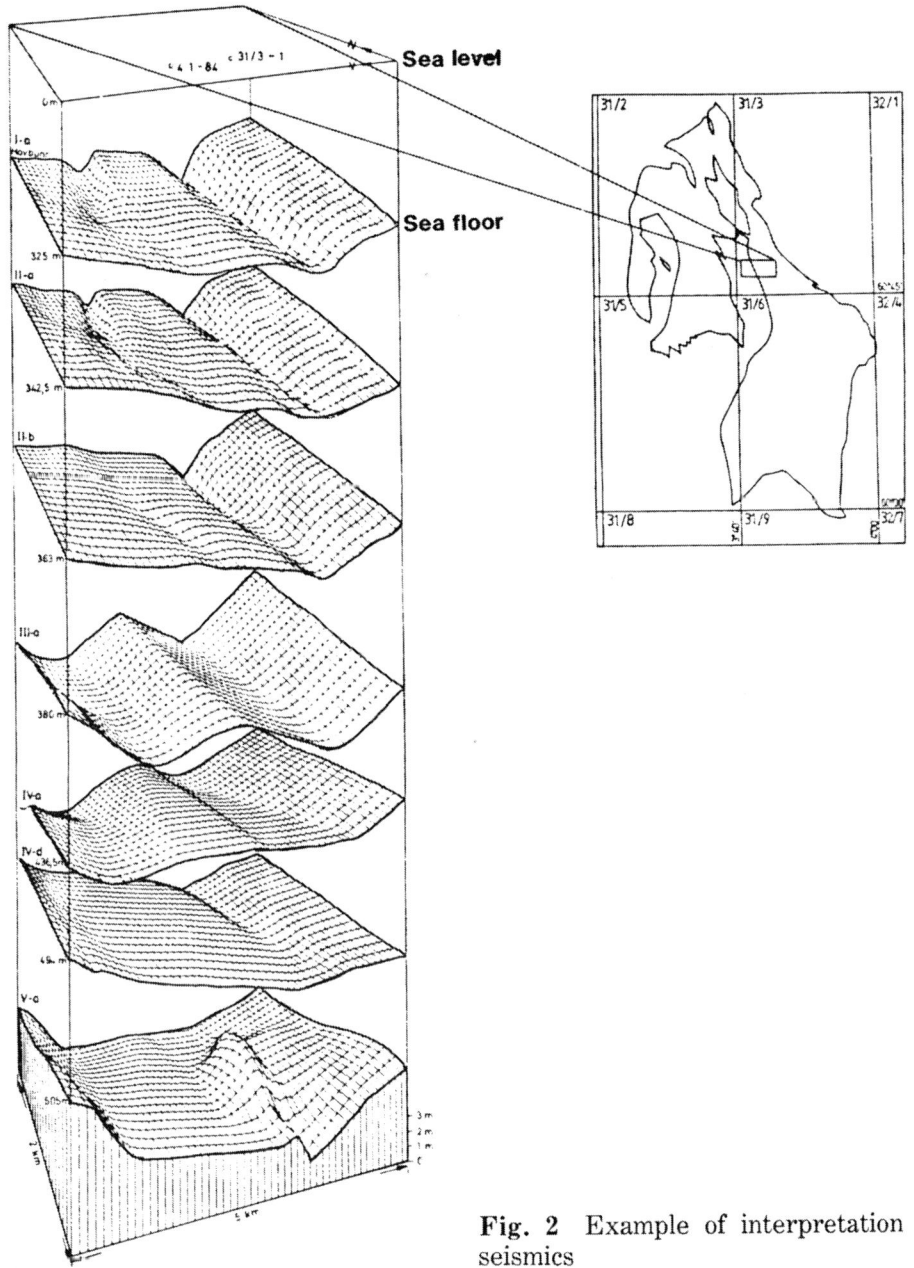

Fig. 2 Example of interpretation of shallow seismics

(2) emphasis on extensive, advanced soil testing on board the survey ship;

(3) the comprehensive use of *in situ* testing with several types of tests for reference and extrapolation purposes.

Quality assurance and control in geotechnical investigations has not been typically ˆrmalized, mainly due to the nature of the investigations, which do not easily provide reference criteria for control. The main contractor's quality assurance procedures were used as a standard throughout the survey. Quality control was established at all levels of the investigation and was performed as a separate function with separate personnel at all important stages.

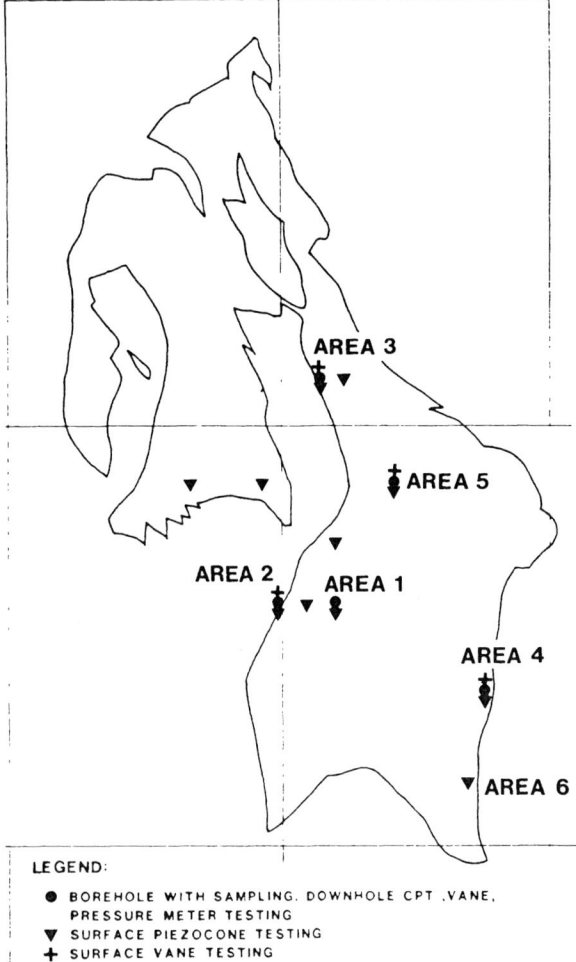

LEGEND:
- ● BOREHOLE WITH SAMPLING. DOWNHOLE CPT .VANE. PRESSURE METER TESTING
- ▼ SURFACE PIEZOCONE TESTING
- ✛ SURFACE VANE TESTING

Fig. 3   Soil investigation programme

Laboratory Testing

Specialist laboratory equipment was taken offshore, including an X-ray used to select high-quality portions of samples for testing. Results from the offshore testing programme were to be used as a basis to quantify possible effects of storage time and transport on sample quality, to provide high-quality data input for designing the extensive onshore laboratory programme, and for selecting preliminary soil design parameters at an early design stage.

In Situ Testing

Detailed technical specifications were estab-

lished for *in situ* soil testing; for example, by specifying testing equipment calibration requirements and reporting routines. *In situ* testing requirements focused on the use of continuous piezocone penetration tests to 40 m penetration together with downhole cone testing. Both surface and downhole vane tests were also performed. In addition, pressuremeter tests were undertaken for reference purposes.

TECHNICAL DESCRIPTION

Technical Specifications

Technical specifications for the geotechnical field work were included in the tender and contract documents. In particular, detailed specifications covered the following equipment used for the Troll East Field soil investigation:

- drillship;
- positioning systems;
- seabed reaction frame (including hydraulic units, TV camera and current meter);
- sampling equipment (hammer, standard push and piston sampling as well as gravity coring);
- cone penetration tests including measurement of pore water pressure (piezocone) both downhole and surface modes;
- vane tests, both downhole and surface modes;
- push-in pressuremeter tests.

The purpose of these specifications was to ensure that sampling and testing were carried out to accepted standards and the relevant data and descriptions reported. Detailed specifications were provided for equipment and testing procedures, instrument calibration prior, during and after field work, and data acquisition and reporting.

Drillship

MV *Bucentaur*, as shown in Fig. 4, was contracted by Norsk Hydro for its maiden voy-

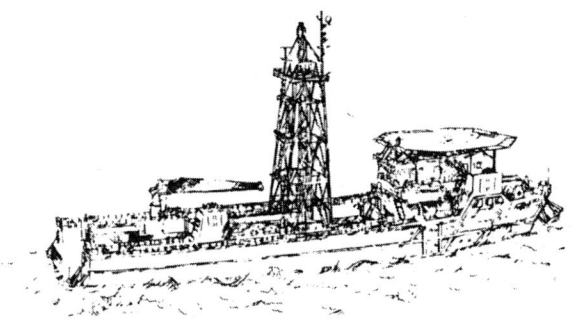

Fig. 4 Geotechnical survey vessel MV *Bucentaur*

age in March 1983. The ship is 77 m long with a 16 m beam. It has a displacement of 4300 tonnes and a deadweight of 2000 tonnes. Four 1200 kW generators drive two 2000 hp azimuth aft thrusters and two 1200 hp transverse tunnel bow thrusters. The GEC Simplex dynamic positioning system is used with positioning reference by deep-water tautwire and Simrad HPR acoustic transponder systems. The HPR system is also used for positioning of the sea-floor jacking unit and vane basket.

A 4.6 m square moonpool is located midships and the drilling derrick has an 80 tonne static hook load capacity. Drill string heave compensation has a maximum stroke of 7 m. The dry-mud system has a total tank capacity of 325 m³ in 5 tanks, with pneumatic fill and discharge, all remotely controlled from the drillfloor. The dry mud is passed through two 4 m³ surge tanks with lime added through a 10 m³ hopper tank. The system is designed for a maximum specific gravity of 2.5 t/m³.

Positioning Systems

Two independent positioning systems were required for continuity of operation if one system broke down. The two chosen systems, Syledis and Hyperfix, were interfaced to a Hewlett–Packard 9826 computer enabling simultaneous and complete utilization of both systems. At each location, simultaneous logging of both systems was carried out and data were collected for comparison.

Stingray Soil Sampling and *In Situ* Testing System

The Stingray drill string control system developed by McClelland (Semple and Johnston, 1979) was used to provide high-quality soil sampling and *in situ* testing. The Stingray system features a remotely controlled, hydraulic sea floor based unit that operates in conjunction with site investigation vessels having a drilling derrick and moonpool.

The basic sea-floor jacking unit has a 3 m square base and is 4.7 m high. It weighs from 15 to 28 tonnes, depending on the amount of ballast used. Figure 5 shows the various components of the Stingray system. Power to the jacking unit is provided by a single electrical umbilical cable connected to a subsea power pack on the sea-floor jacking unit. Hydraulic power and control systems are included in the sea-floor power pack comprising subsea motor and pump, a pressure-compensated hydraulic valve pack and a telemetry pressure can containing all control and telemetry electronics. This sys-

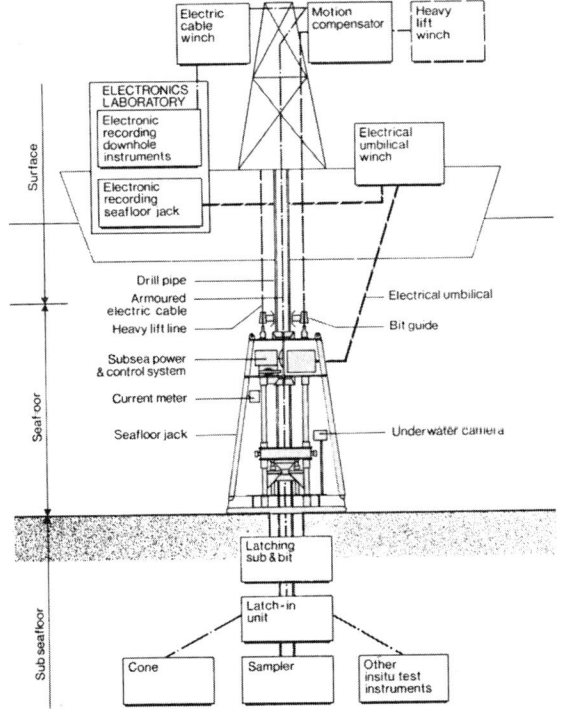

Fig. 5 The Stingray system

tem is capable of operating in water depths up to 900 m.

The sea-floor jack has sensing units to monitor both the roll and the tilt of the unit, the pulldown pressure of the vertical rams and the hydrostatic pressure at the sea floor. Also associated with the jack is a bit guide, seen on top of the jack in Fig. 5, which runs along the two-part heavy lift line to guide the drill string into the jacking unit.

Seafloor Jacking Unit Stability

During the tender stage for the Troll East Field project settlement of the Stingray sea-floor jacking unit was assessed, since very soft soils were known to exist at sea floor. Consequently, a skirt perimeter was added around the base of the jack extending to a depth of about 1.2 m. This skirt increased the soil-bearing capacity by a factor of three, essentially eliminating settlement of the sea-floor unit for the expected duration of soil borings.

Subsea Television Camera

In order to observe the sea-floor jacking unit penetration and also to provide the means to synchronize the timing of the continuous cone penetration machine (described in a subsequent section) a subsea television camera was mounted on the jacking unit. The camera, which incorporates a pan and tilt facility, was mounted to allow viewing of two corners of the sea-floor jacking unit and viewing of the central clamps and drill pipe. Control and video signals were transmitted along the main sea-floor jacking unit umbilical cable. The camera is a silicon intensified target (SIT) underwater television camera capable of operating at low light levels down to approximately quarter moonlight.

Current Meter

To observe current velocity and direction near the seabed and in the water column, a current meter was fitted to the jacking unit, as shown in Fig. 5. Power to the meter, and readout signals from the meter, were transmitted through the existing sea-floor jack umbilical.

Downhole Sampling and Cone Testing

Downhole sampling and cone testing was performed using an updated version of the Stingray system as described by Semple and Johnston (1979). Cone penetrometer and vane testing were undertaken to a maximum of 6 m beyond the base of the borehole. Vane and pressuremeter tools were also pushed into the soil to the required penetration and held motionless during the test. Figure 6 shows the various stages in operating the system for downhole cone testing.

In soft soils, the normal drill-string heave compensation systems will not function, as weight cannot be set off on the bit. In order to 'drive' the drill-string compensator it was

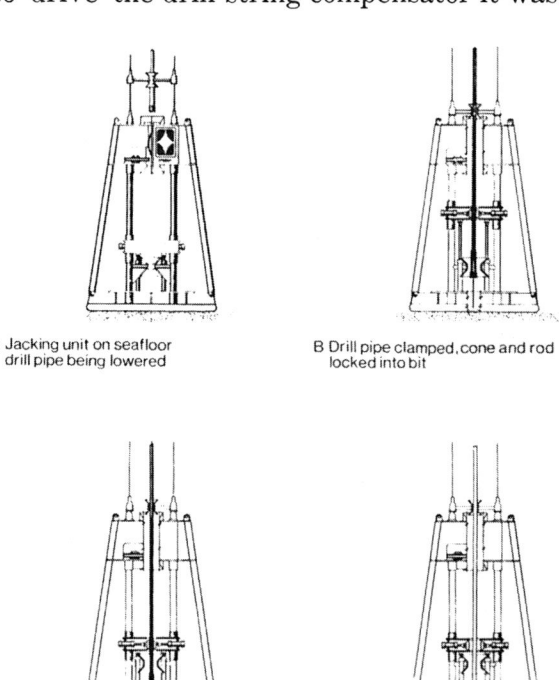

A Jacking unit on seafloor
drill pipe being lowered

B Drill pipe clamped, cone and rod
locked into bit

C Cone pushed by repeated strokes
to maximum penetration

D Cone and rod removed, hole drilled
to next interval

Fig. 6 Downhole testing with Stingray

connected to the sea-floor jacking unit compensator by means of a 'hard tie'. To serve as a deadman and provide cable handling capacity, a modified air tugger with an air-brake system and remote air control permits the driller to operate both the winch and the air brake. The air-brake friction was adjustable, allowing adequate drag to make the sea-floor frame compensator 'drive' the drill-string compensator.

This procedure for drill-string motion compensation accommodates two soil strength conditions, the strong soils that will support the pipe and can be drilled with the motion compensator in the normal mode and the weak soils that can be displaced by washing or jetting. Using the hard-tie system in conjunction with the Dolphin piston sampler, described in a subsequent section of this paper, high-quality samples, up to 2 m in length, were obtained.

Continuous Cone Penetration Tests

In 1984 the Stingray sea-floor jacking unit was modified to accommodate cone testing using a continuous penetration machine (CPM). A land-based CPM, manufactured by Borros of Sweden, was modified by McClelland to operate compatibly with the existing subsea system. This gives the Stingray system the facility to perform continuous push cone penetration tests from the sea floor to refusal. Figure 7 shows the position of the CPM frame within the sea-floor jacking unit. The CPM comprises two hydraulically driven, syncopated chucks vertically aligned one above the other. Each chuck is driven by two pistons, which allows the chucks to move up and down either pushing or pulling the cone rod. The electric and hydraulic controls of the CPM are operated through the existing remotely controlled subsea power pack and associated telemetry signal communications.

Cone Penetrometers

Piezocone electric strain gauge cone penetrometers, designed and manufactured by

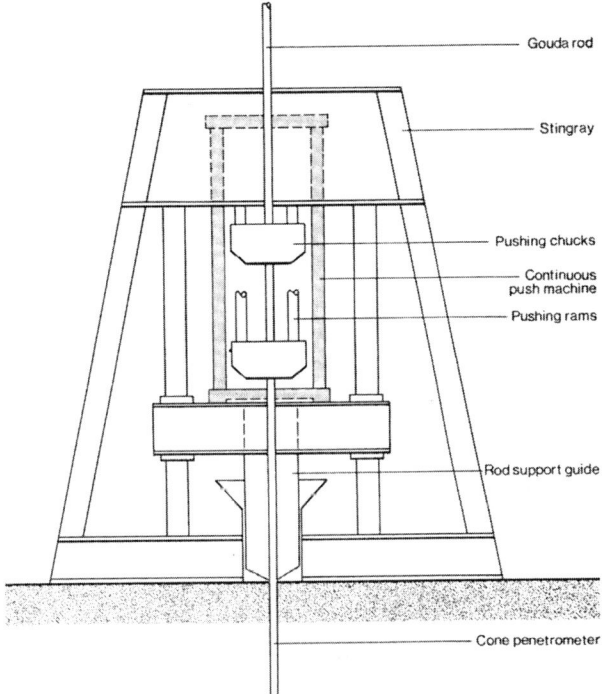

Fig. 7 Stingray continuous push machine

McClelland, were used throughout this project. The dimensions of the piezocones used are given in Fig. 8. All cones were calibrated in McClelland's onshore laboratory before being used offshore. Calibrations were checked after the offshore programme was completed. McClelland's temperature cones also were used for a small number of surface cone penetration tests at the Troll East Field.

Dolphin Piston Sampler

The Dolphin piston sampler is one of a series of four mud-pressure operated tools that have been developed by McClelland to provide state-of-the-art data acquisition with operational simplicity. The other three tools comprise downhole cone penetrometer, remote vane and push sampler. The piston sampler, as shown in Fig. 9, comprises a stroke rate or metering control cylinder, sample tube adapter and fixed piston which allows high-quality samples up to 2 m in length to be taken in very soft soils with

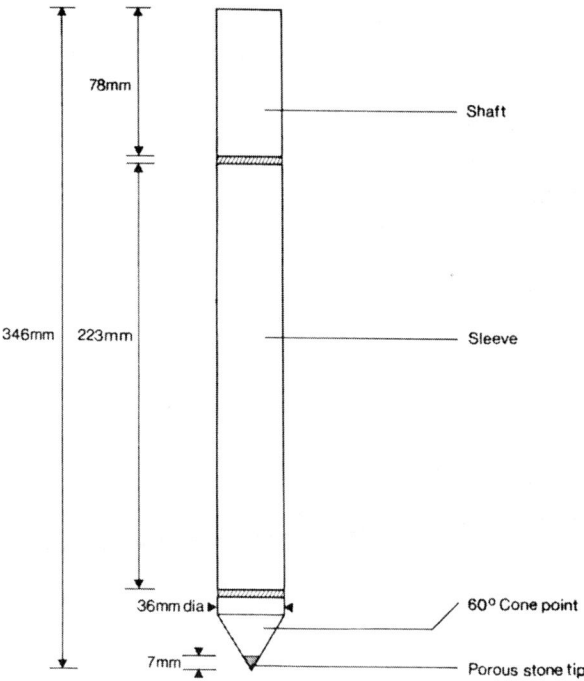

Fig. 8   Dimensions of 10 cm² piezocone

Remote Vane Testing

In situ shear vane testing at the Troll East Field was undertaken using the remote vane downhole and from the seabed surface using the Halibut vane basket, as described by Ehlers *et al.* (1980) and shown in Fig. 10. To determine *in situ* soil strength measurements at shallow penetrations (less than 5 m) the remote vane can be operated effectively without drillpipe by attaching it to the stabilizing ballast collar or Halibut basket.

Halibut vane tests were undertaken simultaneously with downhole *in situ* testing and sampling. For these tests the vane equipment was the same as for the downhole tests, except that it was set with a fixed penetration below the base of the Halibut basket. The system was lowered over the side of the vessel by the *Bucentaur's* crane until the base rested on the seabed and the vane was at the desired penetration. Sufficient wire was spooled off the crane so that no boat motion was transmitted to the frame. The Halibut basket position was monitored on the ship's HPR acoustic trans-

minimal disturbance. The sampler uses a standard 76 mm diameter thin-walled tube and in soft soils provides a sample with less disturbance than a push sampler. The advantage of using this system was that sampling times were decreased significantly. Instead of deployment and recovery with a winch, the sampler was free-fallen to the drill bit. This time saving was particularly important in the deep water of the Troll East Field.

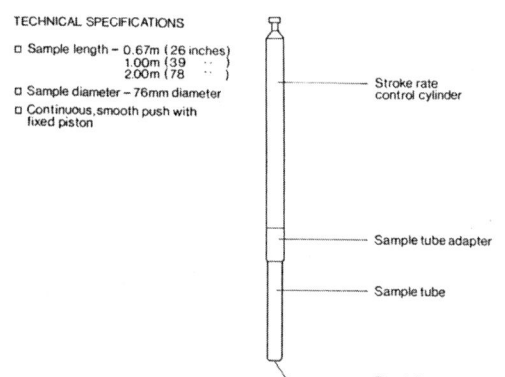

Fig. 9   Dolphin piston sampler

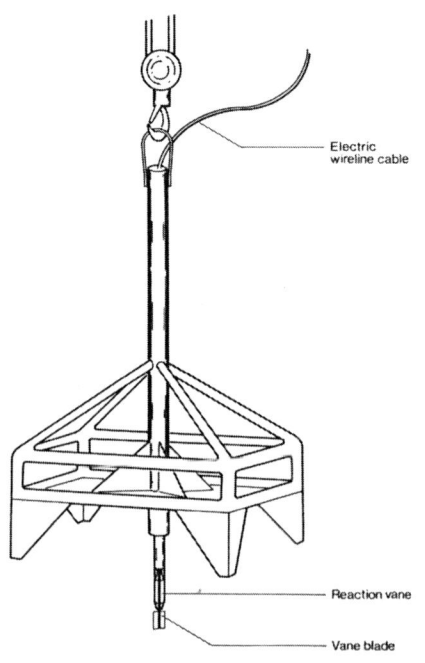

Fig. 10   Remote vane Halibut assembly

ponder system. This enabled control of multiple tests at each penetration. After a test the system was recovered and the vane set at a different distance under the frame so that sea-floor and shallow soil strengths could be measured.

Data Acquisition

A schematic diagram of the data acquisition system used for surface and downhole cone penetration and vane testing is shown in Fig. 11. Signal conditioning at the cone using downhole amplifiers allows the seven-conductor electric wireline to service four transducers in the cone.

Five channels of data, including CPM thrust, were logged during continuous operation. Data is stored in the active computer memory until testing can be interrupted to allow transfer to disc storage. A multichannel strip chart recorder was used to display continuously analogue test data for the operator in adjustment and test-control operations.

Push-in Pressuremeter

The push-in pressurementer (PIP) was ini-

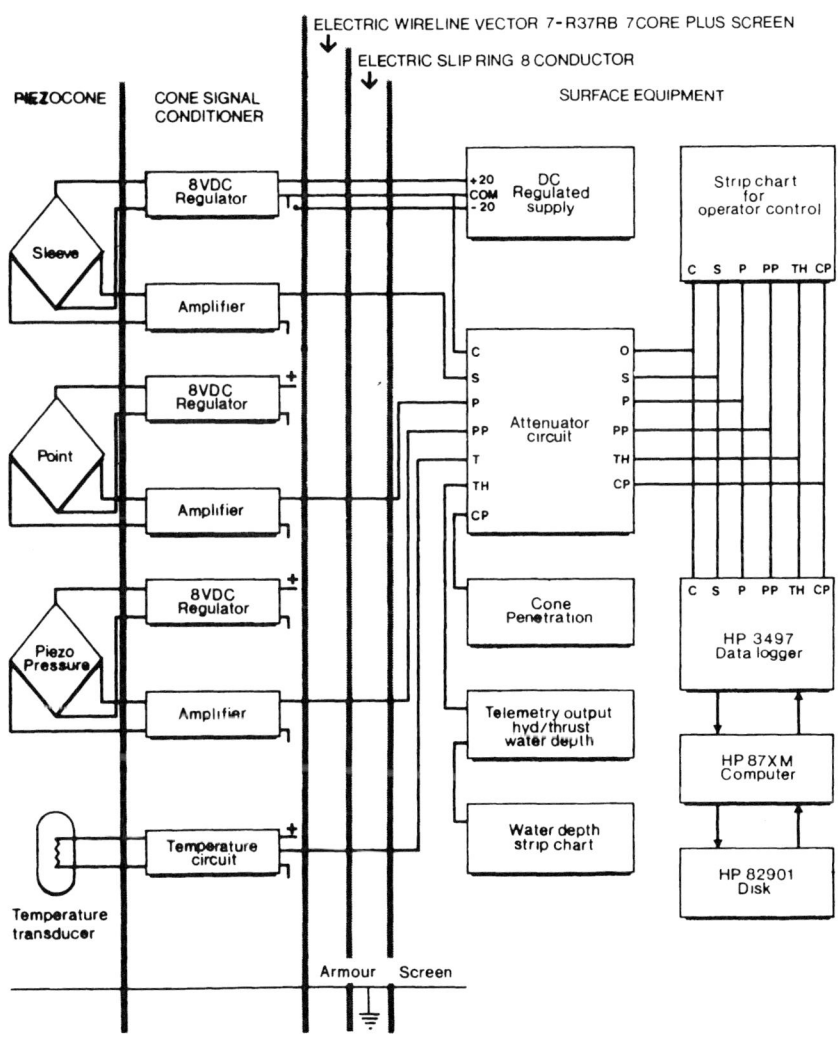

Fig. 11   Cone penetrometer data aquisition system

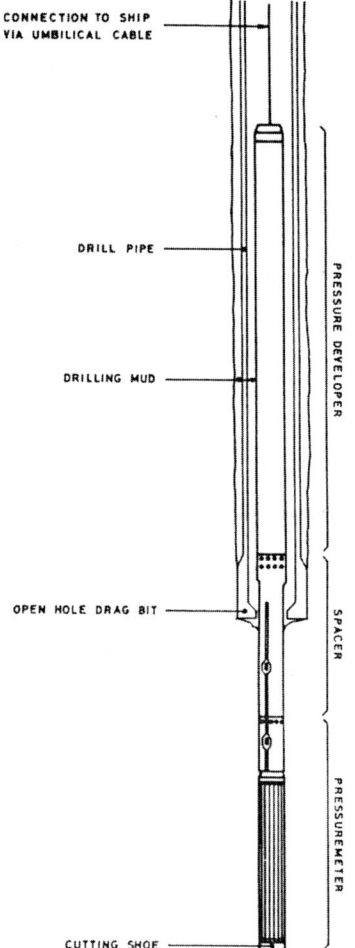

Fig. 12 Push-in pressuremeter

tially developed at the Building Research Station (BRS), and has, since 1980, been operated by Stressprobe Ltd. The PIP is operated downhole using McClelland's latch-in units and wireline vector cable. Details of the push-in pressuremeter system are given by Reid *et al.* (1982). and are shown in Fig. 12.

On-board Laboratory Testing

The laboratory testing undertaken on board the *Bucentaur* during the soil investigation was probably the most extensive that has ever been performed in offshore soil investigations. In addition to the traditional standard field laboratory testing (classification

tests, sample descriptions etc.) test of the following types were performed:

- special geological and geochemical tests;
- radiography of soil sample tubes;
- constant rate-of-strain oedometer tests;
- direct simple shear tests;
- triaxial tests (unconsolidated-undrained and consolidated-undrained).

The various laboratory tests that were performed offshore are described below:

Standard Field Laboratory

Standard laboratory testing was conducted in the permanent 20 m³ NGI-designed main laboratory on the *Bucentaur*. The laboratory contains a specially built hydraulic extruder. Tests performed included sample descriptions, water content determination, soil unit weight, shear-strength estimates and unconfined compression tests. In addition, selected samples were photographed and the majority of the samples were sealed in the sample tubes for transportation to NGI's main laboratory for further testing. A small amount of testing was performed in McClelland's onshore laboratory for comparison and quality assurance.

Geological/Geochemical Tests

The engineering geologist performed the following tests with the purpose of providing input for geological evaluation and for the assessment of corrosion potential:

- detailed description of samples (macro- and microfabric) including use of microscope;
- pH measurement;
- soil resistivity;
- sound velocity by Pundit Device;
- treating and storing samples for special onshore tests, including sulphate-reducing bacteria (SRB) and gas content in pore fluid.

X-ray Examination

To carry out the radiography work offshore,

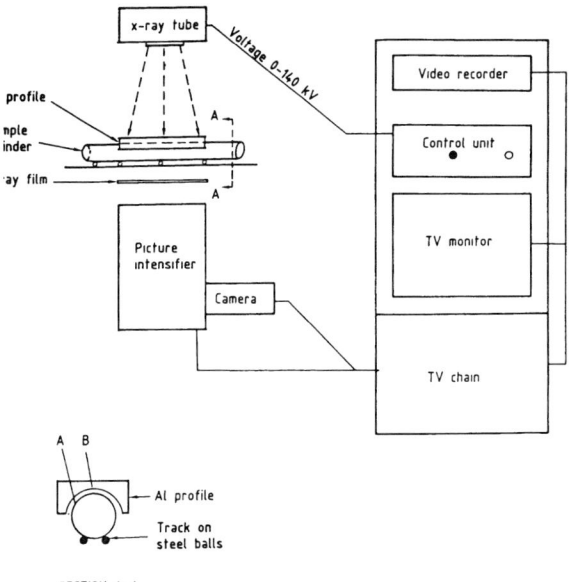

SECTION A-A

Fig. 13   Instrumentation for X-ray examination

TABLE 1
Procedure used on board Bucentaur *during 1984 soil investigation at troll*

1. Sample comes to X-ray container
2. Lead numbers from 0 to 100 (in cm) are attached to sampling tube
3. Sample is exposed to radiation. The sample quality is assessed on TV monitor
4. Message is given to drilling deck or soil laboratory about sample quality and 'usable length'
5. Radiographs of whole sample length are taken (one radiograph gives a picture of about 40 cm of sample)
6. Sample is sent back to soil laboratory
7. Radiographs are developed and soil is later described in detail
8. Description from radiographs is given to soil laboratory for use during visual sample description and allocation of specimens.

NGI provided a 3 m square insulated standard steel X-ray laboratory. The radiography equipment included a cathode ray tube, a high-voltage generator and control unit, a picture intensifier and camera, an X-ray–TV interface, and a TV monitor and video tape recorder. Figure 13 summarizes the instrumentation used. The system was enclosed in a lead-insulated cabinet built on shock and vibration absorbers. All openings were equipped with safety switches.

In addition, the X-ray container included all equipment necessary for the manual development of the X-ray films. The container contains an air-conditioner, sea air filter, coolants for the cathode ray tube, a wash-basin and a wall light board for viewing the radiographs. It is certified to comply with the safety requirements of the Statens Institutt for Stralehygiene of Norway (State Institute for Radiation Safety). X-rays of samples contained in 76 mm diameter steel tubes were taken throughout the project and the procedures used are described in Table 1. Exposure and developing time depend on the physical properties of the soil sample and on the ambient temperature (Lacasse *et al.*, 1984).

CRS Consolidation Test Apparatus and Procedures

The constant rate of strain consolidation (CRSC) test apparatus used offshore was installed in a 3 m insulated standard steel container which was fitted for direct simple shear and CRSC tests. The CRSC test apparatus was designed and constructed by NGI specifically for offshore use. This apparatus differs radically from that used on land and offers very good operating characteristics. The equipment includes two complete automatic CRSC machines, two microcomputer systems, and soil-sample preparation equipment.

Soil specimens were prepared with dry filter stones which were saturated with salt water when the applied axial stress equalled the estimated negative pore pressure in the specimen (Sandbaekken *et al.*, 1985). Specimens were then automatically loaded at a constant rate of strain up to a specified axial stress. Some tests included an unload–reload cycle. If the unload–reload procedure was used, excess pore pressure was allowed to dissipate before unloading to the *in situ* overburden pressure. This procedure was fully automatic after initial manual loading to 25% of the effective overburden stress and saturation of the porous filters. All further steps were regulated by and logged by the microcomputer. At the end of the survey, data were transferred to NGI's main computer for final data plotting.

Direct Simple Shear Test Apparatus and Procedures

The direct simple shear (DSS) test apparatus is based on that in use at the main NGI laboratory in Oslo, with necessary mechanical and data acquisition modifications for offshore work. The DSS test equipment includes a semi-automatic DSS apparatus with associated microcomputer and soil-specimen preparation equipment.

Test procedures offshore include specimen preparation, automatic loading at a constant rate of strain up to the specified axial consolidation stress, and saturation of filter stone as for the CRSC test. The axial consolidation stress is held constant until a sufficient degree of consolidation is achieved; then the specimen is sheared at a shear strain rate of slightly less than 5% per hour. A constant specimen volume is maintained throughout the shearing stage. The test data are logged and processed by the microcomputer in the NGI laboratory container.

Triaxial Testing Laboratory

To undertake triaxial testing offshore simultaneously with boring operations, McClelland supplied their standard containerized offshore soils laboratory. The laboratory is equipped with two total and three effective stress triaxial test cells, capable of testing full-size samples without the need to trim to a smaller diameter. A microcomputer-based data-logging system was incorporated in the offshore laboratory to increase efficiency, particularly on quick, undrained triaxial testing. Test results were plotted on computer-controlled plotters on board the drillship. Unconsolidated undrained and consolidated undrained triaxial tests were performed on selected samples from each

boring to evaluate the undrained shear strength of cohesive samples. Anisotropic consolidated undrained tests were performed with the initial horizontal and vertical stresses equal to the estimated *in situ* stresses. A number of samples were further consolidated to represent the state of stress under the weight of a gravity platform.

SURVEY RESULTS

Field Production

Thirty-six days out of the total contract period of forty days were spent on site undertaking sampling and testing. Nine boreholes were performed with a total of 600 m penetration. Continuous surface piezocone tests were carried out with accumulated penetration of just over 500 m. Average penetration was 35 m. In the boreholes, piston sampling and wireline cone testing were the dominating activities. Standard push sampling was undertaken in harder clays with occasional hammer sampling in very dense morainic sands. Some 121 m of downhole cone testing, 130 piston samples, 173 push samples and 5 hammer samples were performed.

Vane testing was used extensively, particularly in the upper 5 m, due to the significance of this layer in relation to foundation design. Altogether 82 vane tests were performed, 32 of which were downhole tests and the remainder were from the surface Halibut vane basket. Finally, 17 push-in pressuremeter tests were performed. A significant amount of soil testing was carried out on board. In addition to the X-ray of all samples and classification testing on the majority of samples, 33 constant rate of strain oedometer (CRS) tests, 19 undrained triaxial tests and 24 constant volume direct simple shear tests were carried out. All remaining samples were sealed and carefully packed for transportation to the onshore laboratory.

Efficiency, Operational Statistics and Cost

The efficiency of a soil investigation may be reflected in a statistical evaluation of performance for the various types of operations. For comparison purposes the performance units used are hours per meter of boring or meter of cone testing (Amundsen and Lauritzsen, 1982). To enable such an evaluation, however, a detailed breakdown of activities must be made. The activity is given as a percentage of the time spent in the field, exclusive of weather downtime periods.

Some 75% of the field time was spent operating. Actual drilling and sampling and *in situ* testing consumed 51%, pipe and equipment handling 22% and positioning 2%. The remaining 25% was standby time due to the maintenance and testing of state-of-the-art geotechnical equipment and some malfunctioning and repair (19%), drillship (2%), positioning problems (1%) and other factors (3%). Applying the above data to the production figures, 1.05 hours were required per meter of boring and 0.36 hours were required per meter of surface CPT. These figures compare well with similar surveys. The 1983 soil investigation at Troll West Field, on behalf of Norsk Shell, resulted in 1.16 hours per metre and 0.38 hours per metre, respectively.

In terms of cost the statistics also show that, on average, in 1984 each CPT cone lasted 164 m penetration before being lost or replaced. Similar statistics show the 'working' life of drill pipe, cone rods etc. The total cost of the soil investigation was 18 million NOK with approximately three-quarters spent on the offshore operation; the remaining quarter related to onshore laboratory, data interpretation, administration and reporting.

In Situ Test Results

Cone Test Results

Figure 14 presents the results of a continu-ous cone penetration test performed at location CPT 2001 to a depth of 43 m. CPT 2001 is a typical profile for the normally consolidated clay of the Troll East Field with point resistance q_c, generally increasing linearly with depth from 0 MPa at seabed to 2 MPa at 43 m penetration. The dynamic pore-pressure ratio (defined as the excess pore pressure generated during testing divided by the measured or total cone resistance) is typically in the range of 75 to 100%. Figure 15 presents the combined point resistances for CPTs performed in Area 2. Point resistance from three profiles have been superimposed on one figure and show consistency and uniformity in the soil conditions.

Dissipation Test Results

Several dissipation tests were performed during the surface cone tests. Penetration was halted so that pore pressure and cone-point resistance could be recorded for a period of up to 30 minutes. Results of two of these tests performed in CPT 2003 are presented in Fig. 16. When restarting penetration after each dissipation test, an increase in sleeve friction was recorded indicating soil set up around the cone sleeve. This increase in sleeve friction for a number of dissipation tests is shown in Fig. 16.

Temperature Test Results

Two continuous cone penetration tests were undertaken using a temperature cone. Temperature is measured by a thermistor, mounted at the base of the cone shaft, which has an accuracy of about 1°C. The results of two temperature profiles are tabulated in Table 2. The measured temperature increases by about 2°C over a penetration of 40 m.

Remote Vane Test Results

Results of typical surface (Halibut) and downhole *in situ* vane tests are shown in Fig. 17 for tests undertaken in Boring 3.2

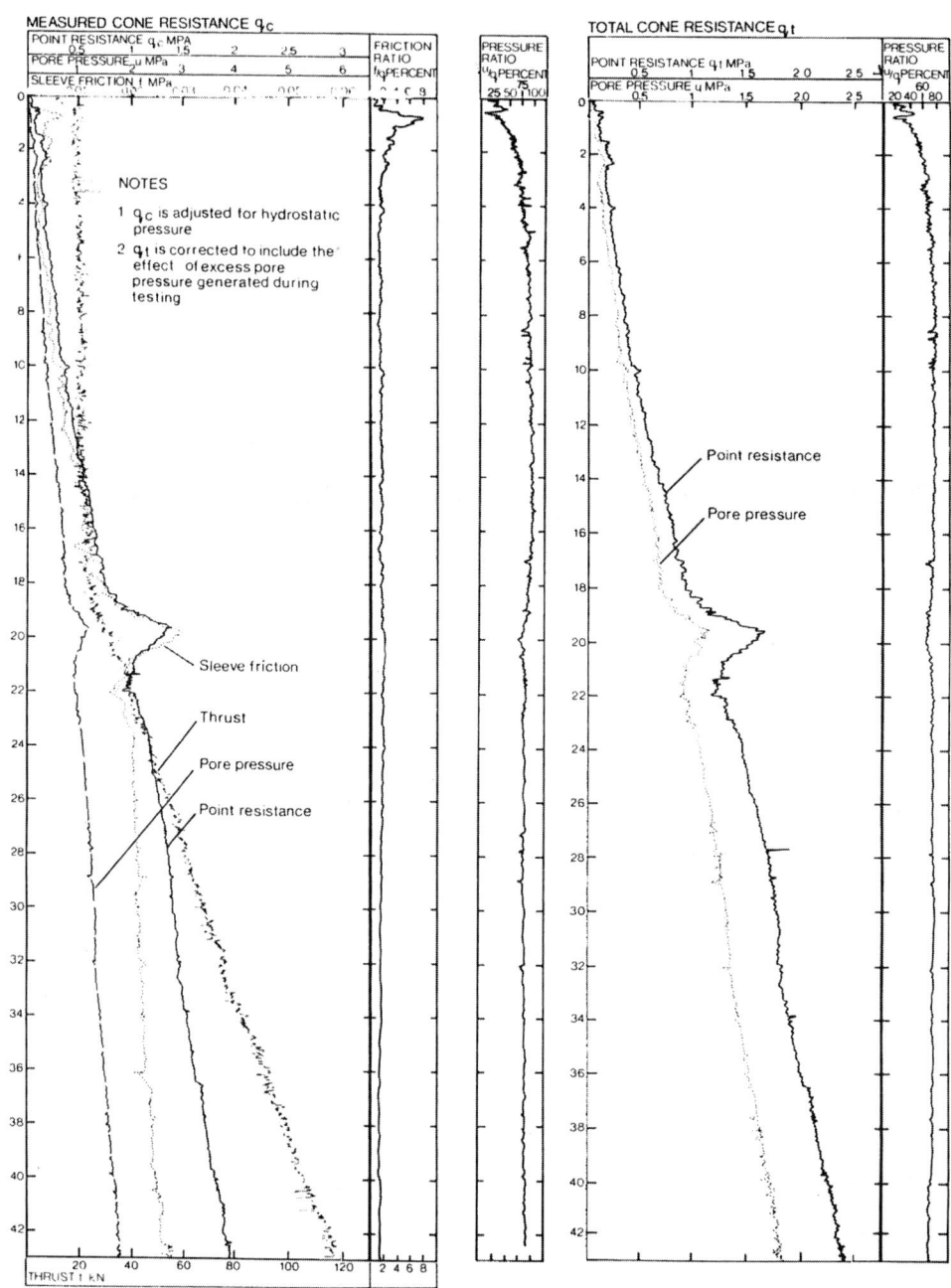

Fig. 14 Result of surface cone penetration test at location CPT 2001

and adjacent to Boring 2.6. The results for both types of test show the mobilized shear stress in kPa with rotation of the vane blades. Peak shear stress occurs at about 7 to 8 degrees of rotation. Due to the required slow speed of rotation of the vane motor,

time was not available to record the residual shear strength.

Results of all downhole tests undertaken in Boring 3.2 and all surface tests adjacent to Boring 2.6 are shown in Fig. 17.

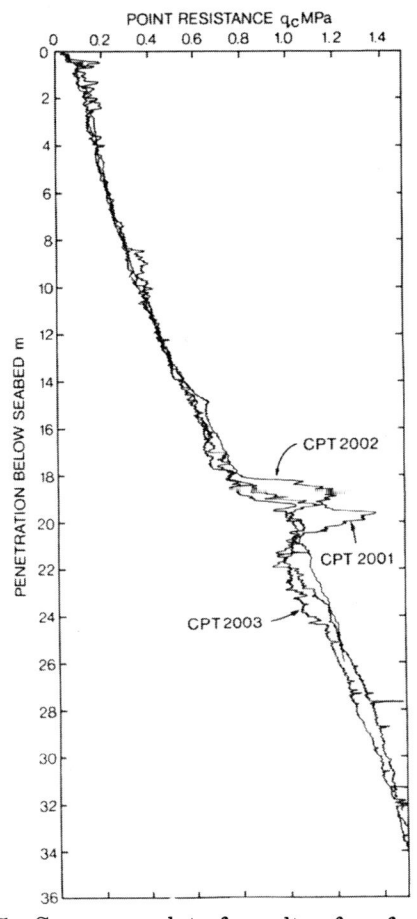

POINT RESISTANCE q$_c$ MPa

Fig. 15 Summary plot of results of surface cone penetration tests in Area 2

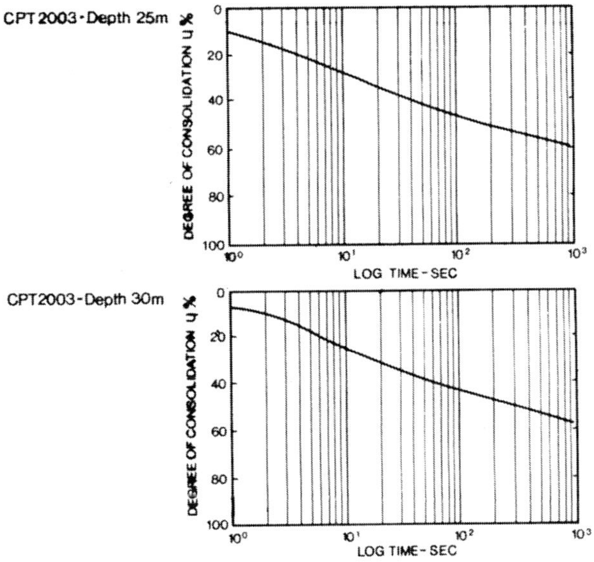

CPT 2003 - Depth 25m

CPT 2003 - Depth 30m

Sleeve friction set up during dissipation tests

CPT	Pen. m.	Sleeve Friction just prior to dissipation test MPa	Peak Sleeve Friction following restart of push MPa
1001	9	0 005	0 007
	21	0 014	0 015
3001	10	0 005	0 011
4001	10	0 006	0 009
	20	0 012	0 015
	30	0 017	0 019
5001	5	0 002	0 005

Fig. 16 Results of two dissipation tests performed in CPT 2033

TABLE 2
Temperature test results

Location	Temperature (°C)	Location	Temperature (°C)
Submerged in moonpool	10	On deck	14
44 m below sea-level	7.4	Submerged in moonpool	10
Sea floor	7.2	44 m below sea-level	6.5
		Sea floor	7.3
Penetration (m)			
1	7.1	*Penetration (m)*	
5	7.1	1	7.3
10	7.3	5	7.1
15	7.6	10	7.5
20	7.9	15	7.8
25	8.3	20	8.2
30	8.6	25	8.5
35	8.9	30	8.8
40	9.2	35	9.1
	CPT 7001	40	9.4
			CPT 8001

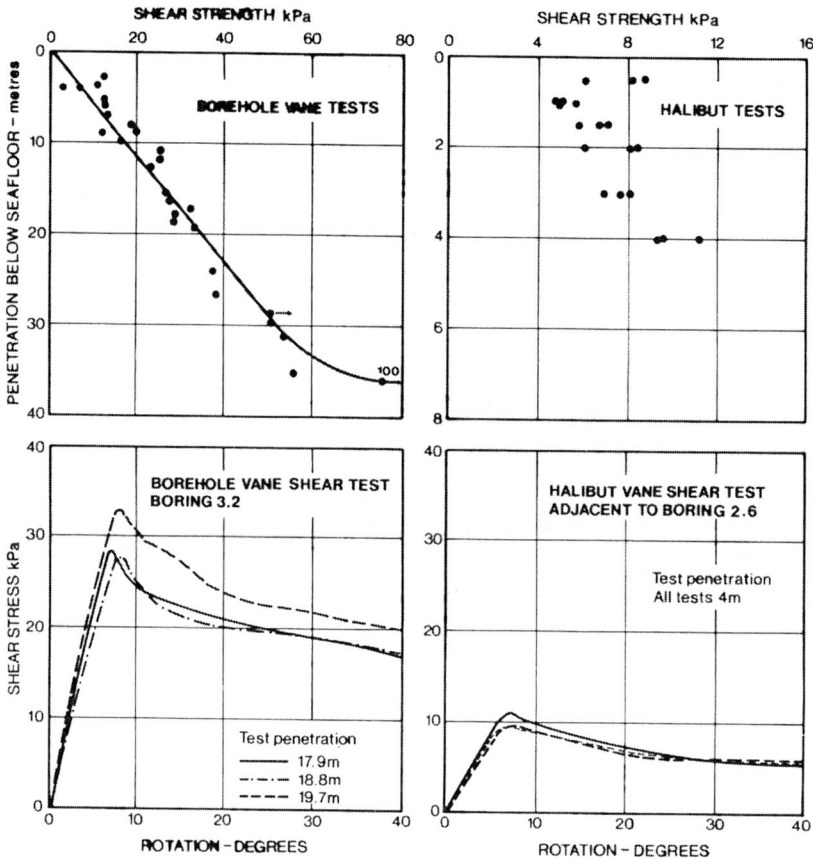

Fig. 17   Results of Halibut and borehole *in situ* vane tests

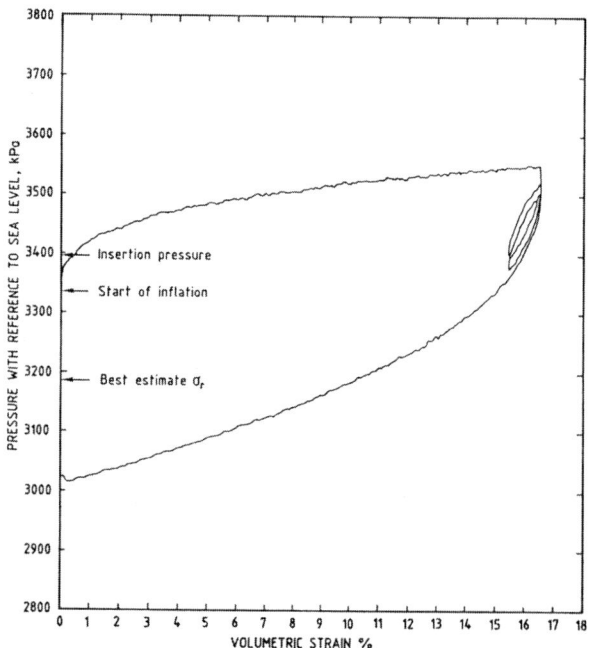

Fig. 18   Results of push-in pressuremeter tests

Push-in Pressuremeter Test Results

The volume expansion pressure curve is shown in Fig. 18 for a test performed at 20 m depth; the curve has been corrected for membrane stiffness. Also shown is the insertion pressure, I, which is somewhat higher than the pressure at the start of the test, indicating that some excess pore pressure has been dissipated. For comparison, the best estimate of the total *in situ* horizontal stress is indicated in Fig. 18.

Laboratory Test Results

As well as the standard index tests, a variety of laboratory tests were performed on samples after X-raying sample tubes and describing the soil. Based on *in situ* and laboratory tests, the soil may be divided into three main strata as shown in Table 3. Figure 19 shows a typical borehole profile from

TABLE 3
Soil properties

Soil layer	Depth interval m	Soil description	Typical values						
			W %	W_L %	I_P %	γ_T kN/m³	s_u kN/m³	% Clay	OCR
1	0 18–26	*Clay*, very soft to firm, high plasticity, slightly overconsolidated	50–60	55–75	35–50	15.0–17.2 increasing with depth	5–60 increasing linearly with depth	35–50	1.3–2.2
2	18–26 44–95	*Clay*, slightly sandy and silty, firm to hard, medium plasticity, Slightly overconsolidated	17–24 with depth	32–37	17–23	20.0–21.0 increasing with depth	50–300 increasing with depth	23–31	1.3–1.8
3	44–95	*Clay*, sandy, silty, hard to very hard, overconsolidated, glacially reworked material	20–28	35–50	25–30	19.0–21.0	400–1000	34–43	2 6

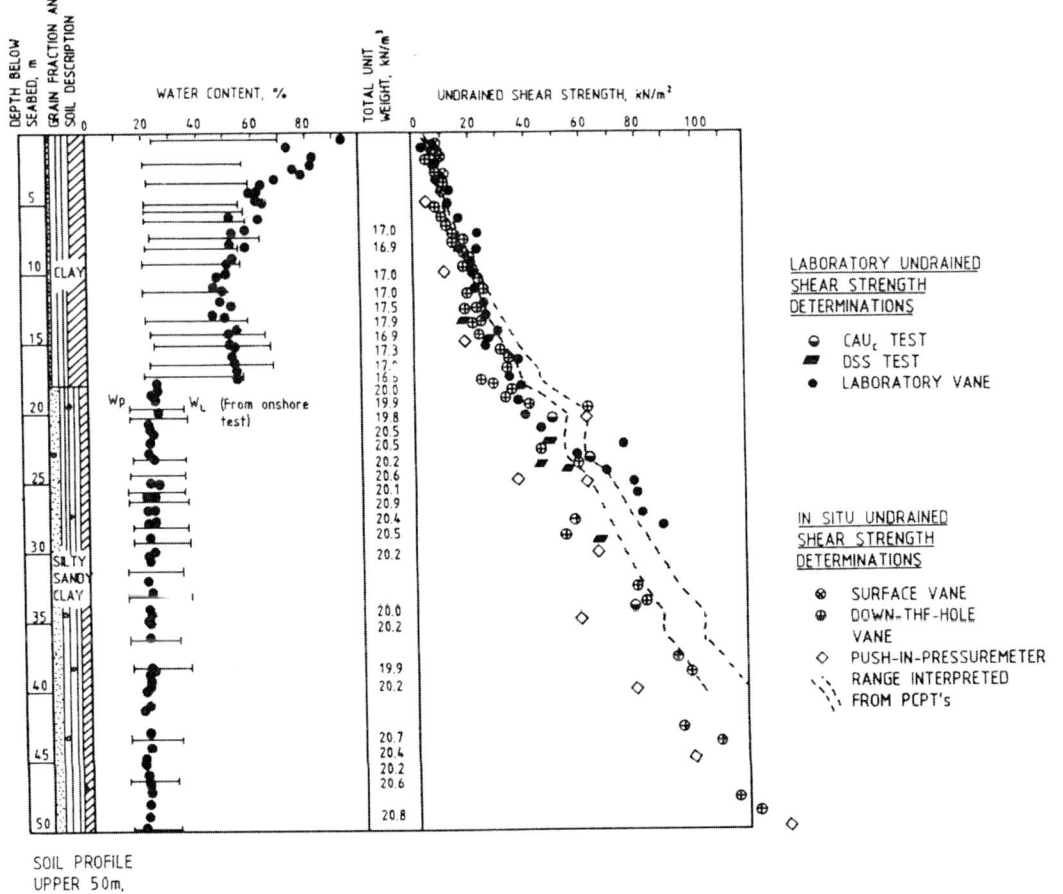

Fig. 19 Typical borehole profile

one of the locations investigated. The results from the undrained shear-strength tests give a well-defined linearly increasing s_u-profile with depth. The s_u-values from the direct simple shear tests (DSS) agree well with the *in situ* vane test results and give somewhat lower values than the consolidated undrained triaxial compression tests (CAUC). The push in pressuremeter (PIP) test results give the lowest s_u-values. Piezocone tests (PCPT) results were correlated with CAUC test results as outlined by Lunne *et al.* (1985). Normalized undrained shear strengths with respect to vertical consolidation stress (s_u/o'_{ac}) are plotted against depth in Fig. 20 giving well-defined relationships. It is interesting to observe that the marked change in unit weight, water content and plasticity at 18 m depth, as shown in Table 3, is not associated with a significant change in undrained shear strength.

Results from CRS consolidation tests are illustrated in Fig. 21a. A clear preconsolidation effect due to ageing is observed. The effective overburden pressure (p'_o), together with estimated effective preconsolidation pressure (p'_c), are presented in Fig. 21b. The overconsolidation ratio (OCR) is well defined in the range 1.4 to 1.8. A systematic comparison between the results of offshore and onshore laboratory tests showed that no significant differences could be found. This indicates that, for the types of tests performed, transportation and storage time do not significantly influence the test results.

OPERATIONAL PROCESS

The offshore operation was a dynamic process, involving sampling and testing, data

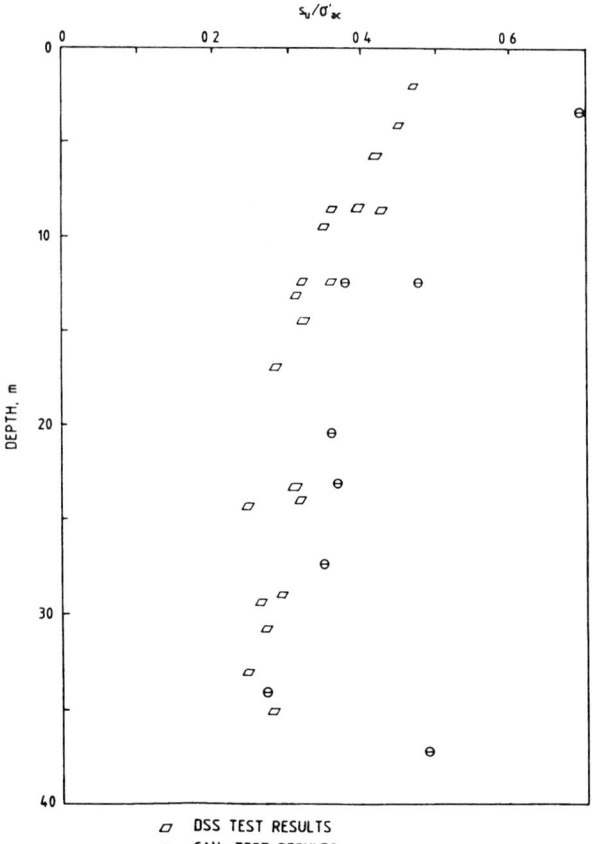

Fig. 20 Normalized undrained shear strength plotted against depth

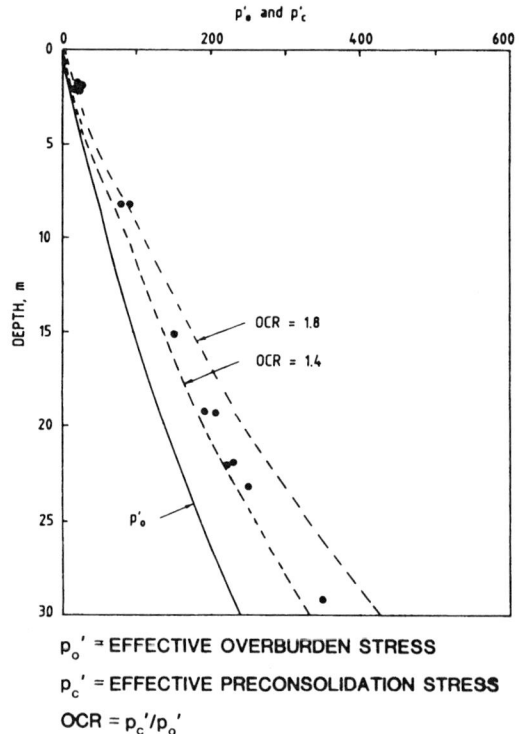

p_o' = EFFECTIVE OVERBURDEN STRESS

p_c' = EFFECTIVE PRECONSOLIDATION STRESS

OCR = p_c'/p_o'

Fig. 21b Effective overburden pressure and estimated effective preconsolidation pressure plotted against depth to give overconsolidation ratio

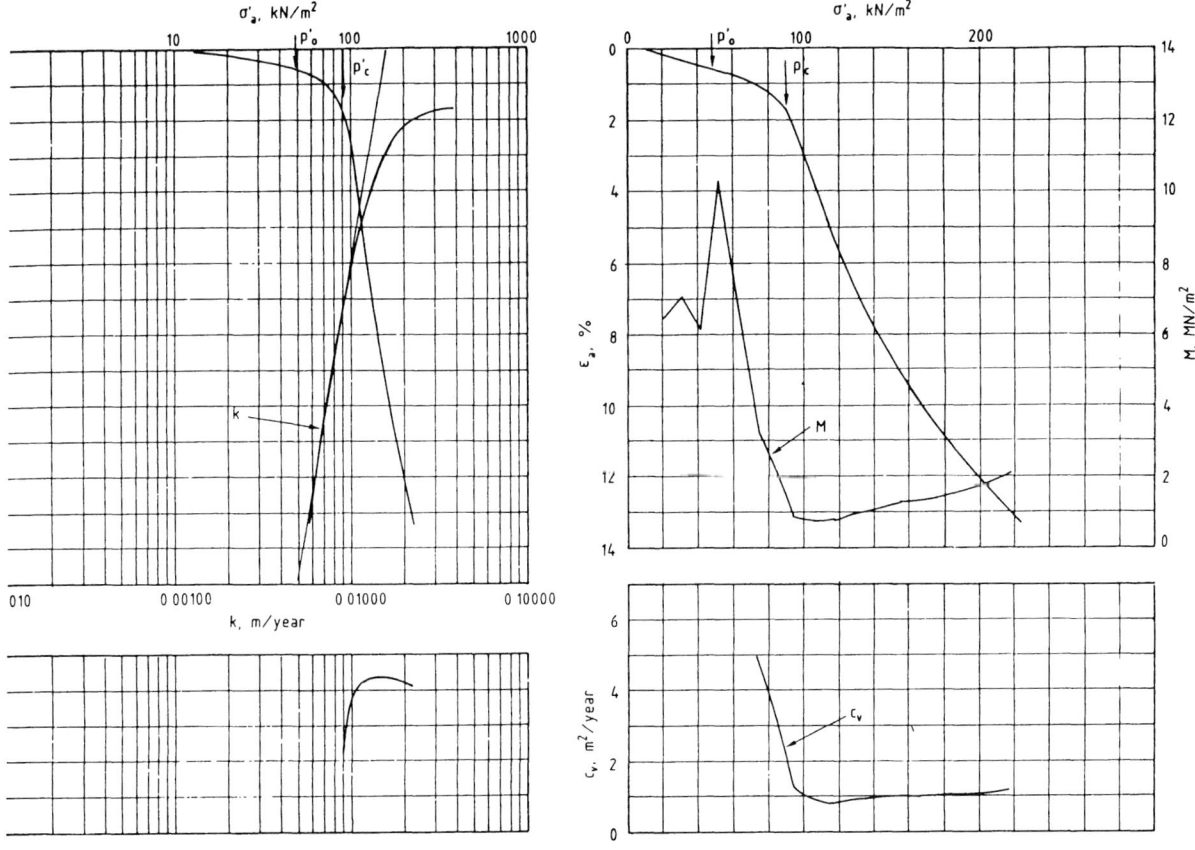

Fig. 21a Constant rate of strain (CRS) consolidation test results

collection, interpretation and feedback, and programme review. On board, laboratory data were collected and reported daily. Similarly all *in situ* test data were edited, plotted and reported within 24 hours of each test. The operation involved frequent communication with the home offices where geotechnical engineers and geologists at Norsk Hydro and NGI participated in the continuous process of comparing all new data from the field laboratory and *in situ* testing with the data used as a basis for the original field programme. Advice for modification of the field programme was relayed to the ship in writing. Communication was made possible by extensive use of telefaximile through the automatic Nordic Mobile Telephone system. This same telephone system was used for telephone conversations and is as easy to operate as the ordinary land-based system. Figure 22 illustrates this process. During the operational process the number of CPTs was reduced due to general uniformity of stratification throughout the Troll East Field in the upper 50 m. The amount of surface vane testing (Halibut) was increased due to some regional variations in strength of the upper 5 m. A number of boreholes were added due to detection of a moraine layer in Area 3, stretching south east from this area. An example of the field programme performed in Area 2 is shown in Fig. 23.

Geotechnical data reporting was part of the administrative reporting system. Here the team leader of NGI reported every 24 hours at midnight to the client's representative. This report consisted of a detailed log of activity, a summary of off-hire periods with descriptions and explanations of inci-

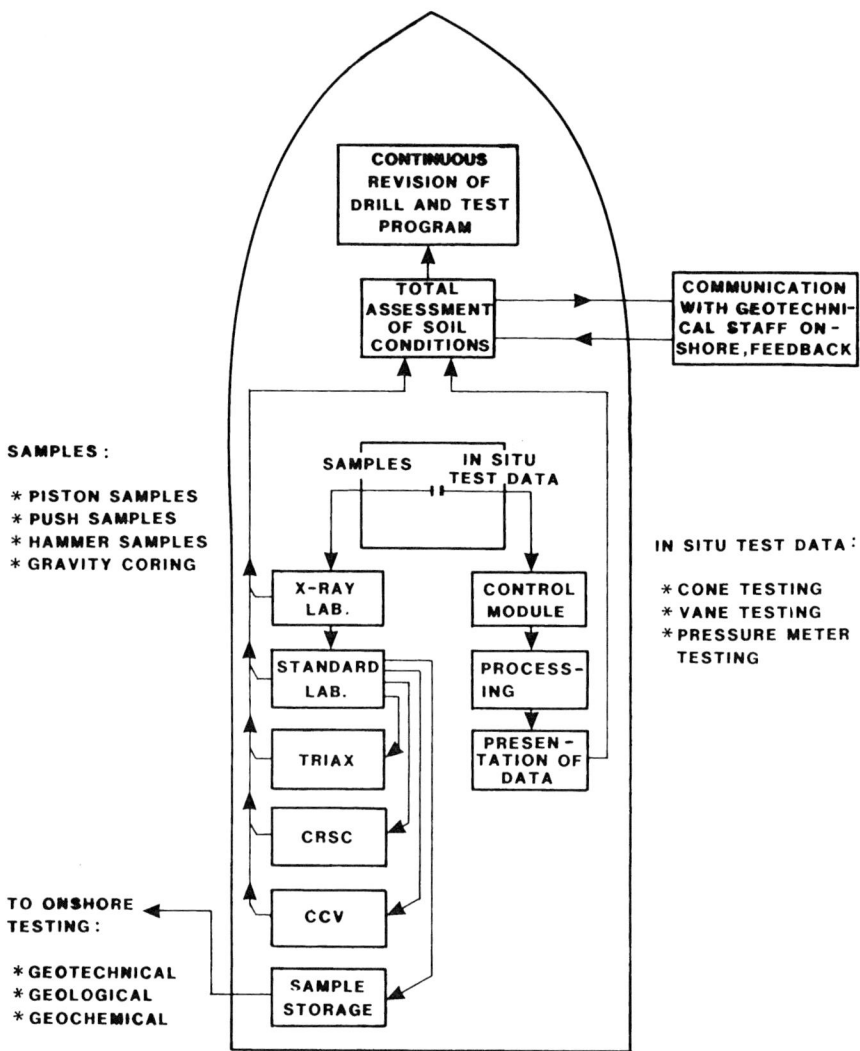

Fig. 22   Process of operation on board geotechnical drillship

dents, a summary of sampling, laboratory testing, *in situ* testing and finally a list of reimbursable and lost and damaged equipment. Similar reports were presented by the ship's captain as well as Racal Survey's surveyor. Based on these reports, a daily telex was prepared which formed, among other things, the basis for payment.

SUMMARY AND CONCLUSION

This paper has described the planning and execution of an advanced deep-water soil investigation which was undertaken to obtain a detailed knowledge of the soil conditions of the Troll East Field and to establish preliminary soil design parameters at potential platform locations. A quality assurance and control system formed an essential part of the fieldwork. For drillship positioning Syledis and Hyperfix navigation systems were used simultaneously. Both systems were found to be very stable with an accuracy of less than 5 m. No significant horizontal movement of the drillship *Bucentaur* was observed when the ship was

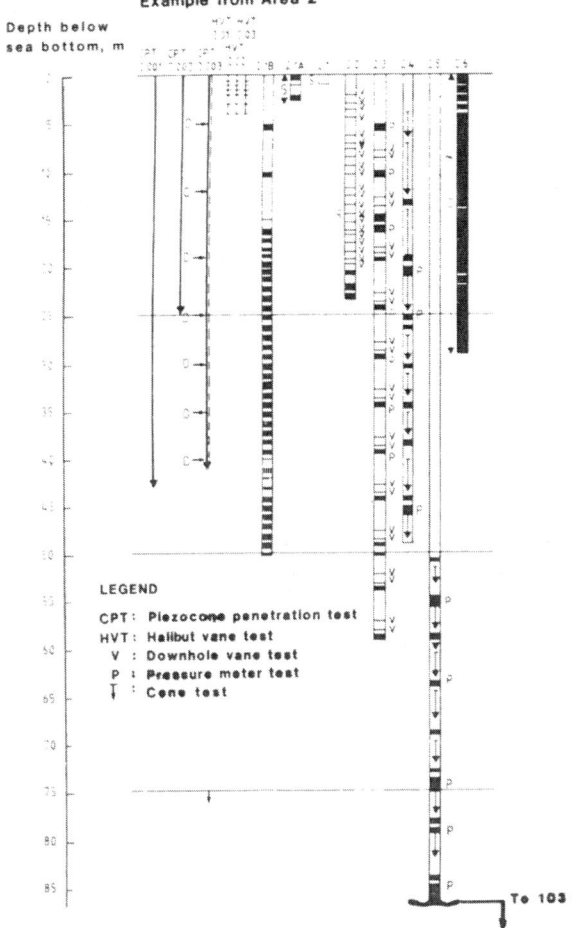

Fig. 23 Soil investigation carried out in Area 2
in 1985

was issued on 10 December 1984 for the Norsk Hydro company's review. The project was successfully completed within the allocated budget of 19.2 million NOK and a time limit of 1 January 1985. Based on the experience gained from this project technical specifications are now proposed for labortory testing, the process of soil design parameter selection and the practical application of quality assurance and control. These specifications will be implemented in future soil investigations.

ACKNOWLEDGEMENTS

We wish to extend our thanks to Norsk Hydro, Saga and Statoil for allowing this paper to be published.

REFERENCES

1. Amundsen, T. and Lauritzsen, R. 1982. Time and cost planning for offshore soil investigations. In *Proceedings, Third International Conference on the Behaviour of Offshore Structures*, Boston, Vol. 2, pp. 417–424.

2. Ehlers, C. J. and Babb, L. V. 1980. *In situ* soil testing: remote vane. McClelland Engineers' Soundings, Spring 1980.

3. Lacasse, S., Iversen, K., Sandbaekken, G. and Morstad, P. 1984. Radiography offshore to assess sample quality. Nordiska Geoteknikermotel, Linkoping, Sweden, June 1984, Vol. 2, pp. 575–584.

4. Lunne, T., Christopherson, H. P. and Tjelta, T. I. 1985. Engineering use of piezocone results in North Sea clays. To be presented at *11th International Conference on Soil Mechanics and Foundation Engineering*, San Francisco.

5. Østmo, S. R. and Kleiven, A. 1984. Shallow seismic investigation Troll Block 31/3, 5 and 6. Report No. ST8393.01, Norsk Hydro, Statoil and Saga Internal Report.

6. Reid, W. M., St John, H. D., Fyffe, S. and Rigden, W. J. 1982. The push-in pressuremeter. *Symposium on the Pressuremeter and its Marine Applications*. Paris, April, pp. 247–261.

positioned on location in the dynamic positioning mode.

Technical specifications were established for sampling and *in situ* testing equipment and procedures. Implementation of these specifications enhanced the quality of results obtained from the investigation. State-of-the-art sampling and testing equipment were used, and a comprehensive soil-testing programme was carried out offshore for the first time. The total cost of this project was 18 million NOK with almost 75% of this amount allocated to fieldwork; the remainder was allocated to laboratory testing, interpretation and reporting. The report describing the soil design parameters

7. Sandbaekken, G., Berre, T. and Lacasse, S. 1985. Oedometer testing at the Norwegian Geotechnical Institute. Presented at *ASTM Symposium on Consolidation of Soils Laboratory Testing*, Ford Lauderdale, Florida, January, 1985.

8. Semple, R. M. and Johnston, J. W. 1979. Performance of Stingray in soil sampling and *in situ* testing. In *Proceedings, Conference on Offshore Site Investigation*, Society of Underwater Technology, London, March, pp. 167–182.

12

Use of the PAM Self-boring Pressuremeter and the STACOR Large-size Fixed-piston Corer for Deep Seabed Surveying

J-B. Faÿ, R. Montargès, P. le Tirant and F. Brucy, Institut Français du Pétrole, France

To meet the needs of geotechnical soil surveying at great water depths, two systems have been specially developed and investigated by the Institut Français du Pétrole in recent years: these are the PAM Self-boring Pressuremeter and the STACOR Large-size Fixed piston Corer.

The PAM pressuremeter consists mainly of a framework set on the seabed, a control cabin on the surface and an electric umbilical cable connecting the cabin and the framework. The self-boring probe of the PAM is derived from the PAF 76 used by the Laboratoires des Ponts et Chaussées for onshore applications. The PAM provides two types of information: self-boring logs recorded in real time, and pressuremeter measurements performed every one or two meters. The PAM has been operated successfully on various offshore sites in the Gulf of Lions and offshore Spain on the Casablanca and Montanazo fields, at depths ranging from 50 to 625 m, with penetrations into the seabed of more than 50 m. The system is currently operational and can be used at water depths of up to about 1000 m.

The STACOR corer was designed in collaboration with Elf-Aquitaine and Total for taking core samples at water depths of several thousand meters. It is made up of a core barrel 20 to 30 m long, a weight-stand of up to 10 tons, and a piston held in a fixed position while the core barrel penetrates into the ground. The fact that the piston is kept fixed enables relatively undisturbed core samples to be taken. These samples are 11 cm in diameter and have a length equivalent to the depth to which the core barrel penetrates the ground. Comparative results have been obtained for core samples taken by the STACOR operating normally (with a fixed piston) and by the same device operating like a standard oceanographic corer. The results show the considerable advantages of the fixed-piston STACOR for taking undisturbed samples.

THE OFFSHORE SELF-BORING PRESSUREMETER (PAM)

Improving knowledge of the mechanical properties of soils is the key condition for better evaluating the safety of offshore structures. Such an objective requires the parallel use of different soil survey techniques — including the pressuremeter technique, which has so far not been widely used offshore. The pressuremeter would incontestably provide further and more reliable data. The wireline method for coring or *in situ* measurements requires the use of a survey drilling ship, which quickly becomes very costly as water depth increases.

The pressuremeter provides full information about the 'stress-deformation' behaviour of the soil up to failure. However, the conditions under which the probe is made to penetrate the ground influence the results obtained considerably.

The PAM uses a self-boring method to implement the operation of the pressuremeter probe (Fig. 1).

Description and Operation of the PAM

Description of the PAM

The Offshore Self-boring Pressuremeter comprises (Le Tirant *et al.*, 1981; Faÿ and Le Tirant, 1982):

- On the seabed, a tripod frame with equipment to enable a self-boring pressuremeter probe to penetrate the soil by pushing.
- On the surface, a control and measurement cabin.
- An electric umbilical cable linking the cabin and the frame.

The main characteristics of the frame (Fig. 2) are as follows:

- height: 8.5 m;
- transportation width: 2.5 m;
- total surface area of the three bedfootings: 20 m²;
- approximate weight: 16 tons in air.

The self-boring pressuremeter probe (Fig.

Fig. 1   The PAM self-boring pressuremeter

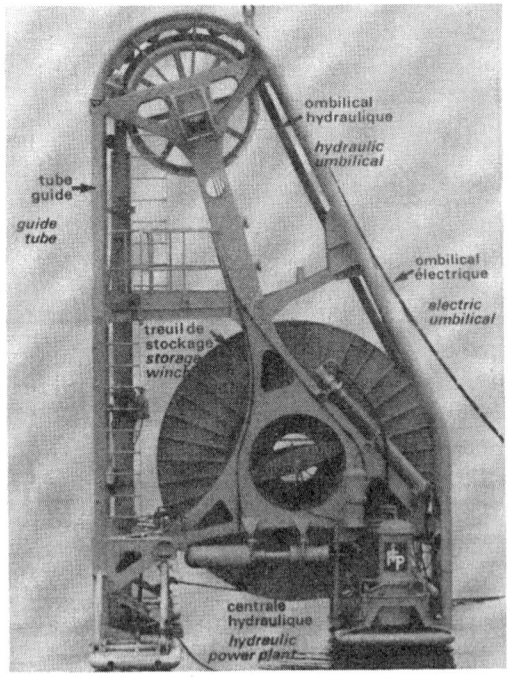

Fig. 2   The PAM: frame and mains equipments

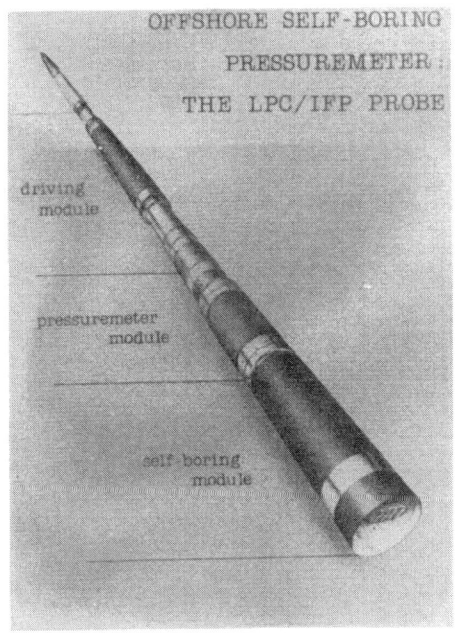

Fig. 3   The self-boring pressuremeter probe

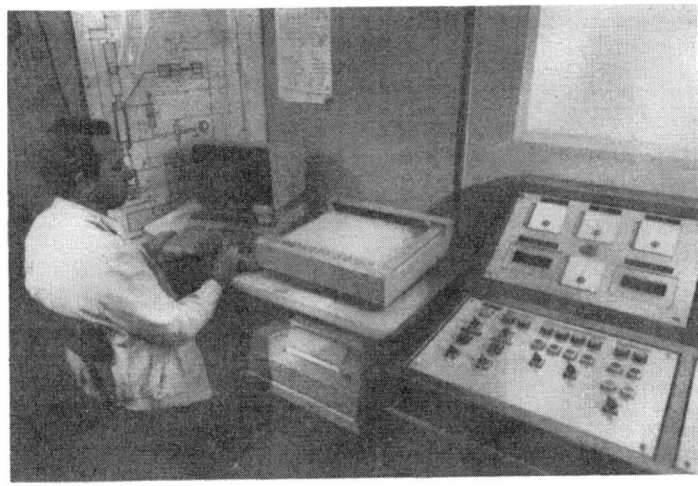

Fig. 4   The control cabin

3) is 6 m long and 160 mm in diameter. It is made up of:

- A self-boring module mainly consisting of a cutting edge and a grinder.
- A driving module with a choice of two models:
  (1) a 10 kN drill collar for soft soils;
  (2) a jack-packer for stiffer soils.
- A pressuremeter made up of a cylindrical cell with a slenderness ratio of 2, which can be expanded to 25% more than its initial volume.

The control cabin (Fig. 4) mainly contains:

- A panel for the manual and visual control of the different functions of the frame and probe.
- A computer for controlling all the probe penetration operations, performing pressuremeter tests and acquiring data which are recorded simultaneously on cassettes and on a printer.

The electric umbilical cable provides the power supply for the electrohydraulic power plants and also transmits commands and measurements (multiplexed) via a coaxial ble (Fig. 5).

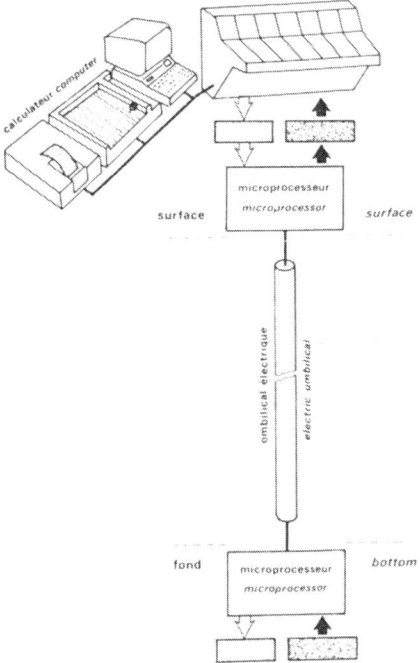

Fig. 5   Sea bottom–surface transmission principle

Operation of the PAM

The PAM is commonly operated from a non-specialized ship (Fig. 6) of the supply-boat type, equipped with:

- An A-frame or gantry crane with a capacity of 20 to 30 tons.

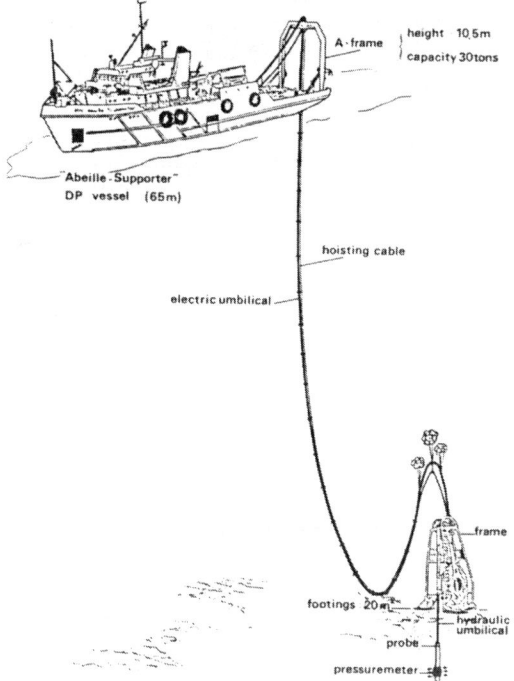

Fig. 6   Implementation of the PAM

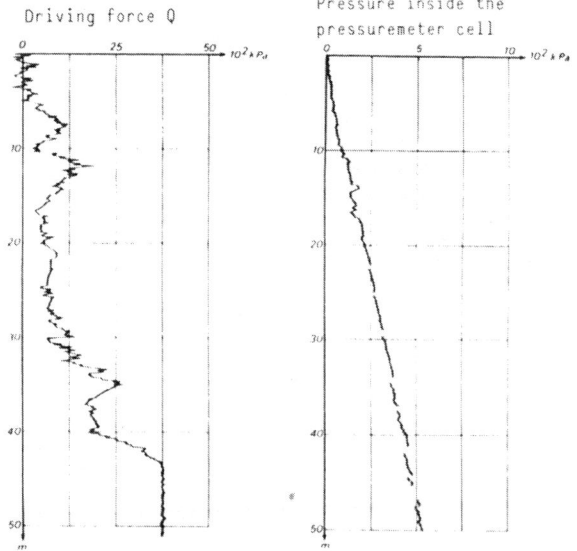

Fig. 7   Self-boring logs Q and p_i

- A winch with a hoisting capacity of about 30 tons so the frame can be hoisted back up (if the bed-footings should become buried into the seabed).

The surface area of the bed-footings (20 m²) provides stability for the frame on very soft soils having a surface cohesion higher than about 1.5 kPa.

PAM Measurements

The penetration of the probe is accompanied by the recording of self-boring logs every 10 cm to provide information on the properties of the formations though which the probe passes (Fig. 7):

- driving force, Q;
- pressure inside the pressuremeter cell at zero dilatation, p_i;
- rate of penetration;
- torque on grinder;
- water circulation pressure.

A pressuremeter test with the simultaneous recording of the 'pressure–volume' curve is carried out at each previously determined level (every 1 or 2 m), or at levels chosen by the operator on the basis of information deduced from self-boring parameters. Pressure and volume corrections are made to the pressuremeter curve in delayed time. The time required to carry out a probe covering about 50 m, with a pressuremeter test every 1 m or every 2 m, is about 24 to 36 hours from the time when the device is positioned on the seabed.

Trials of the Offshore Self-Boring Pressuremeter

The PAM has been systematically tested in clayey and sandy soils at different water depths down to 625 m.

Onshore and Shallow-water PAM Tests

The preliminary test phase took place in 1981. First of all, tests were carried out on a land site with well-known soil properties. This site had consolidatd marine clay with a cohesion of $20 < C_u < $ kPa. Five boring operations were performed to a maximum depth of 17 m (the bottom limit before the gravelly soil). Then tests were done at water depths of about 10 m in sandy and silty formations, in the harbours of Le Havre and Fos-sur-Mer, to a maximum penetration depth of 19 m (the bottom limit before the gravelly formation).

PAM Tests in the Gulf of Lions

In February 1982 the PAM was tested in the Gulf of Lions, 10 km SSE of Sète, at a water depth of 50 m from the dynamically positioned ship *Castor 02*, which is 45 m long and equipped with an A-frame. Several boreholes were executed in recent quaternary silty clay sediments. It took an hour to instal the PAM on the seabed and 17 hours to carry out 23 pressuremeter tests at a penetration depth of 30 m.

PAM Tests in the Casablanca Area (1983)

The PAM was tested offshore from Tarragona in the Casablanca Area from the dynamically positioned ship *Abeille-Supporter*. This 65 m long ship is equipped with a 10.5 m high A-Frame and has a hoisting capacity of 30 tons. Two boreholes were successfully executed. Work on the first, at a water depth of 227 m in a silty clay formation, had to be halted after a penetration of 35 m into the seabed on account of bad weather. The execution of this borehole, including 12 pressuremeter tests, was completed 12 hours after the pressuremeter frame had been set on the seabed. The second borehole was executed at a water depth of 625 m and penetrated to a depth of 51 m, also in a silty clay formation. This borehole with 20 pressuremeter tests was completed in 17 hours. These results confirm the excellent performance and the full operational reliability of the PAM.

Geotechnical PAM Test Results

Efforts are now commonly made to acquire data relating to the geotechnical properties of soils from various different sources (core-sample measurements, cone-penetration tests, borehole logs, etc.). Because the .M can run self-boring logs and acquire

pressuremeter data in the same borehole, it is used to obtain a wide range of data on the geotechnical characteristics of soils. The numerous results acquired on different sites (harbours, Gulf of Lions, Casablanca Area, etc.) and in different types of soils (clay, silt, sand) give a good idea of the wealth of information that the PAM can provide (Le Tirant *et al.*, 1983; Brucy *et al.*, 1984).

Self-boring Logs.

The self-boring parameters recorded every 10 cm of penetration of the probe into the ground are:

- the driving force, Q;
- the pressure inside the undilated pressuremeter cell, p_i;
- the fluid circulation pressure;
- the resistance torque on the grinder.

The values of the resistance torque on the grinder and of seawater circulation pressure are highly indicative of the conditions under which penetration takes place and contribute to the information obtained on the characteristics of the formations through which the probe passes. The pressure, p_i, in the undilated pressuremeter cell during penetration of the probe into the ground, which is between the total horizontal stress in the ground (perfect self-boring) and the hydrostatic pressure (in the case of soil decompression), gives information on the state of stress in the soil.

Driving Force and Cone Penetration Strength

The driving force Q of the probe is a boring parameter which describes the resistance of the soil to the penetration of the probe, in the same way as cone resistance in a CPT. Close correlations between the force Q and the tip strength q_c of a penetrometer have already been established for normally consolidated clays.

Pressuremeter Test Data

The shape of the pressuremeter curve

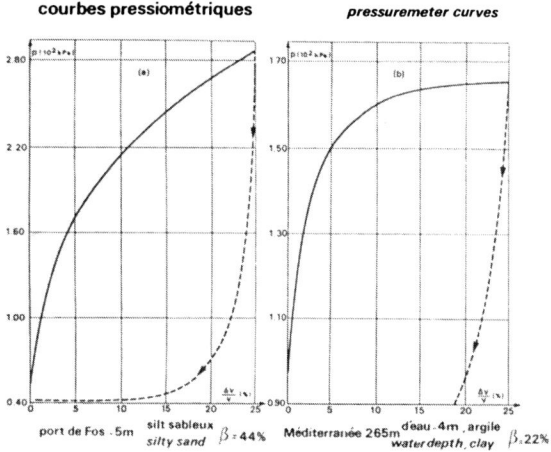

Fig. 8 Pressuremeter curves

$(p, \Delta V/V)$, characterized by the coefficient β, differs considerably according to whether the formation is of a sandy or clayey nature (Fig. 8). The curve can be interpreted to deduce the pressuremeter parameters (Fig. 9) of the soil for use in calculating foundation parameters:

- pressure p_0 in the undilated pressuremeter cell after a period of soil relaxation can be used to deduce an approximate value of the coefficient K_0 of the soil;
- pressures p_2, p_5 and p_{20} characterize the respective states of the soils for 2, 5 and 20% volume dilatations of the pres-

suremeter. Pressure p_{20}, which practically represents the limit pressure of the soil, falls directly within the calculation of the bearing capacity of foundations by the pressuremeter method.

- tangent G_0 and secant Gp_2 moduli (for 2% deformation) are also used directly for calculating foundation design by the pressuremeter method.

Soil Investigation Coefficient

The coefficient $\beta = \dfrac{p_{20} - p_5}{p_{20} - p_0}$

calculated from pressuremeter curves (Baguelin *et al.*, 1978) is very reliable for identifying the nature of soils, results being quite comparable to the identification obtained by the description of core samples (Fig. 10).

Undrained Shear Strength

The shear-strength curve of clayey soils is deduced, by derivation, from the pres-

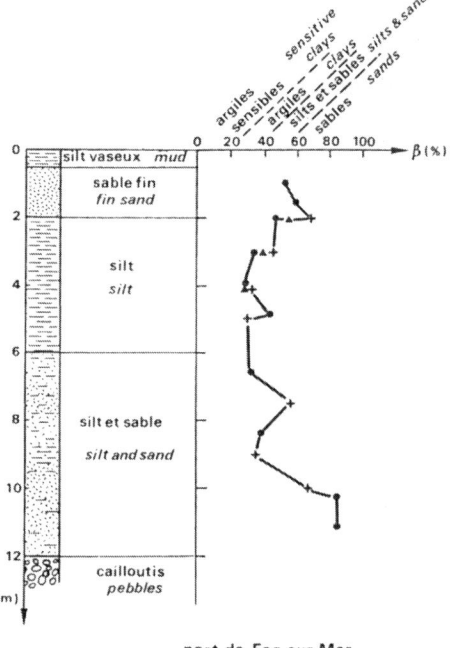

Fig. 10 Example of soil identification from pressuremeter curves

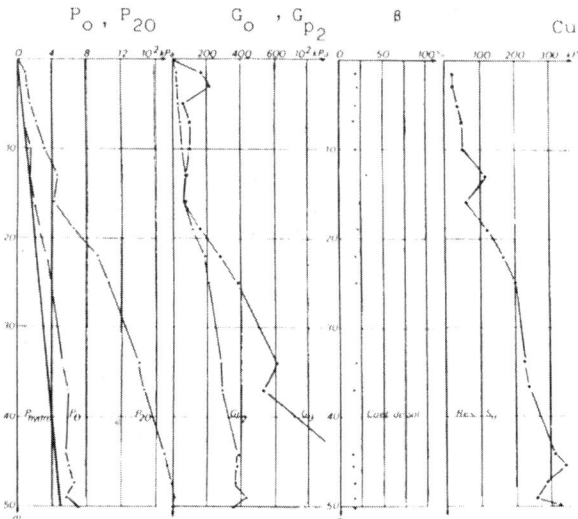

Fig. 9 Pressuremeter results

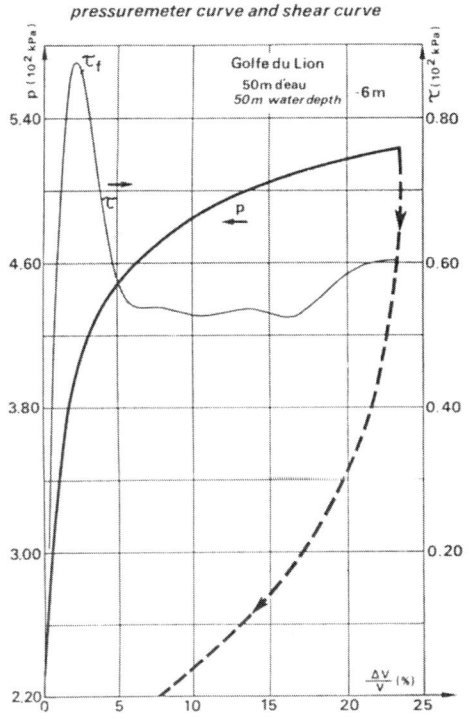

Fig. 11   Pressuremeter curve and shear curve

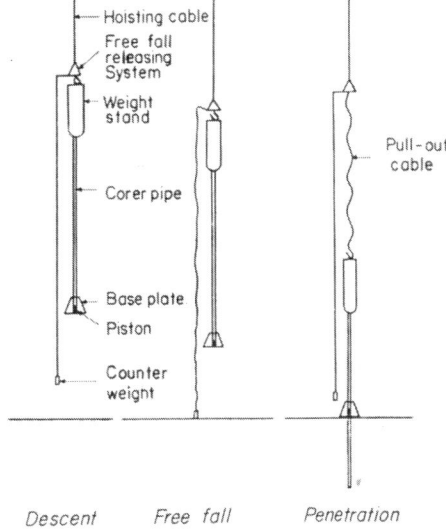

Fig. 12   Principle of the STACOR

suremeter curve. In soft or medium stiff clays, the peak of the shear-strength curve gives the maximum shear strength τ_f generally attained for a volume deformation of 1 to 2% (Fig. 11). For these soils, the C_u value obtained from Ménard's correlation

$$C_u = \frac{p_{20} - p_0}{5.5}$$

is in good agreement with that deduced from the C_u triaxial test.

Application of PAM Measurements

The PAM is particularly well suited for surveying fine sediments (soft or very soft soils with a cohesion $C_u < 50$ kPa, loose carbonate sediments which are extremely sensitive to disturbance, etc.). It thus covers a large portion of seabed soils — especially those beyond the continental shelf — in which conventional techniques of core sampling or *in situ* measurements cannot give

meaningful results. The steady development of petroleum production on the continental slope, at water depths of several hundred meters, raises two types of geotechnical problems linked both to the sizing of new structural foundations (tension-leg platforms, guyed towers, articulated gravity towers) and to the evaluation of the instability risks of seabed slopes and their consequences for seabed installations. The offshore self-boring pressuremeter meets the needs corresponding to the surveying of such soils.

Foundation Design

In a way similar to the pressuremeter rules drawn up by Ménard for the standard pressuremeter test, the Laboratoires des Ponts et Chaussées have developed semi-empirical rules using self-boring pressuremeter results for calculating the limit load of onshore or offshore foundations (Baguelin *et al.*, 1978). These cover:

- lateral friction of piles;
- piles subjected to horizontal loads;
- limit punching pressure at the base of foundations;
- settlement of shallow foundations.

Slope Instability Evaluation

An evaluation of the instability risks of soft soils on the continental slope is indispensable prior to the installation of petroleum structures on the seabed. This evaluation is generally made by using the limit equilibrium approach requiring solely the determining of the intrinsic parameters (C', φ') of the soil or the undrained cohesion C_u. Soil surveying on the continental slope with the PAM thus determines the C_u so as to investigate the instability risks of subsea soils under undrained conditions.

Application of the PAM for Planning Deepwater Production Facilities

Production from hydrocarbon fields on the continental slope, at water depths of several hundred meters, raises the problem of investigating soils which are often soft or very soft, prior to the installation of risers, anchoring piles and flowlines. At such water depths and in this type of soil, the PAM, which is designed for water depths of up to 1000 m, has a great many advantages of both a technical and economic nature compared to other soil investigation methods currently in use.

Technical Advantages of the PAM

The PAM is currently the only device capable of making *in situ* measurements at water depths up to 1000 m. The technical advantages of the PAM involve both its implementation and data acquisition.

(1) From the standpoint of implementation, the PAM does not require a specialized ship (such as a drilling ship) because it is operated from a supply-type ship equipped with dynamic positioning.
(2) From the standpoint of geotechnical data acquisition, the PAM provides a complete range of data which are habitually acquired by different tools or devices:

(a) the probe driving force can be correlated with the penetrometer tip resistance (CPT);
(b) coefficient β deduced from the pressuremeter curve provides a very reliable identification of the soil, equivalent to that obtained from core samples;
(c) pressure p_{20} (close to limit pressure) is directly involved in calculating the bearing capacity by the pressuremeter method;
(d) pressure p_0 can be used to deduce an approximate value of the coefficient K_0 of earth pressure at rest;
(e) maximum shear strength τ_f of the soil is derived from the pressuremeter curves;
(f) *in situ* soil deformation moduli can be obtained solely by the pressuremeter.

Furthermore, the quality of the data acquired by the PAM is independent of the water depth, despite the fact that the disturbing of core samples commonly increases considerable with water depth on account of both sampling conditions and decomposition.

Economic Advantages of the PAM

The cost of operating the PAM from a dynamically positioned ship, of the *Abeille-Supporter* type, is almost independent of the water depth (up to 1000 m) as long as the time required for lowering and hoisting the frame is not considered.

In contrast, the cost of the traditional process of soil surveying by wireline coring and/or *in situ* measurements via the drill string increases very quickly with the water depth because of the tonnage required for the ship used.

THE STACOR LARGE-SIZE FIXED-PSITON CORER

The effective stationariness of the piston is the key condition for taking undisturbed core samples. This condition, which was previously fulfilled by the corer developed by Kermabon and Cortis (1969), was a prime consideration in the STACOR corer.

Description and Operation of STACOR

Principle of the Corer

The STACOR is a free-fall gravity corer with a fixed piston (Fig. 12). It is lowered at the end of a hoisting cable wound around a winch. At a predetermined distance above the seabed, between zero and several meters, a triggering system causes the free fall of the device so that the corer pipe can penetrate into the soil. Then the corer is pulled out of the soil by a cable which had been allowed to uncoil freely during the free fall and penetration of the corer.

The true stationariness of the piston, which is impossible to obtain with conventional corers of the Kullenberg type, requires the positioning reference to be taken in relation to a baseplate falling with the corer pipe and sitting on the seabed while the corer penetrates into the soil. The piston and baseplate are connected (Fig. 13)

by a cable running over two pulleys situated at either end of the core pipe. When the core pipe penetrates into the soil the piston remains at the level of the baseplate.

Description of the Corer

The STACOR corer (Fig. 14) mainly consists of (Montargès et al., 1983):

- the corer pipe with its liner;
- the weight-stand;
- the baseplate and the fixed-piston system;
- the free-fall releasing mechanism.

The corer pipe is made up of modular pieces decreasing in thickness towards the cutting edge. This pipe protects a liner made of plastic or reinforced resin. The adjustable weight enables the weight of the corer to be varied from about 5 to 10 tons with a corer pipe length of 30 m.

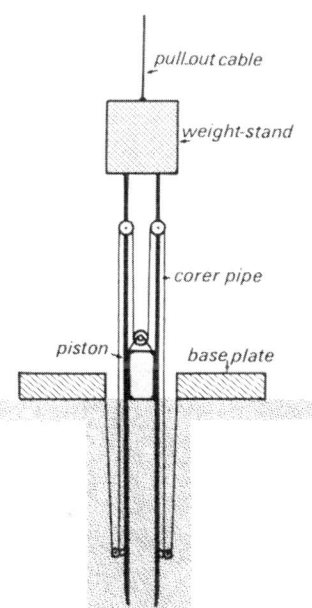

Fig. 13 Principle of the fixed piston

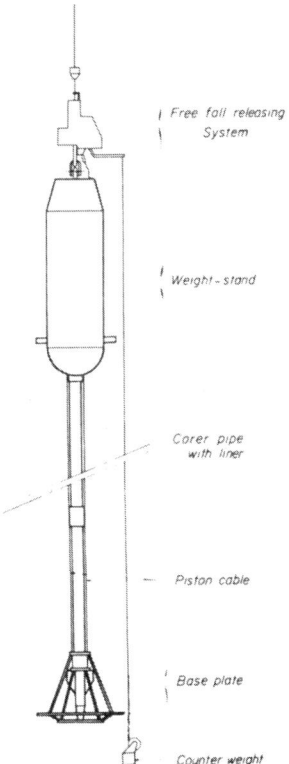

Fig. 14 Description of the STACOR

The piston positioning system consists of:

- a baseplate 1.5 m in diameter made of a tubular frame sitting on the seabed during penetration of the corer pipe into the soil;
- cables connecting the piston and baseplate, running over pulleys situated respectively underneath the weight-stand and on the lower piece of the pipe.

An original releasing device with a cam–links–spring system ensures full safety during the operations of launching the corer over the stern or the side of the ship and lowering it to the seabed. The command is set off when a counterweight strikes the seabed. The pull-out cable, which is coiled in a conical housing situated above the weight-stand, uncoils freely during the free fall and penetration of the corer pipe.

Characteristics of the Corer

The size and weight of STACOR are the result of both the objectives to be achieved and the conditions of implementation:

- Length of corer pipe; up to 30 m
- Length of device, overall; up to 35 m
- Diameter of corer pipe; OD 16 and 17 cm, ID 13 cm
- Diameter of plastic liner; OD 12 cm, ID 11 cm
- Ballast weight; up to 5.5. tons
- Total weight of corer with 30 m pipe; 5 to 10 tons
- Cutting edge; OD 16 cm (maximum), ID 11 cm, cutting angle adapted to nature of soil.

Implementation of the STACOR

Two procedures were investigated for implementing the STACOR: (1) launching it over the stern of the ship, and (2) swinging it out laterally. A trial of the first procedure was carried out with the *Nadir* (Fig. 15), the CGM-IFREMER submarine handling ship. The second procedure will be used in June 1985.

Characteristics of the Ship and Preparation of the Corer

The *Nadir* has a deck 35 m long so that the corer can be set horizontally. This ship is equipped with:

- a self-driven trolley moving on two rails along the axis of the deck;
- an A-frame aft, rated for 60 tons;
- a crane covering the entire deck.

During the set-up and axial movement of the corer, the weight-stand sits horizontally on a cradle attached to the trolley. The device is tiped upright around two trunnions sticking out from the weight-stand. Two lateral upraised catwalks attached to the cradle facilitate the outfitting of the release mechanism situated on top of the weight-stand.

Fig. 15 The STACOR implemented on the *Nadir*

Fig. 16   Core sample recovery

Recovering the Core Sample

After the corer has been laid down horizontally and the cutting edge has been disassembled, the core sample is extracted from the corer pipe (Fig. 16) and preserved in its plastic liner in sections which are carefully closed off by specially fitted plugs.

Experiments of the Corer and Results Obtained

Two STACOR test campaigns have been carried out from the *Nadir*:

- the first, in April 1982, was to test the reliability of the implementation procedures and the operation of the corer;

- the second, in May 1983, was to check its performance after a few modifications of the device.

Coring Operations

During the second test campaign, six coring operations were carried out in normally consolidated silty clays in the Gulf of Lions at water depths between 50 and 1225 m, with the latter being the maximum attainable with the length of cable now available on the *Nadir*. Table 1 sums up the applied experimentation conditions, the penetration depths reached and the lengths of the core samples recovered.

The filling ratio of the corer, defined by the ratio of the length of the core sample to the penetration of the corer pipe, is about 95%, with the exception of test 5 when the STACOR worked like a conventional corer of the Kullenberg type.

Quality of Core Samples Taken

A filling ratio of 95% is the criterion for a good-quality core sample, as checked out, moreover, by a comparison of the core samples taken with a fixed-piston corer with those taken with a Kullenberg-type corer (or acting like one).

A comparison of shear strengths measured on core sample 4, which was correctly taken (filling ratio of 97%), and core 5, which

TABLE 1
STACOR test campaign results in the normally consolidated silty clays of the Gulf of Lions

Test	Water depth (m)	Length of corer pipe (m)	Penetration (m)	Core length (m)	Filling ratio (%)
1	50	15	11.1	10.2	92
2	510	15	14.8	13.9	94
3	510	20	18.3		
4	510	20	18.6	18.0	97
5	510	20	18.8	13.4	71
6	1225	20	17.0	15.3	90

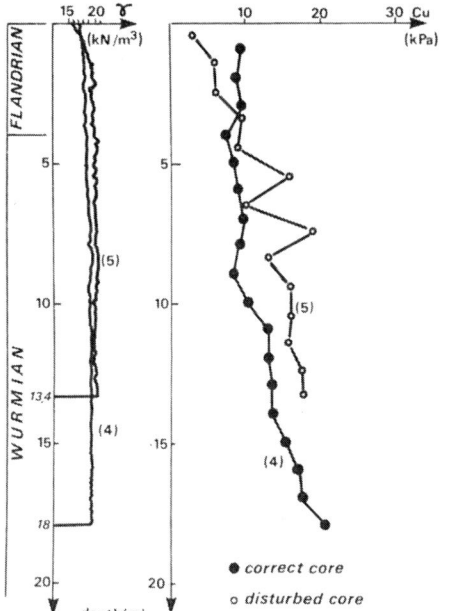

Fig. 17 Comparison of soil characteristics: (4) undisturbed core sampled by fixed-piston corer; (5) disturbed core sampled by conventional piston corer

was disturbed by sampling, shows that (Fig. 17):

- at the level of the Flandrian (down to a depth of about 4 m) a normally constant shear strength was shown by core 4, whereas core 5 indicated low residual cohesion;
- at the level of the Wurmian below 4 m a cohesion gradient was revealed by core 4, whereas core 5 showed shear strength values which were highly erratic and about 40% greater than the preceding ones on account of a packing of the normally consolidated silty clay.

Outlook for STACOR Applications

Because of the size and quality of the core samples taken, the STACOR is a new technique which is well suited for the precise surveying of soils to a penetration of 25 to 30 m under water depths of several thousand meters. The results of the STACOR test campaigns show the perfect reliability of the device, with regard to both its implementa-

tion and its operation; thus its industrial application for geotechnical or sedimentological purposes can already begin to be planned.

Operating Conditions of STACOR

Implementation of the device, experimented aboard the *Nadir*, can be directly transposed to ships of the supply-boat or anchor-handling type equipped with suitable handling equipment such as winches, A-frames and/or cranes. Implementation of the device is very safe because the corer never hangs over the deck, so that operation can proceed even with a heave of the ship's stern of up to 4 m. Under conditions of industrial application of the STACOR, with two devices and crews working continuously it should be possible to take two core samples every day at water depths of 6000 m.

Application of the STACOR for Geotechnical Surveying

In zones undergoing petroleum exploration or where production projects are being planned on the continental slope (under water depths of 1000 to 2000 m) the STACOR is a technique which is perfectly suited for the geotechnical surveying of offshore soils, when planning the size of conductor pipes or anchoring systems, or for analysing any eventual risks of slope instability. Down to 1000 m water depth, it effectively completes the data acquired by the Offshore Self-Boring Pressuremeter (PAM). Beyond 1000 m, it is at the present time the only geotechnical surveying equipment not requiring the use of a drill ship.

Application of the STACOR for Sedimentological Investigations

The STACOR has the advantage of being able to take cores for sedimentological, geochemical and geotechnical investigations on abyssal plains at water depths of 5000 to 6000 m. It can thus be used for deep-sea fan investigation projects as well as for the current needs of previous investigations of sites.

CONCLUSIONS

1. For many years the need to improve man's knowledge of offshore soil for the installation of petroleum structures has been felt very keenly by all experts. The Offshore Self-boring Pressuremeter (PAM) which has been developed and tested by the Institut Français du Pétrole for implementation at great water depth meets this need.
2. The PAM is currently the only device capable of making *in situ* measurements at water depths up to 1000 m. It provides a complete range of soil data which are habitually acquired by different tools or devices. The quality of these data is independent of the water depth.
3. The results of the test campaign carried out successfully in June 1983 in the Mediterranean Sea, off Tarragona, at up to 625 m water depth confirm the excellent performance and the full reliability of the PAM, which is now operational.
4. The characteristics of the PAM (water depths of application, operational remote controls, automatic data acquisition, etc.) make this prototype a new-generation device for soil reconnaissance at great water depths.
5. Test campaigns of the STACOR have illustrated the operational reliability of this fixed-piston corer and the safety of its implementation.
6. Because of the size and quality of the almost undisturbed core samples it takes, the STACOR is a new technique which is suited to surveying normally consolidated silty or clayey soils to a penetration

depth up to 30 m under water depths of up to 6000 m.
7. The STACOR is an economic method for soil investigation for sedimentological or geotechnical purposes.

REFERENCES

1. Baguelin, F., Jézéquel, J.-F. and Shields, D. H. 1978. *The Pressuremeter and Foundation Engineering*. Trans Tech. Publications.
2. Brucy, F., Faÿ, J.-B. and Le Tirant, P. 1984. Three years' experience with the Self-Boring Pressuremeter, *16th Annual Offshore Tech. Conf.*, Houston. Paper OTC 4677.
3. Faÿ, J.-B. and Le Tirant, P. 1982. Offshore Self-Boring Pressuremeter for deep water. *Symposium on the Pressuremeter and its Marine Applications*, Paris. Editions Technip, pp. 305–323.
4. Kermabon, A. and Cortis, V. 1969. A new sphincter corer with a recoiless piston. *Mar. Geol.* 7, 147–159.
5. Le Tirant, P., Faÿ, J.-B., Brucy, F. and Jézéquel, J.-F. 1981. A Self-Boring Pressuremeter for deep sea soil investigation, *13th Annual Offshore Tech. Conf.*, Houston. Paper OTC 4019.
6. Le Tirant, P., Faÿ, J.-B. and Brucy, F. 1983. The Offshore Self-Boring Pressuremeter PAM, its applications at great water depths. In *Proceedings of 2nd Deep Offshore Technology Conference*, Malta, 17–19 October 1983, Vol. 1, 45–74.
7. Montargès, R., Le Tirant, P., Wannesson, J., Valéry, P. and Berthon, J.-L. 1983. Large-size stationary-piston corer. In *Proceedings of 2nd Deep Offshore Technology Conference*, Malta, 17–19 October 1983, Vol. 1, pp. 63–74.

13

A Comparison of Ménard, Self-boring and Push-in Pressuremeter Tests in a Stiff Clay Till

J. J. M. Powell and I. M. Uglow, Building Research Establishment, UK

INTRODUCTION

The Building Research Establishment (BRE) has, for a number of years, been investigating the pressuremeter test, its interpretation and the application of its results for assessing soil properties relevant to design. The work originated with studies on London Clay (Marsland and Randolph, 1977), developed through various research programmes into offshore site investigations (e.g. Henderson *et al.*, 1979; Marsland, 1979) and has continued by returning to the land-based application (Powell *et al.*, 1983).

The idea of the pressuremeter (the expansion of a long cylindrical membrane in the ground) was conceived by Ménard in 1954 as a device installed in a pre-formed oversized hole. The resulting Ménard pressuremeter (MPM) was extensively used in France for foundation design. It was only in the late 1960s that the pressuremeter gained much attention outside France; increased attention was being given to *in situ* testing techniques due to a growing awareness of the

possible problems of laboratory testing — for example, sample size, stress relief etc. The activity resulted in the development both in Britain (Hughes, 1973) and France (Baguelin *et al.*, 1974) of the self-boring pressuremeter (SBP). The requirements for offshore site investigations in the 1970s led to the development by BRE of the push-in pressuremeter (PIP) as a wire-line tool (Henderson *et al.*, 1979). More recently examples can be found of all three types of pressuremeter being used both on land and offshore (see Ghionna *et al.*, 1982). The stage has now been reached where there are numerous developments of pressuremeters with ever-increasing degrees of sophistication; these have resulted from improved instrumentation and improved electronic capabilities, and from a better understanding of the problems involved in pressremeter testing. A review of much of the work, and pressuremeter testing in general, is given by Mair and Wood (1985).

Much has been written about the relative merits of the various types of pressuremeter and their interpretation. The CIRIA document (Mair and Wood, 1985) on the land-based use of pressuremeters comments that there is a need for detailed comparisons of the results from different types of pressuremeter (MPM, SBP, PIP); offshore site investigation practitioners have expressed a similar interest (personal communication). This invited paper aims to compare, for the first time, the results of BRE work on MPM, SBP and PIP pressuremeters. This has necessitated well-documented sites with known large-scale properties. The work has been not only to compare one pressuremeter with another, but also to assess the validity of the parameters measured by comparison with other data. The work at the BRE test-bed site at Cowden is presented here.

LOCATION OF SITE AND DESCRIPTION OF TYPICAL SOIL PROFILE

The tests were carried out at the BRE test bed site at Cowden which is located on the east coast of Britain approximately 23 km north-east of Hull. The soil profile at this site was deposited from ice advancing from the North Sea during the Devensian period (Catt and Penny, 1966), and is composed principally of clay matrix dominated tills separated by layers and/or lenses of silt, sand and fine gravel of varying extent. The presence of numerous erratics (lumps of chalk and harder stones with dimensions generally less than 25 mm), and evidence from microfabric studies, indicate not only

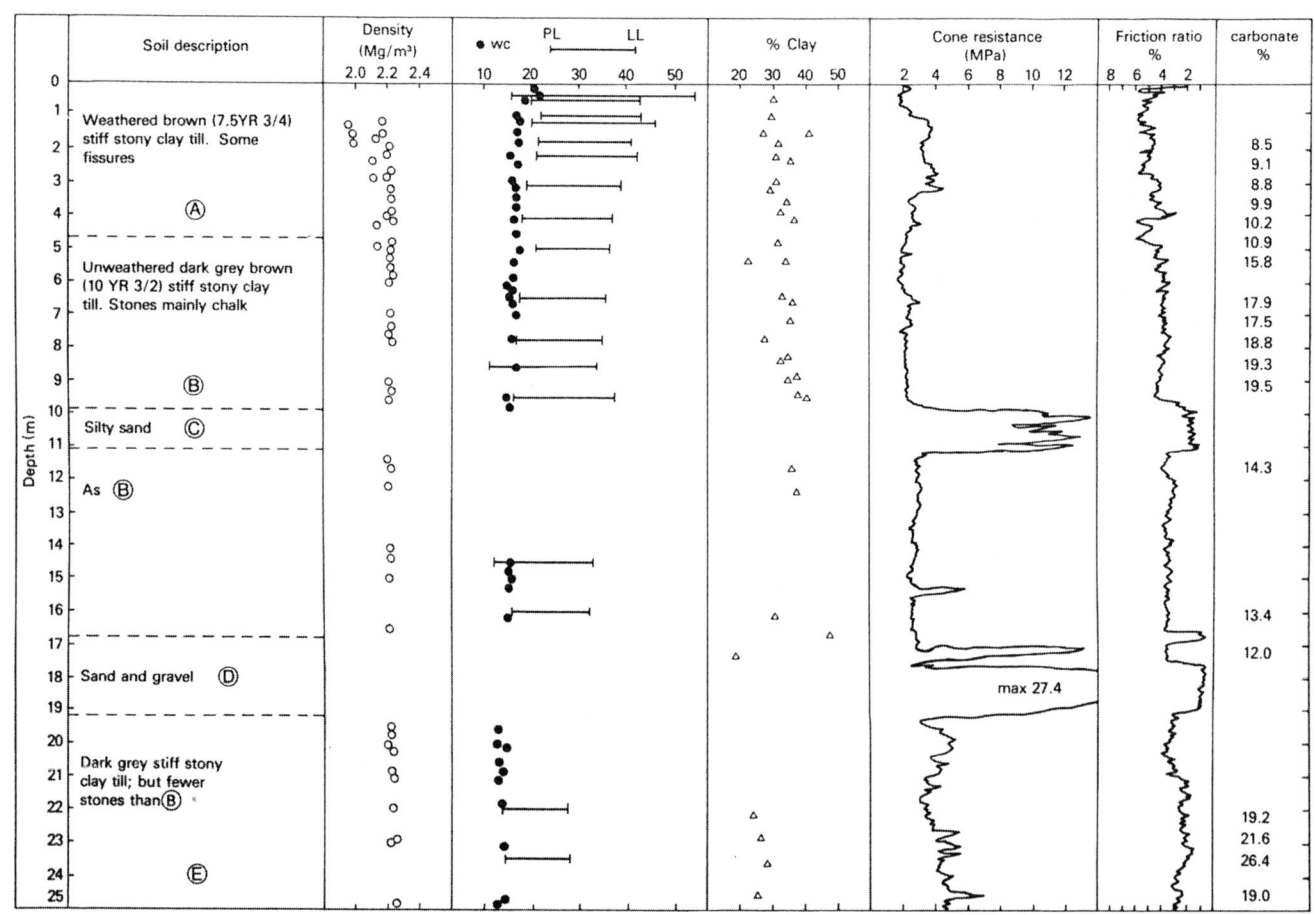

Fig. 1 Typical soil profile

that the tills were deposited directly from, or close to, the ice, mainly subglacially, but also that some flow and melt-out components may be present. The predominance of clay tills and their consistency over the site, as well as the changing depths and thicknesses of sand and gravel, were confirmed by static cone tests.

The relatively uniform nature of the clay tills down to the levels investigated can be seen from the typical profile in Fig. 1. There is a distinct change of colour from brown to very dark greyish brown at a depth of 4–5 m, and this is considered to be due to a zonal weathering of the till. The tills are inorganic, well graded (poorly sorted), fairly dense and of low activity (0.4–0.6). For depths between 3 and 17 m (till B), the bulk densities obtained from good-quality 98 mm diameter pushed samples are almost constant with values between 2.2 and 2.23 mg/m³. Over this depth range, the percentage of clay-sized particles is usually within the range 30–40%. Moisture contents gradually decrease with depth from 16.5% at 3 m to 15% at 16 m and are generally close to, or just below, the plastic limit. Plastic and liquid limits also show a gradual decrease with depth.

A second till (E) is encountered below the gravel layer (D). This till differs from till B by having moisture contents 1–2% lower, a lower clay content at 25% and higher densities, typically 2.25 Mg/m³. A marked increase in cone end bearing resistance is also evident in till E. The carbonate measurements shown in Fig. 1 indicate a depletion in the top 4–5 m together with much higher values in the next 4 m, indicating the transfer of carbonates from the weathered to the underlying unweathered till by leaching. Detailed studies of soil fabric at both macro- and microscales have been made on samples down the profile. Predominantly vertical macro discontinuities have been identified down to 4–5 m; no macro discontinuities have been found below this depth.

The till above the first sand layer, which is encountered at depths between 8 and 13 m, has a mass undrained shear strength

(as determined from 865 mm diameter plate tests) of between 90 and 140 kPa, which is comparable with the mass strengths measured in heavily overconsolidated London Clay. The few tests carried out below a depth of 20 m gave much higher strengths of between 210 and 280 kPa, and this lower till was probably deposited during an earlier advance of the ice.

Oedometer tests indicate that the unweathered clay till down to a depth of 10 m has been subjected to overconsolidation pressures approximately twice the present overburden pressures, but estimates are difficult in these well-graded soils. Porewater pressure measurements made during undrained triaxial tests on samples at their *in situ* moisture content show moderate to strong dilation. Typical curves showing the stress–strain behaviour of the tills are shown in Fig. 2. A fuller description of the

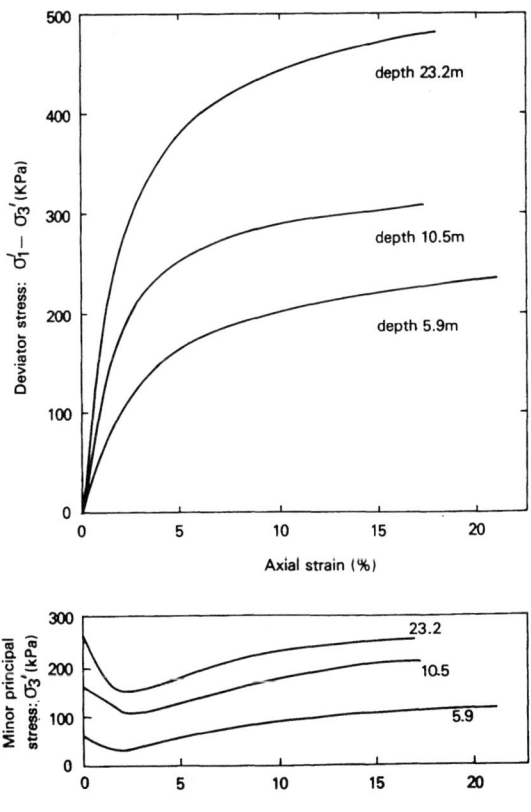

Fig. 2 Typical triaxial stress–strain curves

site geology, soil properties and engineering parameters can be found in Marsland and Powell (1985).

EQUIPMENT AND TEST PROCEDURES

In view of the stony nature of the clay tills at the site, it was considered advisable at the time of the main investigations to make the pressuremeter tests in pre-drilled pockets (the SBP being unable at the time to bore in these soils, and the PIP not having been developed). These first tests were therefore carried out using an MPM pressuremeter.

Ménard Pressuremeter (MPM)

The commercially available type GC Ménard pressuremeter was used. Since detailed descriptions of this equipment are readily available (Baguelin *et al.*, 1978), only the most essential features will be described here. The probe consists of a rubber-sheathed central measuring cell, covered by a longer sheath which forms guard cells at either end of the central cell. In its deflated condition the probe has an external diameter of 58 mm and a central measuring portion 216 mm long.

The central cell (capable of being expanded to twice the original volume) is expanded by water under gas pressure measured on a bourdon gauge, the volume changes being measured by a burette. The guard cells are expanded by gas pressure which is slightly lower than the central cell pressure.

A test pocket approximately 1 m deep with a diameter slightly larger than the probe was freshly drilled for each test using either a hand bucket type auger, or a powered rotary flight auger attached to a drilling rig. The pressuremeter probe was inserted as quickly as possible into the test pocket. The time interval between completion of drilling and reapplication of the initial *in situ* stresses to the walls of the boreholes was about 10–15 min. During a test, the

pressure in the measuring cell was increased in equal increments and maintained at each pressure for a given period of time. In 'standard' tests the pressure was increased in 50 kPa increments which were maintained for 1 min. Volume change readings were recorded at intervals of 15 s, 30 s and 1 min. Pressure corrections for the effects of the resistance of the membranes were obtained by expanding the probe in the unconfined state. Detailed calibration tests in which pressure increments of 20 kPa were maintained for various intervals of time showed that holding each increment for periods greater than 5 min had little effect on the corrections. All the tests described in this paper have been corrected using membrane calibration curves obtained by applying 20 kPa increments for 5 min. This gave a total calibration time of 20–30 min, which was approximately the same as the time required to carry out the 'standard' tests. In addition to the detailed calibrations, quick calibrations (50 kPa pressure increments held for 2 min) were made prior to and following each test in order to check that no significant changes had occurred in the behaviour of the probe. Checks were made for possible volume changes due to the expansion of the connecting tubes and for compression of small air bubbles trapped in the probe. Errors in the final pressures in the stiff soils are considered to be less than 2%.

Self-boring Pressuremeter (SBP)

The self-boring pressuremeter (SBP) used and its method of operation were similar to that described by Windle and Wroth (1977). The SBP is an instrument that should be capable of being inserted into the ground with minimum disturbance to the surrounding soil. It consists of a rubber-sheathed pressuremeter section placed behind a drilling head. The device has a constant diameter of 83.8 mm and is 1.05 m long, the pressuremeter section being 0.5 m long with its central expanding section 0.4 m back from the leading edge of the instrument. Gas pressure is used to expand the membrane and is monitored by a total pressure transducer mounted within the instrument, while the movement of the membrane is monitored by three displacement transducers.

The method of insertion is similar to that of a tunnelling machine in that, as the instrument is advanced, the soil within the SBP is chipped by a rotating cutter and the chippings flushed to the surface. The SBP is pushed into the ground using hydraulic rams which react against kentledge, casing, or ground anchors. The power to drive the rams and the rotating cutter is controlled through a series of valves using a hydraulic power pack. The cuttings are flushed to the surface by pumping water down the inner rotating rods to the cutter and then back up the 50 mm diameter outer rods. After insertion the SBP was left for an average of 30 min before the membrane was expanded at a constant rate of strain of about 1% per minute. This rate was controlled by means of a feedback control unit, which also limited the maximum expansion to about 20% of the original volume. An unload–reload cycle was included when about 2–3% cavity strain had been reached. The data from the test were logged by a microcomputer and stored on floppy disc. Total testing time generally amounted to 20–30 minutes.

Frequent calibrations for membrane stiffness and equipment compliance were made and applied to the data (maximum membrane correction was approximately 50 kPa). Small stones tended to collect at the base of the hole and after several advances had to be removed by withdrawing the SBP and boring past the stones using flight augers on a Minute Man drilling machine.

Push-in Pressuremeter (PIP)

The push-in pressuremeter (PIP), now known as the 'Stressprobe', may be inserted into undersized pre-cored holes or into the bottom of the borehole without pre-coring. The equipment is fully explained by Henderson et al. (1979) and only a brief description of the pressuremeter is given here. The Stressprobe consists of three basic units; the pressuremeter, the pressure developer and the control and data acquisition systems.

The pressuremeter comprises a hollow 78 mm diameter stainless steel cylinder 695 mm long onto which is mounted the 330 mm long inflatable membrane. At the lower end, the cylinder was fitted with a cutting shoe which incorporated a stress-reducing step so that soil could pass unrestricted up the inside of the cylinder. When used in very stiff soils or soils that had a tendency to plug before they were fully sampled, a 50 mm diameter pre-core of soil was first removed from the base of the hole. Testing reported in this paper has been performed both with and without pre-coring.

The inflatable membrane was protected by long stainless steel strips attached at the lower end to the cutting shoe. At the upper end, the strips were attached to a split ring, which was free to move axially in order to accommodate the expansion of the membrane. The membrane was inflated at a constant rate of 2% volume increase per minute by oil pumped by the pressure developer to a maximum expansion of 20% of the original volume. The pressure and volume of oil delivered were monitored electrically. The volume was measured by recording the displacement of a ram within the pressure developer, and the pressure by two piezoelectric pressure transducers, one to measure total pressure, the other to measure differential pressure.

Membrane calibrations were carried out periodically throughout the test programme (maximum correction at full expansion being about 80–90 kPa). During testing, the membrane was expanded under strain-controlled conditions (stress-controlled tests are possible). The test procedure was to expand the membrane to a maximum volume increase of 20% of the original volume and to include at least one unload–reload loop over a range of up to 3% volumetric strain. The reload loop was usually performed at the end of the test before final deflation.

INTERPRETATION OF PRESSUREMETER TESTS

The 'ideal' pressuremeter test in clays should be capable of giving three of the parameters of interest in foundation design, namely:

(a) initial *in situ* horizontal stress (P_{ho});
(b) *in situ* shear strength and possibly a stress–strain curve;
(c) deformation parameters over the ranges of stresses or strains applied to various elements in the ground by the foundation loading, e.g. shear modulus.

Methods of assessing these parameters follow. (See Wroth, 1984, for summary of other possible parameters.)

Estimation of Initial Horizontal Stress (P_{ho})

Good estimates of the total horizontal *in situ* stresses are important both as a starting point for detailed analysis of geotechnical problems (e.g. finite elements) and in the analysis of pressuremeter tests. Reviews of the various methods of estimating P_{ho} from the pressuremeter test have been given by Lacasse and Lunne (1982) and Mair and Wood (1985). In the present work, values of

P_{ho} have been deduced for MPM tests using both the original Ménard approach based on the 'creep' curve (Fig. 3a) and the graphical interactive approach of Marsland and Randolph (1977) (Fig. 3b). For the SBP tests the 'lift-off' method has been adopted. This method takes the pressure at the point at which the first significant movement of the membrane occurs (after compliance corrections) as P_{ho} (Fig. 4). Assessment of P_{ho} from PIP tests is of course impossible because of the significant overstress that can occur in the surrounding soil during insertion.

Shear Strength (c_u)

Undrained shear strengths have been obtained from pressuremeter test results by various workers in a number of different ways. The work has been reviewed by Mair and Wood (1985) (see also Ghionna *et al.*, 1982). In this present investigation two approaches have been adopted, the 'stress–strain' and the 'limit-pressure' methods; both are fully described elsewhere (Marsland and Randolph, 1977; Mair and Wood, 1985) and only a brief outline will be given here.

Stress—strain

Essentially similar analyses were presented

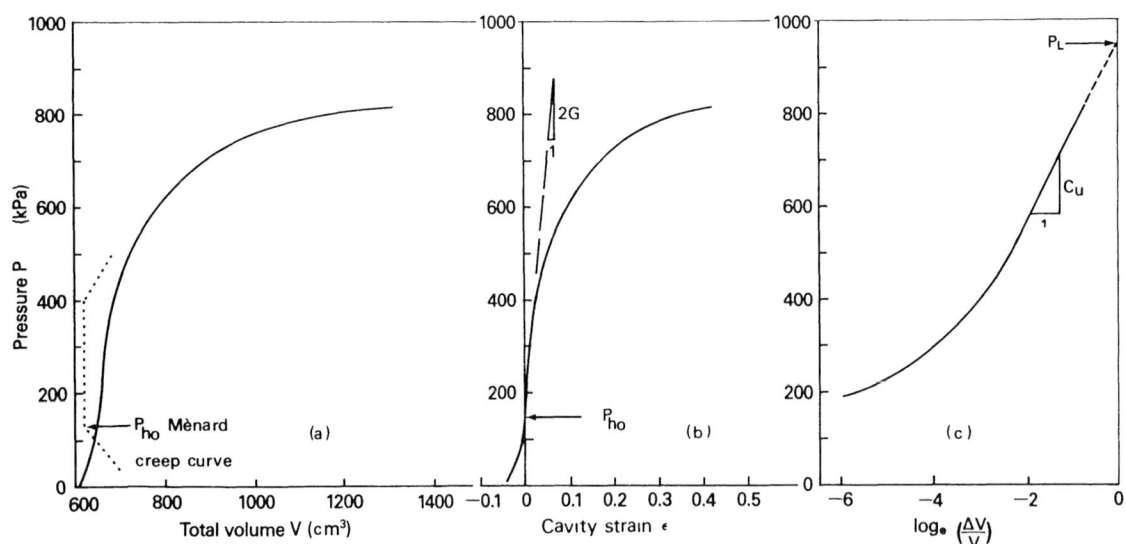

Fig. 3 Typical pressure *vs* volume curve and deduced curves

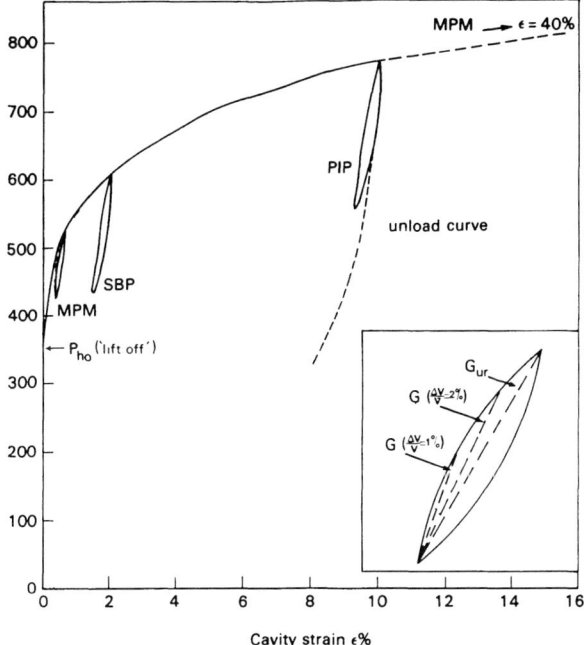

Fig. 4 Position of reload cycles for different pressuremeters for the determination of shear moduli

by Palmer (1972) which could yield full stress–strain curves from the pressuremeter test using the equation

$$\tau = \frac{dp}{d \log_e (\Delta V/V)} \qquad (1)$$

$$\left(\text{slope of the } p \text{ versus } \log_e \frac{\Delta V}{V} \text{ curve}\right)$$

where τ = shear stress; p = current pressure; V = current volume at pressure P; V_0 = volume at reference pressure P_{ho}; $\Delta V = V - V_0$; and $c_u = \tau$ (max).

For the strain-hardening soils reported in this paper, the p versus $\log_e(\Delta V/V)$ curve becomes essentially linear in the later stages and the slope of this part of the curve has been taken for c_u (see Fig. 3c). In later sections of this paper the shear strength from equation (1) will be referred to as the 'Palmer' value.

Limit pressure

Gibson and Anderson (1961) analysed the ᴊse of a pressuremeter test in an elas-

tic–perfectly plastic soil. A limit pressure

$$P_L = P_{ho} + c_u[1 + \log_e(G/c_u)] \qquad (2a)$$

where P_{ho} = *in situ* horizontal stress, G = initial shear modulus, and c_u = shear strength is theoretically reached when the pressuremeter cavity expands indefinitely, i.e. when $\Delta V/V = 1$ (see Fig. 3c). Equation (2a) can be rewritten as

$$c_u = \frac{P_L - P_{ho}}{1 + \log_e (G/c_u)} \qquad (2b)$$

$$\text{or} \quad c_u = \frac{P_L - P_{ho}}{N_p} \qquad (2c)$$

and $N_p = 1 + \log_e(G/c_u)$

where N_p = pressuremeter constant (taken as 6.18 corresponding to a bearing capacity factor $N_c = 9.25$). Equation (2c) yields the 'limit-pressure' value c_u.

Shear Modulus (G)

Values of secant shear moduli have been obtained using the expressions

$$G = \tfrac{1}{2} \frac{\Delta p}{\Delta \epsilon} \qquad \text{or} \qquad G = V_0 \frac{\Delta p}{\Delta V}$$

for initial loading curves (when applicable)

and $G_{ur} = \tfrac{1}{2}(1 + \epsilon) \dfrac{\Delta p}{\Delta \epsilon}$ or $G_{ur} = V \dfrac{\Delta p}{\Delta V}$

for unload–reload loops (see Figs 3c and 4) where ϵ = current cavity strain

$$= \frac{\text{change in radius of borehole due to pressure } (P - P_{ho})}{\text{radius of borehole at reference state } (p = P_{ho})}$$

$\Delta \epsilon$ = change in ϵ due to pressure increase Δp.

TEST RESULTS

In situ stresses measured or calculated for the site are shown in Fig. 5 a and b. The estimates of *in situ* horizontal stress in Fig. 5b were made using results from oedometer

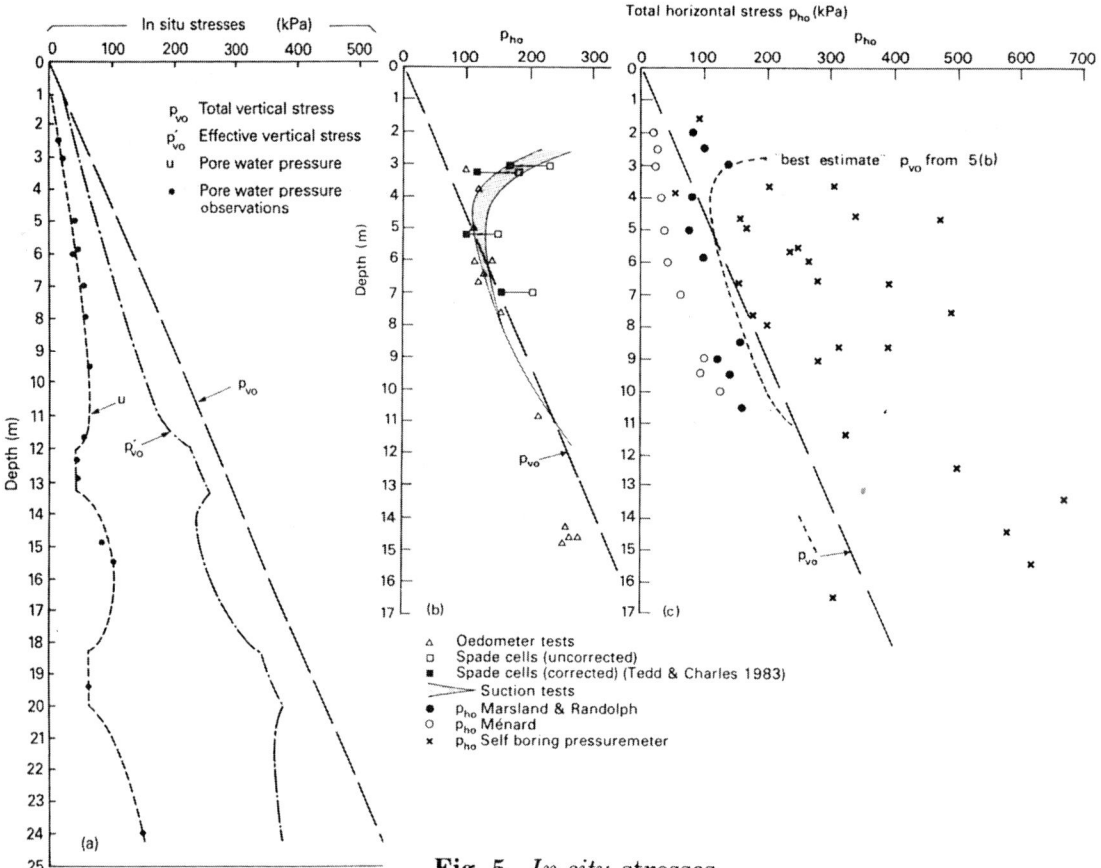

Fig. 5 *In situ* stresses

tests (as in Schmidt, 1966), spade cells (as in Tedd and Charles, 1983), and suction measurements on pushed samples. The results show reasonable agreement and their mean has been considered as a 'best estimate' of P_{ho} for comparison with the pressuremeter results.

Estimates of P_{ho} from the MPM and SBP tests can be seen in Fig. 5c. The 'original' Ménard approach gave values significantly lower than the 'best estimate', whereas the Marsland and Randolph method gave values closer to the 'best estimate' but in general still lower. The application of the Marsland and Randolph method to tests in strain-hardening soils, such as those in this paper, is difficult. The difficulty is due to the fact that the stress–strain relationship in these soils shows greater curvature at relatively low strains than in London Clay, for which the method was developed. If great care is

not taken, this problem can also allow the engineer to weight the interpretation towards a desired answer, if this is known.

The values of P_{ho} assessed from the SBP using the 'lift-off' method, and shown in Fig. 5c, are seen to scatter quite wildly. In general they considerably overestimate the best estimated values. This may indicate the difficulties of the self-boring technique in these soils and the possibility of significant over-stressing during installation (in the extreme tending towards the PIP test). It should be noted that in some cases it was not easy to decide when significant movement of the membrane first occurred on the basis of the average curve of three displacement arms.

However, this average curve in general showed movement when the first arm moved, and the assessments therefore had a tendency to a lower bound estimate.

Typical corrected curves of pressure (p)

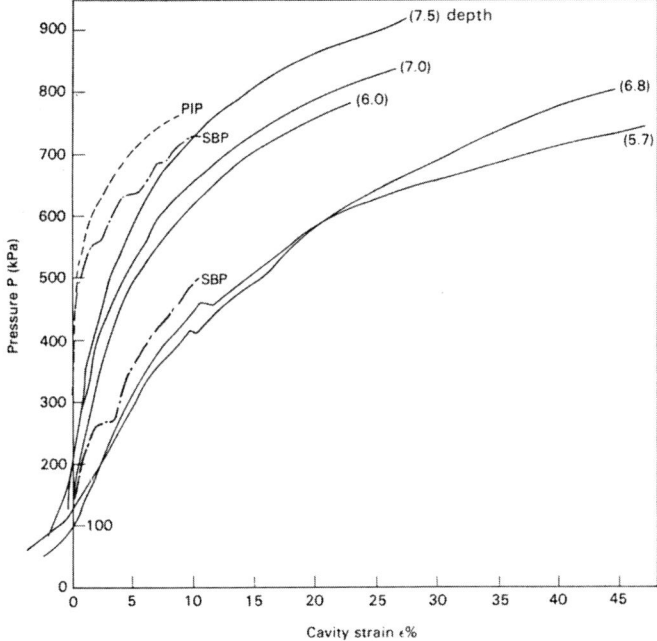

Fig. 6a Pressure *vs* cavity strain for Ménard pressuremeter

against cavity strain (ϵ) are given for MPM, SBP and PIP tests, over the depth range 6–8 m, in Fig. 6 (unloaded–reload loops have been omitted for clarity). Scatter is evident in all the plots, although the SBP tests show the greatest variation. It should be noted that the SBP curves show shapes ranging from the flatter MPM results through to the steeper, over-stressed, PIP tests (see Fig.6a).

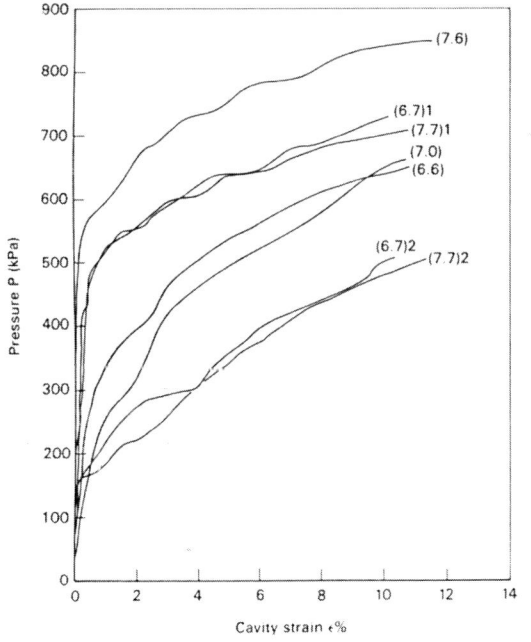

Fig. 6b Pressure *vs* cavity strain for self-boring pressuremeter

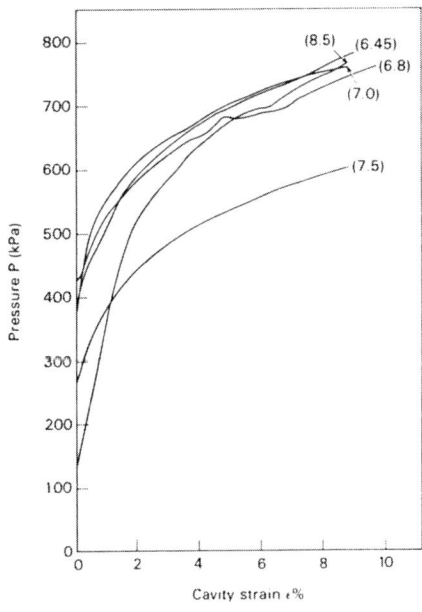

Fig. 6c Pressure *vs* cavity strain for Push-in pressuremeter

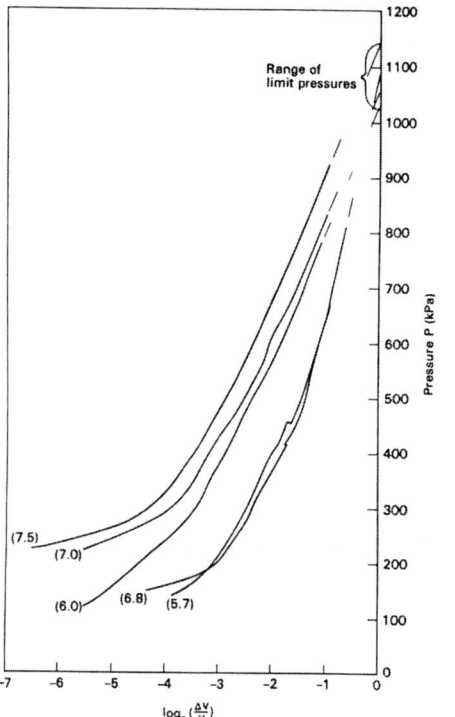

Fig. 7a   Pressure *vs* $\log_e$ volumetric strain for Ménard pressuremeter

In Fig. 7, the data from Fig. 6 are replotted as pressure against $\log_e(\Delta V/V)$. The MPM tests in Fig. 7a give curves with a marked linear portion from which it is quite easy to both assess a value of shear strength from equation (1) (Palmer value), and to extrapolate to a limit pressure. In Fig. 7b, the SBP curves are not quite so smooth as those of the MPM, and often have a less clearly defined linear portion. Additionally, some of the curves exhibit marked changes towards the end of the test. Consequently, assessment of Palmer values of shear strength is less clear and extrapolation to a limit pressure is more open to personal judgement (evidenced by the greater scatter of P_L in Fig. 7b). Results from the PIP tests in Fig. 7c are more consistent than the SBP results, but still show more variation than the MPM results with some difficulties in both assessing Palmer shear strengths and extrapolating to limit pressures.

Shear strengths measured by tests on 865 mm diameter plates in the base of 900 mm diameter boreholes are shown in Fig. 8a (see

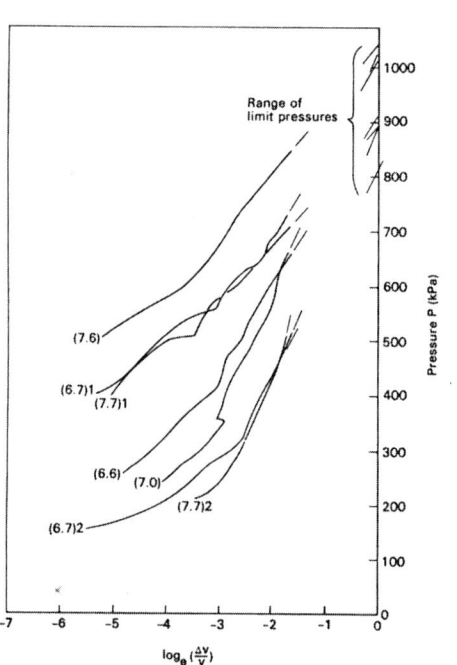

Fig. 7b   Pressure *vs* $\log_e$ volumetric strain for self-boring pressuremeter

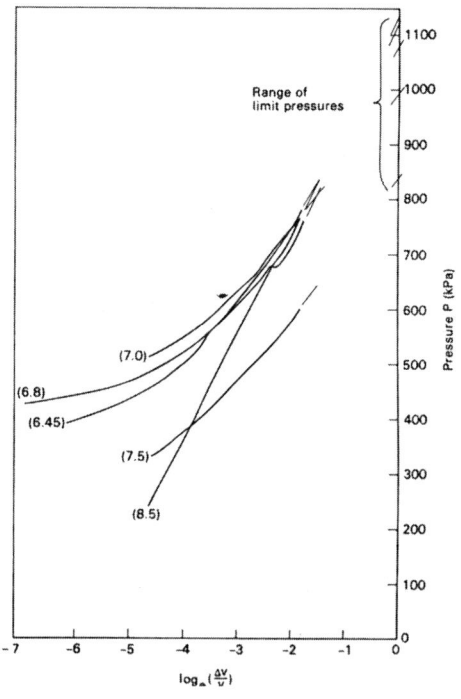

Fig. 7c   Pressure *vs* $\log_e$ volumetric strain for push-in pressuremeter

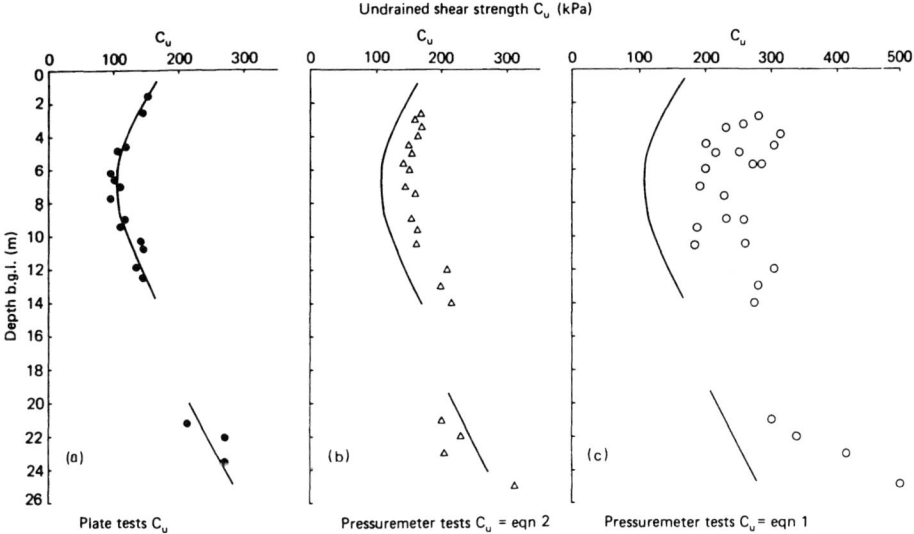

Fig. 8 Comparison of undrained shear strength from pressuremeter and plate tests

Marsland and Powell, 1979, for detailed interpretation). A mean line drawn through this data has been used to define a typical shear strength profile and has been used in subsequent figures as a marker.

It is not suggested that the plate test values of shear strength are 'the values', as it must be remembered that shear strength has no unique value and is test dependent. Possible reasons for differences between the plate and pressuremeter values of shear strength are discussed elsewhere (Powell *et al.*, 1983).

When the results for all the pressuremeter tests down the profile are plotted in terms of shear strength, assessed using equations (1) and (2c), the preceding comments on Fig. 7 are again relevant (Figs 8, 9 and 10). Figs 8 b and c show that for MPM tests much more consistent results are obtained from the limit pressure shear strength than from Palmer values. Factors affecting the Palmer interpretation are discussed elsewhere (Bagelin *et al.*, 1978; Powell *et al.*, 1983), and it is sufficient here to say that the method is particularly sensitive to assumptions about initial *in situ* stresses, and to the degree of disturbance around the borehole (the limit pressure is relatively insensitive to these factors).

Figures 9 and 10 show plots corresponding to those in Fig. 8 for the SBP and PIP results. The limit-pressure approach again gives the more consistent results. However, the scatter of shear strengths from limit pressures is now greater, partly as a result of the difficulty of extrapolating from the smaller final volume expansions. There is also some reduced scatter in the Palmer values, especially for the PIP tests. This is felt to be due either to the reduced or more 'consistent' disturbance that possibly occurs

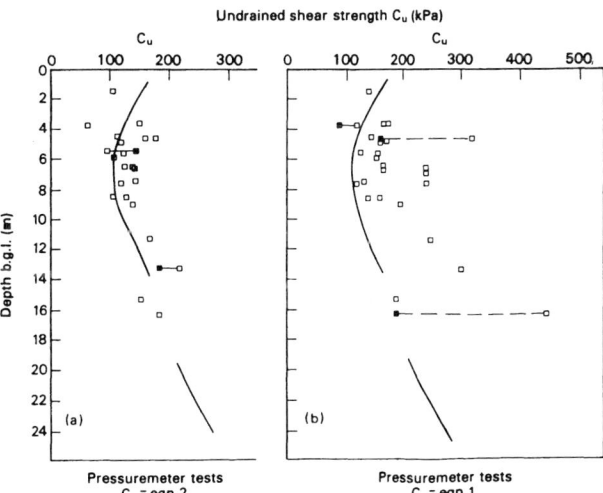

Fig. 9 Comparison of undrained shear strength from self-boring pressuremeter

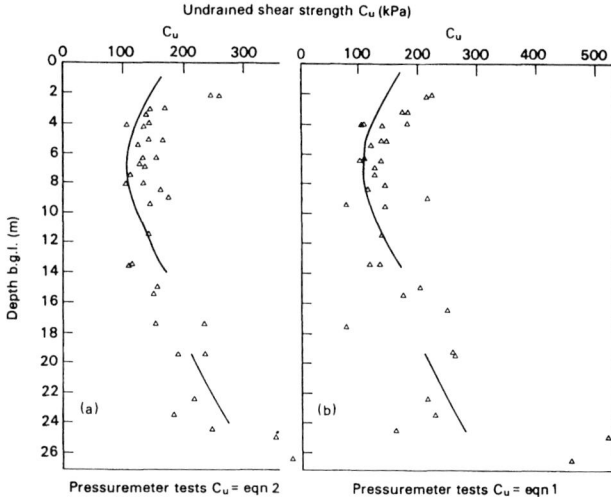

Fig. 10 Comparison of undrained shear strength from push-in pressuremeter and plate tests

with the SBP and PIP tests as compared to the MPM tests.

Shear moduli from all the pressuremeter tests are presented in Fig. 11. The MPM results for initial and reload moduli were calculated over a stress range of P_{ho} to $P_{ho} + c_u$. The SBP and PIP moduli were calculated over their full unload–reload loop, irrespective of stress or strain range. A similar basic pattern is evident with depth, the MPM initial values forming a lower bound with MPM reloads as an upper bound and PIP and SBP scattering in between. It has been shown (Reid, 1982) that in general the values of reload shear moduli are unaffected by the test stage at which the reload loop is performed, provided the loops are over similar stress or strain ranges. However, the moduli values are affected by variations in the strain range considered (e.g. Fig. 4). For these reasons it would seem acceptable to compare reload moduli values from the three types of pressuremeters, irrespective of location within a test (Fig. 4), provided similar strain ranges are used. It should be noted also that in all tests the reload loops were kept within a pressure range of $2c_u$, as suggested by Wroth (1982).

Secant reload shear moduli for 1, 2 or 3% volumetric strain ranges are given in Fig.

12. A certain degree of agreement is evident between the different devices. The SBP results, however, show the greatest scatter. The MPM tests had only a few points on their reload loops and so any hysteresis effect was less clearly defined, giving little difference between the 1 and 2% values. However, Fig. 12 seems to indicate that some of the differences shown in Fig. 11 can be explained when shear moduli are compared over similar ranges. It would be interesting to see if even better agreement could be obtained if all reload loops were done towards the end of the loading stage of the test, especially in the MPM and SBP tests when initial disturbance might have less effect.

In Fig. 13 reload shear moduli from the pressuremeter tests (the PIP range from Fig. 12) are compared with values measured from unload–reload cycles on 865 mm diameter plates (Marsland and Powell, 1985). The values from the plate tests are in general higher than those from the pressuremeter. However, if the average values for shear strains under a plate test (Simpson *et al.*, 1979) are considered, then the plate test reload loops are over about 0.5–0.75% shear strain. In Fig. 13 additional values of shear moduli for PIP and SBP tests have been plotted for 0.5% volumetric strain (which is approximately 0.5% shear strain at small strains). Considering the limitations of the above approach, the agreement with plate test values is somewhat surprising and interesting; it must be stressed that this is a very tentative comparison. Recent work (Jardine *et al.*, 1985) has shown remarkable agreement in shear moduli values from laboratory tests using internal strain measurements on triaxial samples, and from plate tests when comparisons are made over similar strain ranges.

Data from the SBP tests is replotted in Fig. 14 as the ratio of Palmer c_u to 'limit-pressure' c_u against the 'overstress' ('lift-off' pressure minus 'best estimate' of P_{ho}). A marked trend is evident with the results tending to those deduced from PIP tests at higher 'overstresses'.

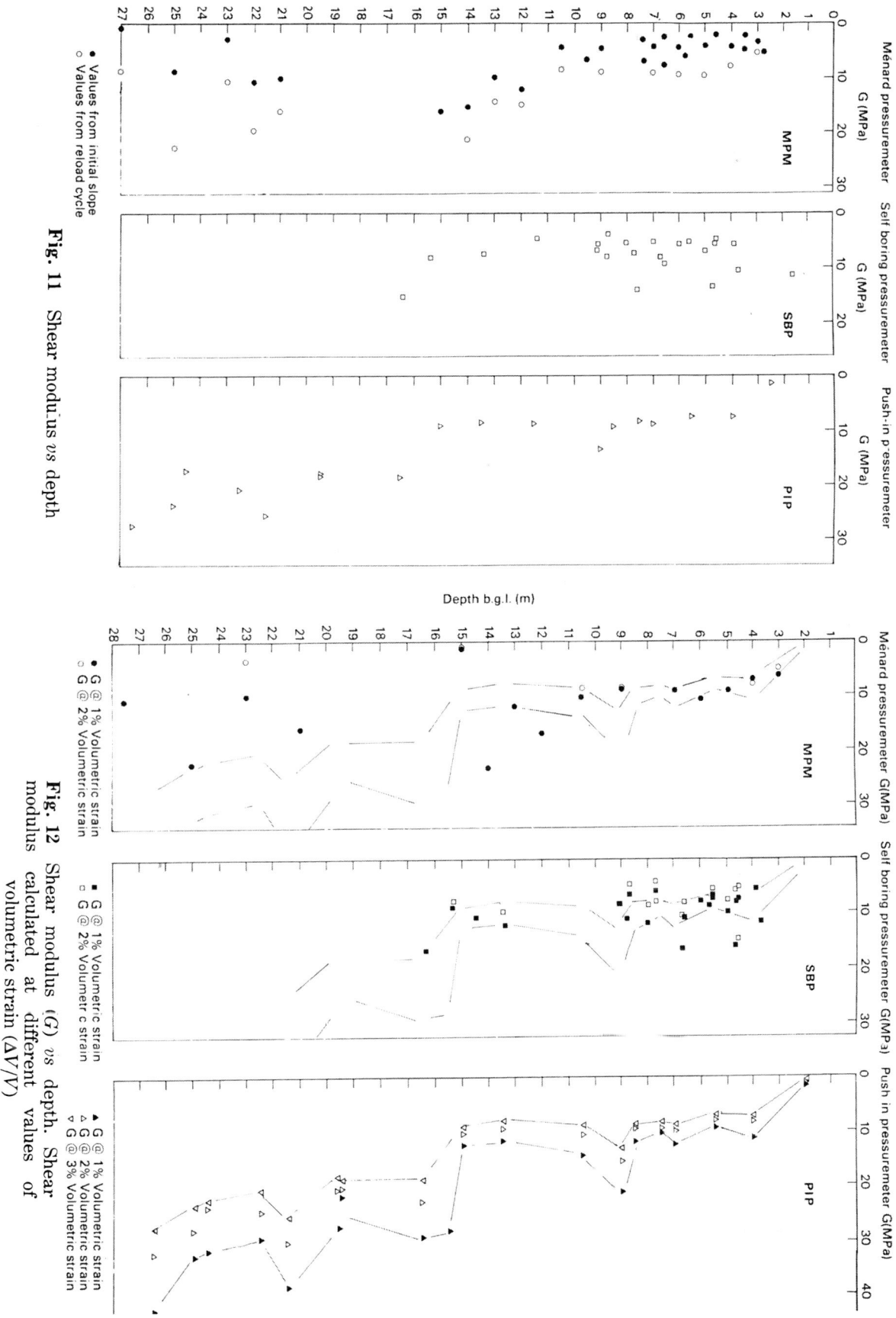

Fig. 11 Shear modulus *vs* depth

- Values from initial slope
○ Values from reload cycle

MPM

SBP

PIP

Ménard pressuremeter

Self boring pressuremeter

Push-in pressuremeter

G (MPa)

Depth b.g.l. (m)

Fig. 12 Shear modulus (*G*) *vs* depth. Shear modulus calculated at different values of volumetric strain (ΔV/V)

- G @ 1% Volumetric strain
○ G @ 2% Volumetric strain

■ G @ 1% Volumetric strain
□ G @ 2% Volumetric strain

▲ G @ 1% Volumetric strain
△ G @ 2% Volumetric strain
▽ G @ 3% Volumetric strain

MPM

SBP

PIP

Ménard pressuremeter G(MPa)

Self boring pressuremeter G(MPa)

Push in pressuremeter G(MPa)

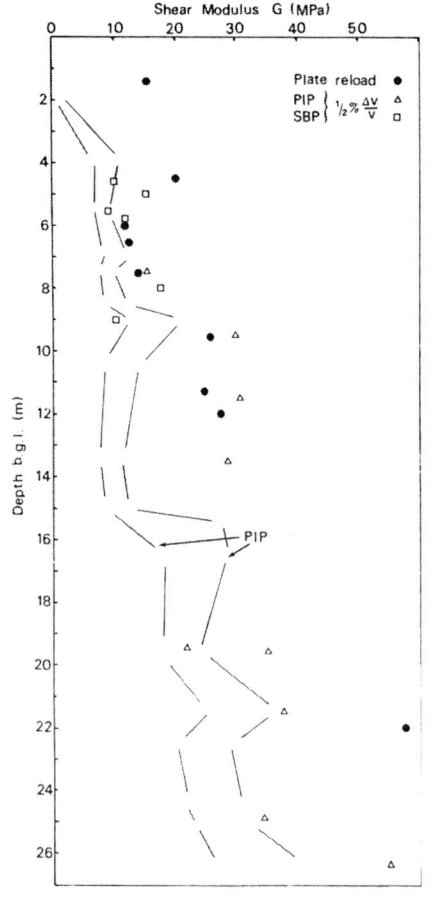

Fig. 13 Shear moduli from plate and pressuremeter tests (reloads)

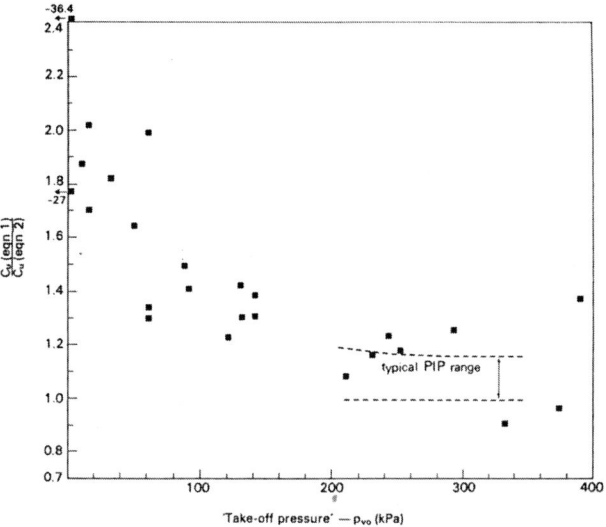

Fig. 14 Overstress *vs* shear strength ratio

At low values of overstress, it becomes possible to estimate initial shear moduli values in some SBP tests. In Fig. 15, values of shear strength using equations (1) and (2c) for the SBP test are plotted against initial shear moduli along with MPM data from Powell *et al.* (1983). It will be seen that there is close agreement between the MPM

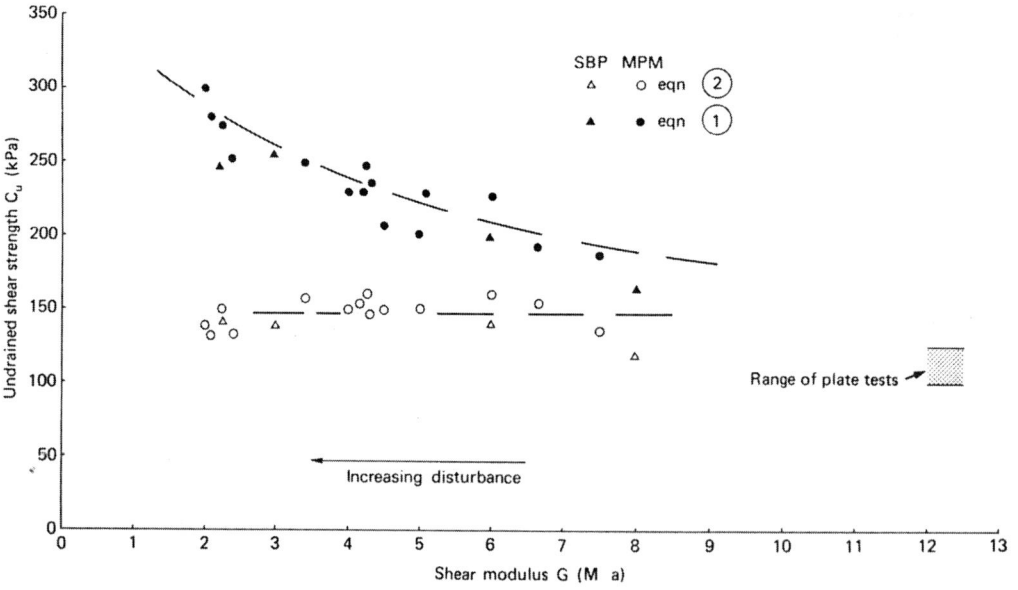

Fig. 15 Variation of shear strength with shear modulus (depth 5–8 m)

and SBP tests, which both show the susceptibility of the Palmer approach (equation 1) to the value of initial shear modulus which itself is an indication of the degree of disturbance.

A similarity between Figs 14 and 15 is evident. PIP tests with their marked overstress or 'pseudo' high initial shear modulus would fit into both patterns plotted to the right. They further highlight the susceptibility of the Palmer type of analysis to test quality.

CONCLUSIONS

Pressuremeter tests have been performed using three different types of pressuremeter which were installed in the ground in markedly different ways. The test-bed site used was a glacial clay till, which over the upper 15 m was a relatively insensitive, well graded, strain-hardening deposit with K_0 values close to 1.

Rates of testing were similar in terms of total test time for all three devices, so minimizing the problems of rate effect which otherwise could have confused the comparisons (Anderson, 1979; Powell et al., 1983).

Estimating the in situ horizontal stress (P_{ho}) from MPM and SBP test data was far from easy. While the Marsland and Randolph approach for MPM tests gave values tending to the 'best estimated', the procedure could become operator sensitive owing to the significant curvature of the stress–strain curves. Estimates of P_{ho} from the SBP tests using the lift-off method tended to overstimate in situ P_{ho}, although the lower bound values came close to the in situ values.

The test results have shown that, irrespective of device, method of installation or testing procedure, all three pressuremeters can yield results which tend towards the same ultimate or 'limit' pressure. The smaller maximum volume expansions of the SBP and PIP devices as compared to the MPM led to slightly greater scatter in values of limit pressure. The consistency of limit pressure values meant that shear strengths deduced from them, using a pressuremeter factor N_p, were also consistent. In using this limit-pressure approach for shear strength, it will be necessary to consider the assessment of N_p and what G/c_u ratio is relevant for a particular deposit (also at what strain level the shear modulus G should be assessed).

All pressuremeter types led to values of shear strength from the Palmer approach which were sensitive to disturbance and showed considerably more scatter than the limit-pressure approach.

Shear strengths were in general higher than those obtained from plate tests, which were similar to those assessed from laboratory tests (see Marsland and Powell, 1985, for discussion of laboratory testing) even though K_0 is close to 1. This behaviour has been found elsewhere (Ghionna et al., 1983; Wroth, 1984). Reload shear moduli from the three devices were in reasonable agreement when compared over similar strain ranges. In general, only the MPM tests could be used to give initial shear moduli and these were significantly lower than the reload values. A tentative comparison with plate test shear moduli showed the possibility of better agreements.

The difficulty of self-boring a pressuremeter into a stony deposit of this type is almost certainly the reason for the scatter of results for P_{ho} and shear modulus with the SBP test. Behaviour ranging from that similar to the MPM test to that similar to the PIP test appears to have been possible with the SBP test. The PIP test appears to produce repeatable results which may be the result of more repeatable installation; however, estimation of P_{ho} was impossible. MPM tests were susceptible to the variable disturbance possible during installation.

Finally, it should be stated that these findings will be very much soil-type dependent. Similar comparisons are needed in other deposits and work is currently in hand at both BRE and the Norwegian Geotechnical Institute (Aas, 1984) to extend the range of soil types.

ACKNOWLEDGEMENTS

The work in this paper forms part of the research programme of the Building Research Establishment and is published by permission of the Director. The authors wish to thank their many colleagues who have contributed to this work over a number of years. The SBP tests were performed for the BRE by P.M. *in situ*; some of the PIP test results came from Stressprobe Ltd (Fyffe *et al.*, 1982).

REFERENCES

1. Aas, G., Lacasse, S., Lunne, T. and Madshus, C. 1984. *In situ* testing: New developments. NGI Report No. 52155–31.
2. Anderson, W.F. 1979. Discussion 'Design parameters for stiff clay'. In *Proc. 7th European Conf. on SMFE*, Brighton, Vol. 4, pp. 149–153.
3. Baguelin, F., Jezequel, J. F. and Le Mehaute, A. 1974. Self-boring placement method of soil characteristics measurement. In *Proc. of Speciality Conf. on Subsurface Exploration for Underground Excavation and Heavy Construction*, Henniker, New Hampshire, pp. 312–332.
4. Baguelin, F., Jezequel, J. F. and Shields, D. H. 1978. *The Pressuremeter and Foundation Engineering*. Trans Tech. Publications, p. 617.
5. Baguelin, F., Jezequel, J. F., Lemee, E. and Le Mehaute, A. 1972. Expansion of cylindrical probes in cohesive soils. *J. Soil Mech. Found. Div. Proc.* ASCE No. 98, SM11, pp. 1129–1142.
6. Catt, J. A. and Penny, L. F. 1966. The Pleistocene deposits of Holderness, East Yorkshire. *Proc. Yorks Geol. Soc.* 35, 375–420.
7. Fyffe, S., Reid, W. M. and St John, H. D. 1982. The use of the Push-in Pressuremeter in offshore site investigation. In *Proc. Oceanology International Conference*, Brighton.
8. Ghionna, V., Jamiokowski, M. Lacasse, S., Lancellota, R. and Lunne, T. 1983. Evaluation of the self-boring pressuremeter. In *Proc. Int. Symp. on In Situ Testing*, Paris, Vol. 2, pp. 294–301.
9. Ghionna, V. N., Jamiolkowski, M. and Lancellotta, R. 1982. Characteristics of saturated clays as obtained from SBP tests. *Symposium on the Pressuremeter and its Marine Applications*, Paris, pp. 165–186.
10. Gibson, R. E. and Anderson, W. F. 1961. *In situ* measurement of soil properties with the pressuremeter. *Civ. Engng Publ. Wk Rev.* **55** (55).
11. Henderson, G., Smith, P. D. K. and St John, H. D. 1979. The development of the Push-in Pressuremeter for offshore site investigation. In *Offshore Site Investigation*, Proceedings of SUT Conference, London, pp. 159–168.
12. Hughes, J. M. O. 1973. An instrument for *in situ* measurement in soft clays. PhD thesis, University of Cambridge.
13. Jardine, R. J., Maswoswe, J., Fourie, A. and Burland, J. B. 1985. Field and laboratory measurements of soil stiffness. Submitted for *XI ICSMFE*, San Francisco, 1985.
14. Lacasse, S. and Lunne, T. 1982. *In situ* horizontal stress from pressuremeter tests. *Symposium on the Pressuremeter and its Marine Applications*, Paris, pp. 187–208.
15. Ladanyi, B. 1972. *In situ* determination of undrained stress–strain behaviour of sensitive clays with the pressuremeter. *Can. Geotech. J.* 9(3), 313–319.
16. Mair, R. J. and Wood, D. M. 1985. A review of the use of pressuremeters for *in situ* testing. CIRIA Project 335.
17. Marsland, A. 1979. The interpretation of *in situ* tests in glacial clays. In *Offshore Site Investigation*. Proceedings of SUT Conference, London, pp. 217–228.
18. Marsland, A. and Powell, J. J. M. 1979. Evaluating the large-scale properties of glacial clays for foundation design. In *Proc. 2nd Conf. on Behaviour of Offshore Structures*, Vol. 1, pp. 193–214.
19. Marsland, A. and Powell, J. J. M. 1985. Field and laboratory investigations of clay tills at the BRE test-bed site at Cowden Holderness. In *Proc. Int. Conf. on Construction in Glacial Tills and Boulder Clays*, Edinburgh, 12–14 March 1985.
20. Marsland, A. and Randolph, M. F. 1977. Comparisons of the results from pressuremeter tests and large *in situ* plate tests in London clay. *Geotechnique* 27(2).
21. Palmer, A. C. 1972. Undrained plane-strain expansion of a cylindrical cavity in clay: a

simple interpretation for the pressuremeter test. *Geotechnique* **22**(3), 451–457.

22. Powell, J. J. M., Marsland, A. and Al Khafaji, A. N. 1983. Pressuremeter testing of glacial clay tills. In *Proc. Int. Symposium on In Situ Testing*, Paris, Vol. 2, pp. 373–378.

23. Reid, W. M., Fyffe, S., St John, H. D. and Rigden, W. J. 1982. The 'Push-in pressuremeter'. In *Proc. Symp. on the Pressuremeter and its Marine Applications*, Paris, pp. 247–262.

24. Schmidt, B. 1966. Discussion of 'Earth pressure at rest related to stress history'. *Can. Geo. J.* **3**(4), 239 242.

25. Simpson, B., Calabresi, G., Sommer, H. and Wallays, M. 1979. Design parameters for stiff clays. In *Proc. 7th European Conf. on SMFE*, Brighton, Vol. 5, pp. 91–126.

26. Tedd, P. and Charles, J. A. 1983. Evaluation of push-in pressure cell results in stiff clay. *Int. Symp. on Soil and Rock Investigations by In Situ Testing*, Paris, Vol. 2, pp. 579–584.

27. Windle, D. and Wroth, C. P. 1977. The use of a self-boring pressuremeter to determine the undrained properties of clays. *Ground Engng* **10**(6), 37–46.

28. Wroth, C. P. 1982. British experience with the self boring pressuremeter. In *Symposium on the Pressuremeter and its Marine Applications*, Paris, pp. 143–164.

29. Wroth, C. P. 1984. The twenty-fourth Rankine Lecture: the interpretation of *in situ* soil tests. *Geotechnique* **34**(4), 474–492.

14

Comparison of In Situ and Ship-Board Vane Measurements on a Deep-Sea Clay

Armand J. Silva, University of Rhode Island, Kingston, Rhode Island

NTRODUCTION

Analysis of most marine geotechnical engineering problems requires information on the shear strength of the natural sediment. One approach is to test samples from borings or cores using conventional geotechnical procedures, such as miniature vane, triaxial compression or direct shear tests. However, the disturbance caused by sampling, pressure release, handling, transportation, etc. can be severe, or at least undetermined, and hence it is often difficult to assess the *in situ* properties from the results. The problems are probably greater in deep water where sampling equipment is usually tethered on a long cable and the dynamics of coring are difficult to control. In addition, the volume expansion on bringing the sample to an atmospheric condition is 2 to 3% in the deep ocean basins and this would presumably cause changes in effective stresses and microstructural alterations. The expansion of gases greatly increases this problem and in many cases can render samples essentially useless for strength determinations. For these reasons, there has been a growing interest in conducting more *in situ* geotechnical tests. The vane shear method is only one of several which can be used to determine the *in situ* strength of an ocean sediment. Some others that are being used are the cone penetrometer, the pressuremeter and even the standard penetration test. There are instances where a particular method or device may not be well suited to the situation. For example, the vane shear method is intended for use with fine-grained (cohesive) material and it cannot be expected to give meaningful results for anything coarser than a very fine sand or silt. Yet the vane method is used almost routinely in testing oceanographic core samples without regard for the texture of the material. Conversely, the cone penetrometer method is probably not as reliable with clay as it is with sand.

One of the advantages of the vane method is that it is a fairly direct measurement of strength, since the sediment is sheared on a well-defined surface, and, with reasonable assumptions regarding stress distribution, it is a simple matter to obtain a relationship between the torque required to produce failure and the shear strength. If torque versus rotation angle is monitored through the full test range, it is possible to estimate the shear modulus and to determine residual strength. The remoulded strength and sensitivity can also be determined if another test is conducted at the same site, after full rotation to assure complete remoulding.

The main purpose of this paper is to present and compare some data on *in situ* and laboratory vane measurements made on a homogeneous illitic deep-sea clay in the north central Pacific. The *in situ* system used in this study was designed specifically for use in deep water (6000 m) or on a bottom-supported platform. In the present configuration it can take a series of approximately 25 separate measurements within a depth of about 1.5 m to produce a profile of strength versus depth. For the first deployment, in 5800 m water depth, the test sequence was reduced to 10 measurements. Two large-diameter piston cores were processed on board the ship within 10 hours from recovery. In addition to physical property data and subsampling, shear-strength measurements were made with a laboratory miniature vane device. This ocean test provided an opportunity to compare the *in situ* measurements, conducted under carefully controlled conditions, with those on very good quality core samples.

BACKGROUND ON VANE SHEAR METHOD

There are several methods and instruments to measure *in situ* strength and deformability of marine sediments, including the vane, cone, and pressuremeter. Depending on the water depth and sediment type, *in situ* tests can be performed from a sea-floor vehicle, from a sea-bottom platform, through the drill string, from a submarine, by a diver, by the impact penetration method or by vibro-

driving. Briaud (1980) gives a detailed review of available methods and instruments.

In this paper, a short review of vane shear testing systems that are presently available will be presented. It is anticipated that references will be made to the work of the following authors: Aas (1965), Briad (1980), Doyle *et al.* (1971), Flaate, (1966), Lee (1979), Monney (1973), Perlow and Richards (1972), Richards *et al.* (1972), ASCE (1975).

DEVELOPMENTAL HISTORY OF URI/ISV SYSTEM

General Requirements

A more detailed description of the development history can be found in recent papers (Babb, 1982; Babb and Silva, 1983; Silva *et al.*, in press). The motivation for the *in situ* vane (ISV) system described here came from the US Subseabed Disposal Program (SDP) which is a study to determine the feasibility of burying solidified high-level nuclear wastes within certain geologically stable deep-sea sediments (Hollister *et al.*, 1981). A similar but smaller system had been designed for another project (Silva and Pekin, 1981) and the experience gained was applied in the development of the ISV system. The first long-term *in situ* experiment of the SDP will be the *In Situ* Heat Transfer Experiment (ISHTE) to determine the thermal, geochemical, and geotechnical response to a heat source in the sediment. As currently planned, this experiment will be carried out over a one-year period in 5800 m water depth in the north central Pacific ocean. Various components (thermal sensors, piezometers, pore-water sampler, etc.) will be mounted on a tubular structure approximately 4.3 m long and 2.3 m high and a 400 W isotopic heat source will be implanted into the sediment to a depth of 1 m below the sea floor. The heater and vane will be situated in such a way as to be outside the zones of significant stress influence from the support pads. The vane shear measurements will be made at the end of the one-year experiment with the vane passing within 20 mm of the heater. Comparisons of *in situ* measured responses with the predictions of numerical models for thermal, mechanical, and chemical behaviour will be used to evaluate the applicability of the techniques being developed in the SDP (Percival, 1983).

The present configuration of the ISV system has in part been dictated by the needs of the ISHTE, but the same basic design can be used for other geotechnical applications. Based on preliminary studies, the basic design requirements for the ISHTE version of the ISV system were defined as follows:

Water depth (pressure)	6000 m (600 bar)
Sediment strength	0.5–70 kPa
Profile depth	1.5 m
Time on bottom	1 year
Sediment temperature	1.5–300°C

It was decided to make the ISV system autonomous, with its own power, controller and data acquisition system, so that it could be used in other sea-floor studies. For the ISHTE application, the system will be activated by the ISHTE platform master controller and a subset of the vane shear data will be sent to this controller for eventual telemetry on a surface ship through an acoustic link (Backes *et al.*, 1981). In the present configuration, the ISV system can be mounted on a bottom-supported platform or used from a submersible.

Large-scale laboratory Experiment and ISV-Model A

Because of the complexity of ISHTE and the fact that most of the instrumentation is entirely new, a short-term (30 days) scale model (0.287 scale) laboratory experiment was conducted in the 3 m diameter pressure vessel at the Naval Ship Research and Development Center, Annapolis, MD. The test bed for this ISHTE Simulation Experiment consisted of a 1 m diameter by 1 m deep tank of saturated, reconstituted, reconsolidated north Pacific illite (Babb and

Silva, 1983). A tank of seawater was mounted above the test bed and the entire apparatus was pressurized to 550 bars and maintained at a temperature of 4°C for a period of 30 days.

The ISV configuration was modified for use in this simulation experiment. Because of space limitations, this prototype (Model A) was shortened to a 0.6 m penetration but was designed for operation in the physical and geotechnical environment of the actual seabed experiment. Therefore the full-scale ISV system (Model B) is very similar in concept. The Model A ISV system consists of three major integrated packages:

- vane shear probe mechanical system (with internal volume/pressure compensator);
- electronic control and data acquisition system;
- power supply.

The mechanical system is contained in a pressure-compensated housing with pressure equalization accomplished by two compensator tubes equipped with 'floating' pistons. One end of each tube remains exposed to the seawater, allowing pressure changes to displace a teflon piston and compress the mineral oil compensating fluid in order to equalize internal and external pressures.

For the simulation experiment, power was supplied from a source external to the pressure vessel. The Model A system was used successfully during the 30-day simulation experiment while at a pressure of 550 bar, and three strength profiles were obtained:

(1) unheated profile prior to sediment heating and pressurization;
(2) heated profile near the heater (within 20 mm) after 30 days' heating under 550 bar pressure;
(3) unheated profile (within 70 mm of heater) after sediment cool-down and depressurization.

Comparison of data from the pretest unheated and heated profiles indicates that a ten- to twelve-fold increase in shear strength occurred at the midplane depth of the heater in the temperature zone of approximately 200°C. This strength increase is greater than the six- to ten-fold increase predicted from earlier laboratory experiments (Hadley *et al.*, 1980).

ISV, Model B, Used in Deep Ocean Test

The *in situ* vane shear device (ISV, Model B) that will be used on the ISHTE platform consists of four integrated packages:

- mechanical system;
- electronic controller and data acquisition system;
- volume compensator;
- power supply.

Both the mechanical system and volume compensator are oil filled and designed to operate at deep-ocean pressure. The power supply and controller operate at atmos-

pheric pressure and are thus contained in high-pressure housings on the platform. As in Model A, cabling between these packages is oil filled to provide volume compensation and special marine connectors are used. The controller is programmed with fixed operation sequences for the vane with the appropriate sequence selected by commands transmitted from the ISHTE master controller. In this section, some details are presented on the mechanical system, followed by a discussion of the electronic controller and information on the power supply.

Experience gained from using the *in situ* vane device developed for the simulation experiment (Model A) was employed in designing the device needed for the ISHTE. For example, it was determined that purchased internal components used in Model A were capable of operating at ambient deep-ocean temperature and pressure after a dormant period of one month. It was con-

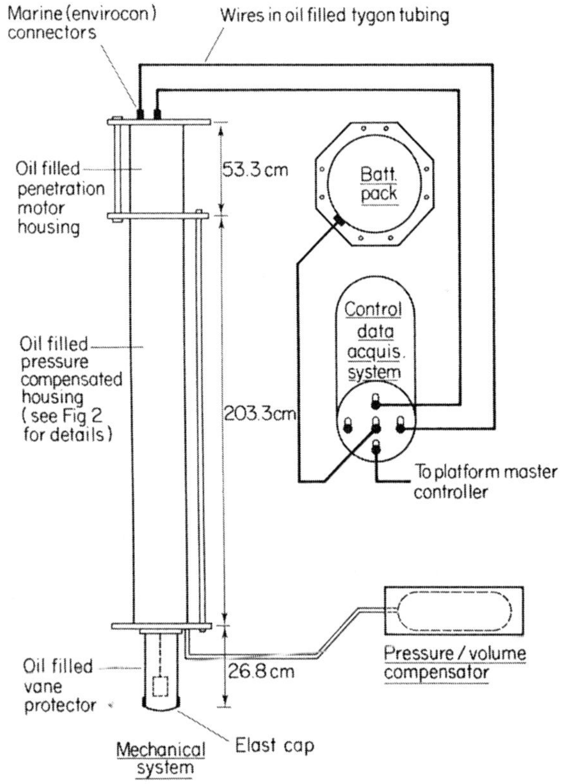

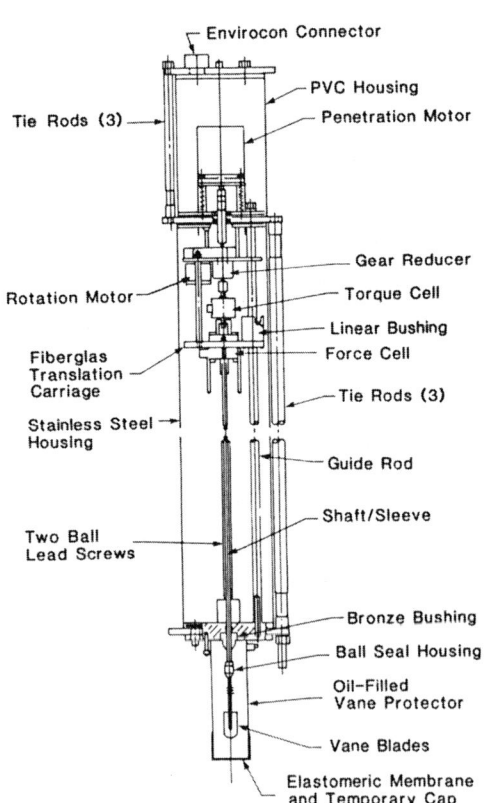

Fig. 1 Schematic diagram of *in situ* vane system

Fig. 2 Assembly drawing of ISV Model B, penetration depth of 1.5 m

cluded that these same components could be used for the ISHTE vane, but some significant design modifications and refinements were incorporated in Model B. The new arrangement is shown schematically in Fig. 1. Figure 2 shows some details of the mechanical unit and Fig. 3 is a photo of the entire system mounted on a platform that was used during deep-water tests in the autumn of 1984. Non-metallic components are incorporated in less critical areas to reduce the weight, and the volume-compensating bladder is mounted externally. Another new feature of the design is a self-compensating vane protector housing with an elastomeric cap to isolate the vane and seals from seawater until just before penetration.

The heart of the electronic system is an Intel 8751 microcontroller which has 4k of Eprom program memory, 128 bytes at RAM, a full duplex serial port, two 16-bit counter/timers, and two external interrupts. System configuration characteristics, such as the number of sample sites, their locations, gear ratios, output data format, etc., are stored in a standard 2716 Eprom which allows the user to easily modify the system configuration to meet changing requirements. Interfaced to the 8751 microcontroller chip are a Sensotec model 41, a force transducer to measure the penetration force, a Lebow Model 2120, a torque transducer to measure the shear torque, a Datel LPS-16 cassette tape deck, and four Airpax stepper motor-drive cards. Sensors of different ranges can be substituted to match the system's sensitivity to the sediment conditions.

Power for the entire system is provided from batteries housed in a 0.43 m diameter Benthos glass sphere. There are three separate battery packs within the sphere, with 10 Ah capacity for the vane controller electronics, and a 12 V pack with 40 Ah capacity for the two motors.

The system electronics are normally in a powered-down state, except for the computer interface board which remains powered up to wait for commands from the master computer system. After some initial readings, the microprocessor computes and issues a number of individual commands to the stepper motor-drive card that controls the penetration motor. Each step of the penetration force is averaged to yield one force sample per 3 mm of penetration. The averaged data is stored on magnetic tape and placed in RAM for transmission to the master computer on the ISHTE platform.

After reaching the depth of the next station, the microprocessor selects a rotation direction and rotates the vane at a rate of 1°/s. The direction of the vane is alternated

Power Supply (in glass sphere)

Vane Mechanical System →

Controller Pressure Housing

← Volume Compensator

Fig. 3 ISV Model B mounted on component test cruise platform

at each successive station, keeping the vane blade rotation within the same quadrant. The output of the torque cell is read and the raw measurement stored on magnetic tape while the RAM receives averaged data. The data is averaged so that 100 data points are placed in memory for each series of torque measurements. The first 80 data points represent the first 45° of rotation, and the last 20 points represent the final 45° of rotation. This averaging technique was chosen to highlight the area where the sediment is expected to fail, usually within the first 10 to 20° of rotation. At the end of the rotation, just before the penetration to the next station, a final torque measurement is made to determine if a residual torque is being exerted upon the vane shaft. If a residual torque is sensed, the vane rotation direction is reversed to relieve the torque before penetration to the next station. However, a maximum reversal of 15° is imposed as a limit. The amount of reversal is subtracted from the next 90° rotation sequence so as to keep the test rotation within the same quadrant. The microprocessor uses the same 8-bit A/D converter to process the torque data. The torque measurement resolution is equal to the rated torque of the torque cell divided by 128. For a 1.412 N m (200 inf oz.) torque cell this yields a sensitivity of 0.011 N m (1.5625 inf oz.) per bit. Finer sensitivities than this can be achieved by changing torque cells.

Data from a pair of rotation limit switches mounted upon the motor shafts are collected and stored on the tape and in RAM. These limit switches allow verification of proper vane extension and rotation during post-mission data analysis.

Once the vane has completed its full extension and rotation sequence, it signals the master computer. The master computer system then requests that the data collected in the RAM be telemetered over a 2400 band serial link. After verification of the data, the master computer issues a command to retract the vane fully. The microprocessor then retracts the vane, and, upon full retraction, shuts the power off to all boards except the computer interface board.

RESULTS OF THE AUTUMN 1984 DEEP-WATER TEST

Cruise Plan and Site Characteristics

The main objective of the 1984 cruise was to test all the components — except for the heater that will be incorporated in the ISHTE — near the deep-water site selected for the one-year experiment. Almost all the instrumentation has been developed especially for the ISHTE. The main instruments are the following:

- thermal sensors, on heater and in sediment;
- thermal conductivity probes (line sources);
- piezometers for pore water pressure monitoring;
- pore-water sampler for geochemical analysis;
- vane shear system for geotechnical analysis;
- ion migration experiment with overcorer;
- hydrostatically activated corers;
- cameras to monitor conditions at site and around heater.

In addition, a special acoustic tracking system with a transponder network and a master controller and data acquisition system have been developed (Olson et al., in press). For the 1984 cruise, the instrumentation was mounted on two separate platforms and a heat source was not provided.

The vane shear measurements reported here were taken in a water depth of 5845 m in an abysall hill region at 30°20.827′N, 157°50.921′W within a SDP study site designated as MPG-I (MPG-I lies between latitudes 30° and 31°30′N and longitudes 157° and 159° and 159°W). The sediments in this area have been studied quite extensively and the upper few meters are generally characterized as being fine-grained

TABLE 1
Summary of geotechnical properties in MPG-I (Upper 4 m, 26 cores)

Average $w(\%)$	e_0	$w_L(\%)$	$I_p(\%)$	Silt (%)	Clay (%)	C_c	OCR
110	3.05	90	50	33	66	0.85	10.6
98–137	2.56–3.33	81–101	42–71	—	—	0.46–1.50	1.2–48.00

*From profile plots, corrected for 35% salt

Note on data format: $\dfrac{\text{average}}{\text{min.–max.}}$

illite-rich clays of medium sensitivity, low strength, low permeability and high compressibility (Table 1).

Coring, Sediment Sampling and Geotechnical Characteristics

The hydrostatically actuated corers (HLC) used on the September 1984 cruise (ATLAS-1-84) were designed especially for the ISHTE and use the ambient water pressure to drive the core tube into the sediment while preventing movement of the piston (Percival *et al.*, 1984). The core has an inside diameter of 102 mm, a smooth outside barrel, and tapered nose cone. Because of the very controlled way in which the corer is taken, it should be expected to recover an excellent quality core with minimal disturbance. However, at the site, the manganese nodule cover is estimated to be 30–40% and it is possible that nodules could be dragged down into the sediment.

There were two lowerings of the main platform. On the first lowering, one full suite of ISV measurements was made and then a core (HLC-1) was taken at a horizontal distance of 370 mm from the vane. On the second lowering only a few measurements were taken with the ISV and two cores were taken (HLC-2, 3). HLC-2 was taken adjacent to the 82 mm heater implant rod and is therefore disturbed, but HLC-3 was at a distance of 742 mm from this rod. The horizontal distance between the two lowerings was 66 m. The core samples were inverted vertically and hydraulically extruded incrementally by using the corer piston and ram. A motorized laboratory miniature vane (Wykeham–Farrance) was used with a torque transducer, 12.5 × 12.5 mm vane and 60°/min rotation rate. The vane apparatus was rigidly attached to the core barrel to minimize relative movement between the two. Samples were taken for water content determinations and several types of subsamples were obtained for detailed laboratory analysis. At the time of writing this paper, only the water content and vane shear data were available.

Approximtely thirty large-diameter (102 mm) gravity cores were taken in MPG-I with the closest (DS1208-81) being within 1 km of the ATLAS-1-84 site. The upper few meters of sediment in MPG-I exhibit very uniform physical property characteristics and it seems reasonable to make comparisons between these two sites. Therefore data from two gravity cores are also presented here.

Some typical physical property data for the upper four meters of the illite-rich clay in MPG-I are shown in Table 1. Throughout the region the upper 3–4 m shows high 'apparent' overconsolidation with OCR values of more than 3 down to 1 m depth (Silva and Jordan, 1984). The water content profile for HLC-1 (Fig. 4) shows that there are variations downcore. Comparison with other cores in MPG-I indicate that this variability is fairly typical with a rapid decrease from over 130% at the surface to less than 110%

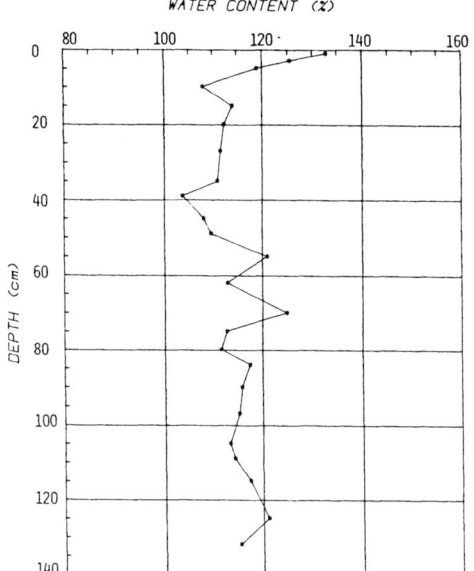

Fig. 4 Water content versus depth, ATLAS
1–84, HLC-1

within 0.1–0.2 m depth. The water content increases somewhat within the 0.2–0.4 m depth (in HLC-3 the increase was from 110 to 117%) and then decreases again to less than 105% within the 0.4 to 0.7 m depth. There is a zone of higher water content (over 120% in both HLC-1 and HLC-3 down to about 1.4 m and then another significant decrease to about 110% at 1.7 m). In summary, the water content profile below the surface for the upper 2 m consists of three zones of low water content (including the zone below 1.7 m) separated by two zones of higher water content.

ISV Preparations, Calibrations and Data Conversion

Because of delays in fabrication and the need to meet the ship schedule, only preliminary calibrations and environmental effects studies were made prior to the 1984 cruise. Since that time a great deal of effort has been put into the determination of all possible effects on the ISV measurements. Following is a summary of the major items that were checked out and an indication of

how the results were corrected (if necessary).

1. Vane size

Since the vane used in the ISV tests (30 × 45 mm) was considerably larger than the one used on the cores, a study was made using laboratory remolded illite from MPG-I to determine how this might affect the results. The results from several measurements with each vane size showed no significant size effect, with the larger vane showing slightly lower values (within 2%).

2. System calibration

Several sets of calibration tests, using a pulley and weight apparatus, were conducted with the assembled mechanical system in both the horizontal and vertical orientations. All the tests, including one set done at 9°C, were quite consistent. The factors used in converting controller output data to torques were as follows:

clockwise rotation   34.18 ozf in/V
anticlockwise rotation   34.07 ozf in/V

3. Pressure and temperature effects on torque sensor

A special variable load system was developed to apply known torques to the transducer in an oil-filled, cooled pressure chamber. Tests were conducted in both rotation directions at room temperature (22°C) and atmospheric pressure, and at 2°C and 58.6 MPa (8500 psi) pressure. Compared to the factory calibration the calibration factor at high pressure and low temperature was only 1.9% greater, and the room-temperature atmospheric pressure calibration was less than 1% greater. The final adjustment made to account for the combined pressure and temperature effect was to increase the calibration factors of the system by 1.8% in the clockwise direction and 2.1% in the anti-clockwise direction. The end result was that the calibration factor was the same in each direction, i.e. 34.79 ozf in/V.

4. *Mechanical friction effects*

The mechanical friction in the system was evaluated at several different times by rotation of the vane with no applied torque and with the unit filled with oil. The readings were always very consistent and within 8 ± 1 controller readout units, N, or bytes of the set central position of 128. Therefore a voltage of ± 0.313 V, corresponding to 8 units, was subtracted from each peak voltage output. It should be noted that controller readings, N, are converted to voltage, V, using the following relationship:

$$V = \frac{5N}{128} - 5 \quad \text{(in volts)}$$

5. *Voltage correction*

It was recognized that there was excessive power drain just prior to the cruise, but repairs were not attempted at that time. During the penetration sequence, there was a slight decrease in voltage (that was monitored) from 10 to 9.76 V. Therefore a correction was made by multiplying the readout voltage by the ratio of the assumed voltage (10 V) to the actual excitation voltage.

6. *Exposed shaft correction*

The length and diameter of the exposed shaft above the vane are constant at 96 mm and 6.35 mm respectively. A correction for the torque on this portion was made by assuming a remoulded strength of 50% of the actual strength. This correction (3.7%) was subtracted from the measured torque to yield the corrected torque used to calculate shear strength around the vane blades.

7. *Conversion to shear strength*

The equation for the conversion of corrected torque to strength includes the vane dimensions of $D = 30$ mm, $H = 45$ mm, and

$R = 17.3$ mm, where R is the radius of a circular arc at the vane bottom (see Fig. 2), and assumes a triangular stress distribution on the top and bottom surfaces.

$$S_u = \frac{T}{[(\pi/2)D^2H + (\pi/16)D^3 + (\pi^4/54)R^3]}$$

$$S_u = 0.0901 T^{\text{ozf in}} \text{ kPa}$$

Comparison of results

The results of the ISV measurements and two hydrostatic cores are shown in Fig. 5. The ISV results indicate an almost linear increase in strength within the upper 0.5 m with an intercept of about 3 kPa at the surface and a value of about 6.8 kPa at 0.5 m. Between 0.5 m and 1.05 m there is variability about the general trend which shows a slight decrease in the rate of increase of strength with depth. Below 1.05 m the trend again approaches linearity.

The HLC-1 core was taken on the same lowering as the ISV measurements. The core appeared to be in very good condition

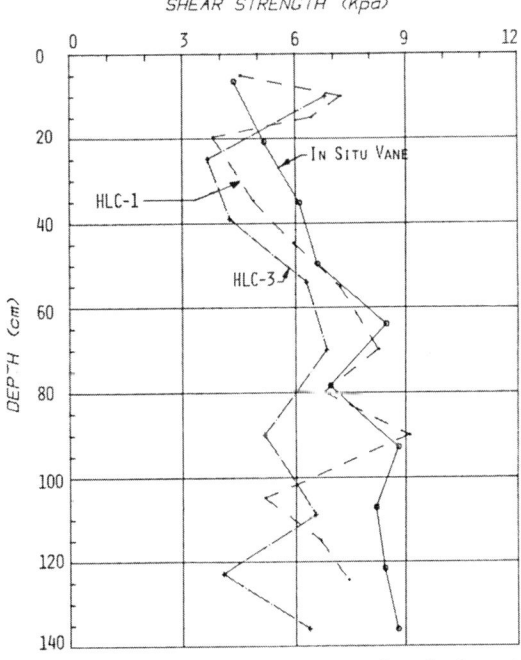

Fig. 5 Shear strength versus depth *in situ* vane versus HLC-1, ATLAS 1–84, HLC-3

during extrusion. A few points are worth noting. There is extreme variability in the upper 0.2 m with some points higher and one lower than the ISV results. This may have been caused by compression during the extrusion process but is more likely to have been due to some difficulties caused by ship vibrations. Below 0.2 m the HLC-1 and ISV trends are remarkably similar, although below 0.9 m the decrease shown in the core results is much greater.

The HLC-3 core was taken on a different lowering (66 m away). Visual observations indicated some disturbance — possibly by intrusion of nodules — but overall the core seemed to be in good condition. Except for one point at 0.1 m, the shear strength of this core was considerably lower than both the *in situ* results and the HLC-1 results. It should be noted that this corer was hauled aboard manually and therefore may have been subjected to more disturbance than HLC-1. However, relative to normal piston coring or gravity coring, the mechanical dis-

turbance imparted to these two cores was probably much less.

Results from two large-diameter gravity cores from the DS 1208-81 cruise (less than 1 km away) are shown in Fig. 6, along with the ISV results. Except for the lower portion, GC-12 shows strengths very similar to the ISV results. The results for GC-13 are considerably less than the ISV strengths.

SUMMARY

The results of a deep-water test with a new *in situ* vane shear system have been described. A few summary comments follow.

(a) All subsystems of the ISV system operated successfully, there were no major malfunctions or failures and the system design appears to be sound.

(b) The strength profiles for HLC-1, HLC-3, and the ISV measurements follow very similar trends for most of the 1.4 m depth. However, the ISV profile generally shows slightly higher strengths than HLC-1 and considerably higher strengths than HLC-3. Some of the extreme variability in the lower parts of the two cores does not show up in the ISV profile. It is likely that the lower strengths in the cores are caused by disturbance rather than actual sedimentary conditions.

(c) The strength profile for one nearby (<1 km) gravity core is similar to the ISV profile, but the other is considerably lower.

(d) In order to quantify the differences between core and ISV measurements, a numerical integration was made between the depth intervals of 2.00 to 1.15 m to determine the areas of the respective curves. Ratios were then calculated to characterize the differences; the results of these analyses are summarized in Table 2.

The most direct and reliable comparison is for core HLC-1, since this was

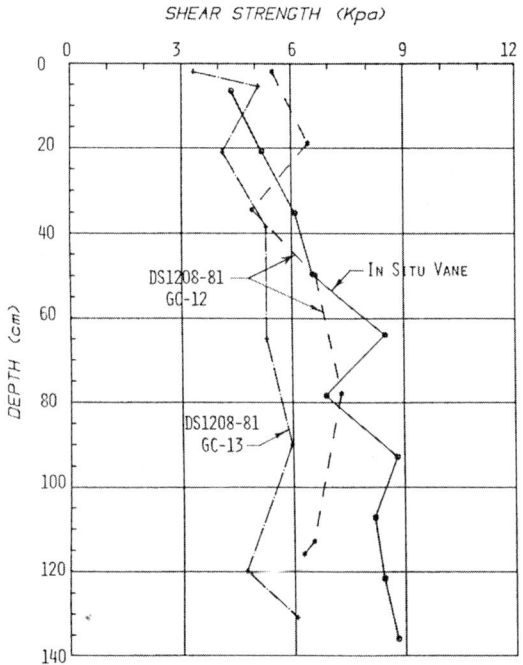

Fig. 6 Shear strength versus depth *in situ* vane versus DS 1208-81 GC-12, GC-13 MPG-I area

TABLE 2
Shear strength comparisons (average from 0.20 to 1.15 m depth)

Core no.	% Reduction ISV to core	Ratio ISV/core	Comments
HLC-1	13	1.15	Hydrostatic corer: same lowering
HLC-3	25	1.33	Hydrostatic corer: 66 m away
GC-12	12	1.13	Gravity corer: <1 km away
GC-13	31	1.44	Gravity corer: <1 km away
Minimum	12	1.13	
Maximum	31	1.44	
Average	20	1.26	

taken on the same lowering as the ISV. As shown in Fig. 5, the agreement between ISV and core measurements between 0.5 and 0.9 m is excellent, but the disparity above and below this zone is such that there is an average of 13% reduction. Based on this one core, the average core strengths would need to be increased by 15% to obtain *in situ* strengths. The corresponding values for HLC-3 are much higher (25 and 33%). There is a greater difference between the two gravity cores (12 and 31% reduction).

The average percentage reduction for all cores is 20%. Therefore the core results would need to be increased by 26% to obtain the *in situ* strengths.

(e) The cores used in the comparisons are considered to be of very good quality, since two of them (HLC cores) are large-diameter piston cores taken in a controlled manner from a bottom-supported platform, and two (GC cores) are large-diameter, thin-walled gravity cores. Smaller diameter cores, especially standard piston cores, used in deep water probably produce greater sample disturbance and the corrections suggested above may not be applicable.

REFERENCES

1. Aas, G. A. 1965. A study of the effect of vane shape and rate of strain on the measured values of *in situ* shear strength of clays. In *Proc. 6th Int. Conf. of Soil Mech. and Found. Engng*, Montreal, Vol. 1, Div. 3, pp. 141–145.

2. ASCE, 1975. *In Situ Measurement of Soil Properties*, Vols I and II. *Proc. Geot. Eng. Division Specialty Conference, 1975*.

3. ASTM, 1972. Field vane shear test in cohesive soil. Designation D2573-72.

4. Babb, J. D. 1982. Development of an *in situ* vane for strength measurement of deep sea sediments. MS thesis, University of Rhode Island.

5. Babb, J. D. and Silva, A. J. 1983. An *in situ* vane system for measuring deep sea sediment shear strength. *IEEE/MTS Proceedings, Oceans '83*, Vol. 1, pp. 598–602.

6. Backes, J. L., Bell, B. M. and Olson, L. O. 1981. Long-baseline deep ocean acoustic tracking and telemetry system. *IEEE/MTS Proceedings, Oceans '81*, Vol. 1, pp. 1–8.

7. Briaud, J. L. 1980. *In-situ* tests to measure soil strength and deformability for offshore engineering. Research Report, Texas A&M Research Foundation, College Station, TX.

8. Doyle, E. H., McClelland, B. and Ferguson, G. H. 1971. Wire-line vane probe for deep penetration measurements of ocean sediment strength. *Offshore Technology Conference*, Paper No. OTC 1327.

9. Flaate, K. 1966. Factors influencing the results of vane tests. *Can. Geotech. J.* 3(1).

10. Hadley, G. R., McVey, F. F. and Morin, R. 1980. Thermophysical properties of deep ocean sediments. In *Marine Technology '80*. Marine Technology Society.

11. Hollister, C. D., Anderson, D. R. and Heath, G. R. 1981. Subseabed disposal of nuclear waste? *Science* 213, 1321–1326.

12. Lee, H. J. 1979. Offshore soil sampling and geotechnical parameter determination. *Offshore Technology Conference*, Paper No. OTC 3524.

13. Olson, L. O., Backes, J. L. and Miller, J. B. In press. Communication, control and data acquisition systems on the ISHTE Lander. *IEEE*.

14. Monney, 1973. Analysis of sediment shear strength at various rates of shear. In *The Physical and Engineering Properties of Deep Sea Sediments* (Ed. A. L. Inderbitzen), Virginia.

15. Percival, C. M. 1983. The subseabed disposal program *In Situ* Heat Transfer Experiment (ISHTE). SAND80-0202, Sandia National Laboratories, Albuquerque, N. Mex.

16. Percival, C. M., McVey, D. F., Olson, L. O. and Silva, A. J. 1984. *In Situ* Heat Transfer Experiment (ISHTE). *Mar. Geotech.* 5(3/4), 361–378.

17. Perlow, M. and Richards, A. F. 1972. In-place geotechnical measurements from submersible Alvin in Gulf of Maine soils. In *Offshore Technology Conference*, 1972.

18. Richards, A. F., McDonald, V. J., Olson, R. E. and Keller, G. H. 1972. In-place measurement of deep sea soil shear strength. ASTM, STP, pp. 55–68.

19. Seabed Programs Division, 1983. The subseabed disposal program: 1983 Status Report. SAND83-1367, Sandia National Laboratories, Albuquerque, N. Mex.

20. Silva, A. J. and Jordan, S. A. 1984. Consolidation properties and stress history of some deep sea sediments. In *Seabed Mechanics*. Edited Proceedings of IUTAM Symposium, University of Newcastle Upon Tyne, September 1983, pp. 25–40.

21. Silva, A. J. and Pekin, O. 1981. An *in situ* geotechnical measurement system for deep sea surficial sediments. *Trans. AGU* 62(45). (Abstract No. 01-2-C-5.)

22. Silva, A. J., Babb, J. D., Lipkin, J., Pietryka, P. and Butler, D. In press. *In situ* vane system for seafloor strength investigations. *IEEE*.

23. Silva, A. J., Criscenzo, S. J., Jordan, S. A. and Babb, J. D. 1983. URI Geotechnical Program of the *In Situ* Heat Transfer Experiment. ISHTE Annual Report No. 2, University of Rhode Island.

15

The Influence of Geological Processes and Test Procedures on Measured and Evaluated Parameters

A. Marsland, Building Research Establishment, UK

INTRODUCTION

The best possible evaluation of the relevant geotechnical parameters is essential for the adequate and economic design of all major structures. The large-scale *in situ* properties of the ground depend on the nature of the source material, the effects of geological processes, environmental conditions, and the stresses to which the ground has been subjected. All these factors are reflected in the *in situ* geotechnical properties and the soil fabric features at scales from the micro to the massive which can be present in the ground. Parameters relevant in a particular foundation design are also a function of the nature, level and orientation of the loadings and the length of time for which they are applied.

Fabric features can vary in nature, scale and orientation both down individual soil profiles and with location. As well as providing evidence of past environmental and stress conditions, fabric features often play dominant role in large-scale geotechnical

behaviour (Marsland, 1971a; Rowe, 1972; McGown and Radwan, 1975).

The results of laboratory tests also depend on the degree of disturbance, the degree of saturation and the effective stresses developed within the samples prior to testing. While *in situ* tests may avoid major relief of *in situ* stresses, the results obtained are affected to varying degrees by the mechanical disturbance and changes of stress which occur during insertion. Non-uniform stress fields and the degree of control of the drainage which occurs during insertion and testing also present problems of interpretation where these differ significantly from those acting on soil elements affected by the foundation. The rate and duration of loading affect both field and laboratory test results and vary with both the type of test and the soil. All the above factors significantly affect the correlations between full-scale behaviour and test results. It follows that measured parameters generally differ from the full-scale parameters which control the performance of foundations. The traditional method of attempting to overcome this problem uses empirical correlations between data obtained from specific standard tests and the behaviour of particular types of foundations or elements of foundations. However, many of the available correlations have either been derived for a particular local situation or show such a large scatter that their use is limited to the design of foundations where the overall factor of safety is high. The purpose of this paper is to highlight some of the outstanding problems in assessing relevant parameters for offshore design and to suggest ways in which they may be improved.

GENESIS OF NORTH SEA SOILS

Most of the soils in the North Sea and the adjacent parts of the continental shelf which are of interest to the geotechnical engineer were deposited during the Quaternary. During this period the area was exposed to large variations in climate, which resulted in glaciations and intervening warmer periods. The continual changing positions of ice fronts, sea levels and climatic conditions led to sequences of errosion, deposition, consolidation, and post-depositional modifications. Studies of land borings and exposures, combined with more recent data from offshore borings and geophysical surveys, show that ice-sheets which developed in the highlands of Scotland and Norway extended well into the North Sea on several occasions (Løken, 1976; Boulton et al., 1977; Oele and Schuttenhelm, 1979). At its maximum (200 000–300 000 years BP), the ice covered the whole of the North Sea area, including the UK, Scandinavia and the northern parts of Germany and the Netherlands. During this period the sediments of the North Sea were subjected to extensive erosion, redeposition and high total overburden pressures. The extent of the ice during the last major glacial advances during the Devensian (10 000–75 000 years BP) were less extensive and much of the southern North

Sea was free from ice (Boulton et al., 1977; Behre et al., 1979). Whether the Scandinavian and British ice coalesced during part of the Devensian is still a matter for speculation, but there seems to be general agreement that both reached the Dogger Bank in the central North Sea. Superimposed on the local glacial conditions were the changes in sea-level produced by the increases and decreases of the global ice masses. Various sources suggest that the sea-levels, when the Devensian ice caps surrounding the North Sea reached their maximum (about 18 000 years BP), were probably 130 m lower than at present (Jardin, 1979; Jelgersma, 1979). Even neglecting the continuing settlement due to techtonic movements, there is little doubt that the southern and central North Sea were above sea-level for appreciable periods after the commencement of the Anglian glaciation about 300 000 year ago. The situation in the northern North Sea is more debatable, but even there the sea would have been shallow for substantial periods with much of the area covered with grounded ice and the possibility of some ground exposure. From the above discussion it is evident that a full range of glacial, glaciomarine, fresh-water depositional environments occurred within the North Sea Basin at some time. The location of the ice fronts, shorelines and glacial lakes at any one time are not known with any degree of precision and are the subject of much ongoing discussion (Aarseth and Sejup, 1984). There is, however, general agreement that a situation similar to that portrayed in Fig. 1 probably occurred on more than one occasion during the various glacial periods.

Much of the material deposited during earlier glacial advances was removed or reworked by later advances. In the southern parts of the North Sea deposition was mainly from sheets of land ice, but in the northern parts of the North Sea there were more periods of marine deposition with the soil being supplied by subglacial streams and ice calving. Evidence from studies on present-day land areas adjacent to the North Sea indicate that many of the ice-free

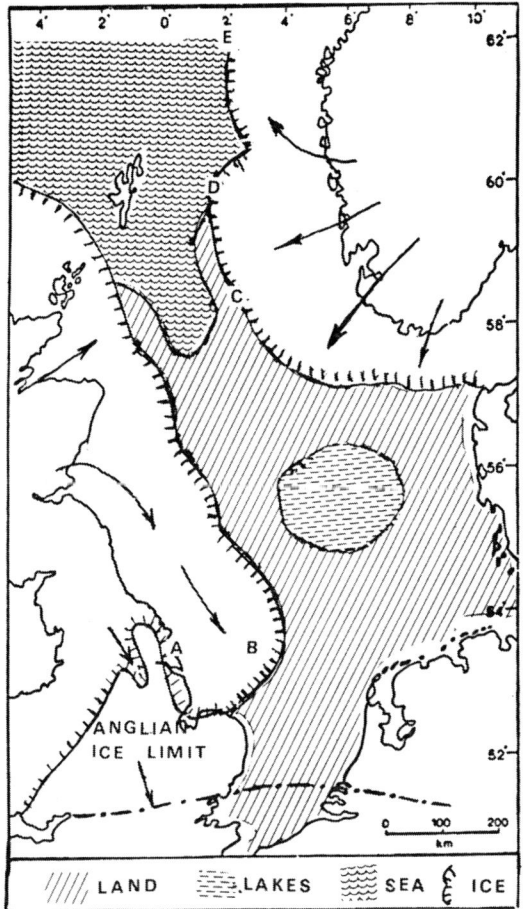

Fig. 1 Generalized map of possible conditions in the North Sea during part of Devensian Glaciation; based on data published by Boulton *et al.*, 1977; Jardine, 1979; Jelgersma, 1979 and Løken, 1976

observations around present-day ice-sheets (Boulton, 1972), suggest that subpolar and thick polar glaciers, while frozen to their beds in their outer zones, are underlain by thin films of liquid water elsewhere. The transition from one subglacial regime to the other provides the mechanism for the incorporation of debris by basal freezing with the concentration of material becoming progressively greater towards the ice margins as a consequence of the downwasting of the ice. Deposition of this debris is affected by a variety of processes, including melt-out, flow and lodgement. The complex sedimentary sequences produced by these processes around modern polar and subpolar glaciers bear a striking resemblance to the sequences commonly seen in British glacial deposits. This led Boulton (1972) to propose that at their maximum extents the British ice-sheets were of polar or subpolar type.

The wide range and complexity of the deposits formed in land-ice and glaciomarine environments are discussed by Boulton and Deynoux (1981). Further aspects of glaciomarine sedimentation are summarized in a recent paper by Powell (1984).

Maximum deposition occurs near the ice fronts present at the particular time. The amount and mode of deposition depends on the activity and fluctuations of the ice front at the particular locality. For example, during the late Devensian advances an extensive surging lobe developed in the western part of the southern North Sea adjacent to the east coast of England (Boulton *et al.*, 1977). This deposited deep layers of mainly subglacial till that are found on land in Holderness (north of Hull) in eastern England and extend well out into the North Sea within the area which was covered by this lobe.

In addition to the wide range of primary depositional processes, the soils have often been subjected to post-depositional changes other than simple loading and unloading. Grounded ice causes appreciable remoulding and shearing of subglacial deposits. Similarly, advancing ice produces substantial disturbance of the local bedrock and mater-

land areas within the North Sea Basin were subjected to very cold conditions which would have caused deep freezing to occur in some areas. This ties in with the concept that the main streams of warm, moist air were further south during the glacial periods than they are today (Liljequist, 1974) and as a consequence cold, dry continental conditions existed, particularly in the mid and northern parts of the North Sea.

The dynamic conditions of the ice masses, particularly as influenced by the thermal regime, affect the patterns and processes of glacial erosion and deposition. Models of rift entrainment and deposition, based on

ial deposited during previous glacial retreats. Shearing also occurs as a result of differential ice loading, slipping or 'flowing' of soft wet deposits, and melting of buried ice during deposition. The degree of consolidation that occurred during ice loading would have depended both on the thickness of the ice and the ease or otherwise with which drainage could occur. Drainage could have been prevented by ground freezing and the long drainage paths from ground below thick ice areas would have restricted the degree of pore water pressure dissipation. Freezing and drying caused further disruption which substantially modified the effects of deposition and post-depositional consolidation.

SOIL FABRIC FEATURES IN NORTH SEA AND ADJACENT LAND DEPOSITS

'Soil fabric' as used in this paper refers to the nature, spacing, and directional properties of all types of fabric features at all scales. In order to obtain a reasonably complete picture it is necessary visually to examine 'broken' surfaces of as many samples as possible. The use of suitable illumination and a large low-powered magnifier are essential. Careful progressive separation of the samples along visible discontinuities provides data on their nature and spacing and often reveals the presence of further smaller-scale discontinuities. Selected areas of a typical sample should be examined at increasing magnifications, such as 25, 50, 100, 250, 500 and 1000 times, with occasional higher magnifications being used to inspect individual particles. Such multiscale inspections are necessary because features that can be easily observed at one particular scale can be easily missed at other scales. In addition there is often a heirarchy of important features. Such studies can be readily made using a scanning electron microscope, and the examples given in this paper were obtained as part of a BRE research contract undertaken at Keele University. Further details of some aspects of these studies are given by Marsland et al. (1982), Derbyshire and Love (1985), and Love and Derbyshire (1985). The fabric features found in a particular deposit are a reflection of the composition of the parent rocks, the degree of comminution, depositional processes and environments, as well as post-depositional changes due to environmental and stress conditions.

While the presence of soil fabric features and their importance for the engineering properties of land-based soils was becoming increasingly recognized in the mid-1960s and early 1970s (Ward et al., 1965; Marsland and Butler, 1967; Bishop and Little, 1967; Rowe, 1972; Marsland, 1971b, c, 1975), surprisingly little detailed attention has been paid to fabric features in offshore deposits. The potential importance of fabric studies in the evaluation of the geotechnical properties of offshore soils was stressed by the author (Marsland, 1977), and some typical examples of macrofabric in samples obtained from the North Sea Frigg field (near location D, Fig. 1) were given. Schjetne and Brylawski (1979) gave an example of a finely laminated clay from the east Shetland basin and described the use of X-ray inspection to discover fabric features and possible disturbance. Schjetne et al. (1979) mentioned fissuring of clays and the possible importance of freezing and chemical cementation. An illustration of a clay containing polished, striated discontinuities (obtained from the Aberdeen ground beds just north of the Forties field by the Institute of Geological Sciences) was published by Redding (1976).

As part of the overall studies of the geotechnical properties of both land and North Sea deposits, the BRE contracted the geography department of Keele University to make detailed studies of the sedimentological and fabric properties of soils from typical locations. In the early stages of the offshore investigations one of the main difficulties was to obtain sufficient good-quality samples in order to give a reasonable coverage of the variable fabrics at specific sites and to relate these to geotechnical properties. In spite of the difficulties of

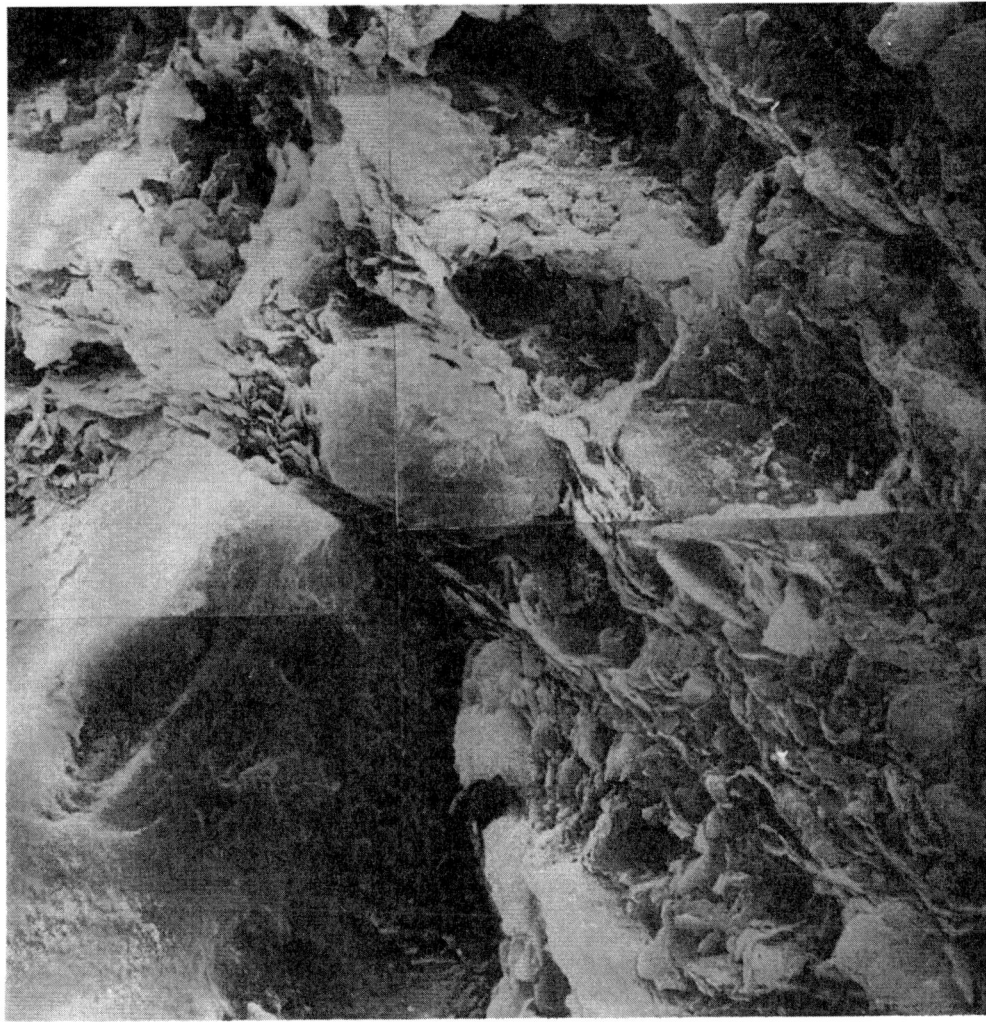

Fig. 2   Photomicrograph of till from a depth of 14.5 m at the B.R.E. test site at Cowden (location A
Fig. 1); photograph width represents approximately 0.1 mm

obtaining sufficient good-quality samples, significant progress was made, as indicated by Marsland *et al.* (1982). Late in 1981 the author and M. A. Love of Keele University were invited by the Norwegian Geotechnical Institute and Stat Oil to make fabric studies of soils from the Statjord field in the Norwegian sector. The great number of fresh samples which were available for inspection, together with the extensive geotechnical data which had been obtained, made it possible to use the soil fabric studies in the detailed evaluation of design parameters. As the studies have progressed a widening range of fabric features have been found and examples of some of these, related to probable depositional environments, are given in the following section.

The presence over a considerable area of the southern North Sea and adjacent areas of eastern England of deep till sequences deposited from an extensive surging ice lobe during the Devensian has already been mentioned. The land-based BRE test-bed site at Cowden north-east of Hull is located near the coast in this area (location A, Fig. 1) and the geotechnical properties measured at this site have been given by Marsland and Powell (1985). Streaked-out clasts on chalk found in the coastal exposures indicate appreciable

TABLE 1
Some references on macrofabric features in glacical clays observed in land areas around the North Sea

Author	Features studied	Area of study
Richter, K. (1929)	Divisional planes, shear layers, compaction	Germany
Bulow, K. V. (1939)	Latent and visible shear layers, ground moraine	Germany
Penny, L. F. and Catt, J. A. (1967)	Joints, stone orientation, fold	England
Boulton, G. S. (1970)	Deposition, fracture, tectonic shear	Norway
Rowe, P. W. (1970)	Fissures, permeability	England
Kazi, A. and Knill, J. L. (1973)	Fissures, orientation, stress conditions	England
Pusch, R. (1973)	Macrofabric, Microfabric, fissures, permeability	Sweden
McGown, A. *et al.* (1974)	Macrofabric, fissures, strength, anisotropy, orientation	Scotland
McGown, A. and Radwan, A. M. (1974)	Slope stability, fissuring	Scotland
Marsland, A. (1975)	Fissures and bedding planes, strength and moduli	North-East England
Marsland, A. (1977)	Fissures and macroped fabric, strengths and moduli	South-East England

subglacial shearing. However, detailed visual inspection of the tills in 100 mm diameter samples and in exposed sides of large diameter boreholes taken to a depth of 22 m showed that observable macro discontinuities were restricted to the weathered zone in the upper 4 to 5 m. This is contrary to the general findings in subglacial tills found in many other parts of Britain, as can be seen from Table 1. Even though the Cowden tills contained few macro discontinuities, both the onshore and offshore tills in this area (locations A and B, Fig. 1) contained numerous microshears which are typically 0.05 to 0.2 mm long, as can be clearly seen in Fig. 2. The highly mixed nature of the tills both at Cowden and at West Sole (locations A and B, Fig. 1) is evident from the micrographs in Figs 2 and 3 where particular note should be made of the clay wrapped round the silt and sand particles. These tills contain about 30% of clay-sized particles and have densities between 2.2 and 2.24 mg/m². In view of these relatively high densities the evaluated over consolidation

pressures and the measured *in situ* horizontal stresses were surprisingly low. High densities can, however, be achieved by thorough mixing of well-graded soils and this, together with the general absence of observable macro discontinuities, suggests that these tills were probably deposited in a relatively wet state.

Further north (location C, Fig. 1) the deposits in the upper 100 m become more variable due to periods of deposition in glacial lakes and under glaciomarine conditions, which gave rise to more laminated deposits such as that shown in Fig. 4. Dense subglacial tills are, however, still present as illustrated by the micrograph in Fig. 5. Other tills which have a more open structure (see Fig. 6) are also present. There is also evidence in this area of substantial post-depositional shearing, as shown by the multiple shears in Fig. 7 and the highly polished microshears in the photograph published by Redding (1976).

Further north still (around locations D and E) the deposits show more evidence of

Fig. 3 Photomicrograph of till from a depth of 11.4 m at West Sole near B in Fig. 1; photograph width represents approximately 0.2 mm

glaciomarine deposition. Although many of these deposits are distinctly laminated, others are relatively uniform. Due to the lack of clasts, these more uniform deposits are usually considered to have been deposited under glaciomarine conditions, but some have microfabric features characteristic of well-mixed subglacial tills. These features could be due to glacial remoulding of previously formed glaciomarine deposits. Some dense subglacial tills are still present, as was shown by the presence of a very dense horizon in the Frigg profile (location D) published by Marsland (1977). Highly polished

shear discontinuities such as that shown in Fig. 8 have been found at several sites near locations C, D, and E (Fig. 1).

In the areas around D and E the clays show a variety of foliated, platy, and lenticular features which often incorporate horizontal bedding features. Typical examples are shown in Figs 9 to 12. Many of these features are similar to those associated with present-day frozen ground such as shown by Mackay (1974) and Boulton and Paul (1976). Similar features (Fig. 12) have been produced by artificially freezing (a single freeze–thaw cycle) a sample of soft, intact

Fig. 4 Laminated clay with silt and fine sand partings near C in Fig. 1; sample width 72 mm

previously unfrozen lake clay from Winnipeg. Some of the marcofabrics caused by freeze–thaw processes have been categorized by Van Vliet and Langohr (1981). These authors have made extensive studies of the soil fabrics in the fragipan horizons in northern France and Belgium and some of these features are illustrated in Fig. 13. Although other much larger features having dimensions of up to many metres occur in frozen ground, the ones illustrated in Fig. 13 are those most likely to be found in samples. In fact a complete heirarchy of fabric features is often formed. During the investigations of the soil fabric in samples obtained from borings in location E in 1982–1983, M. A. Love and the author observed small-scale macropeds at some horizons which had dimen-

sions as small as 1.5–4 mm. These small ped-like features are particularly characteristic of ground in which the intestial water has been frozen *in situ* (Mackay, 1974; Chamberlain and Gow, 1978; Johnson, 1981). Photomicrographs of this soil revealed a pattern of even smaller scale peds with dimensions of 0.2–0.3 mm, as shown in Fig. 14. Similar microped features have been observed in remoulded compacted clay soils subjected to freezing cycles (Smart and Tovey, 1981). More recently a suite of similar features has been found in samples from a second site in the same general area (near E, Fig. 1) by the author's co-workers at Keele University (Derbyshire *et al.*, 1985). It seems highly probable that close inspection will reveal similar features at many more locations.

INFLUENCE OF FABRIC FEATURES ON STRENGTH

Soil fabric features at all scales can significantly affect the large-scale *in situ* strengths, as well as those measured by *in situ* and laboratory tests. The strengths relevant in a design of a particular type of foundation are influenced by the orientations, continuity, surface characteristics and, in some cases, the scale of the soil fabric features. The importance of particular features depends on the direction of shearing induced by the foundation loading and the extent to which the soil is modified by the installation of the foundation. The scale and the restrictions imposed on the directions of the applied stresses in both field and laboratory tests means that each type and size of test has the potential to produce different results. The nature and scale of the macrofabric features in the soil can significantly affect the ratios of the strengths measured by different tests. They can also lead to significantly different ratios being obtained between the measured and large-scale strengths. This can still be the case even when disturbance during sampling, or in the case of *in situ* tests insertion, is minimized and the direc-

Fig. 5 Photomicrograph of dense well-mixed fabric from depth of 22.0 m near C in Fig. 1; photograph width represents approximately 1.6 mm

tions and magnitude of the stresses applied in the tests faithfully reproduce those on typical elements in the ground surrounding the foundation. The influence of the relative scale of a static cone and a soil containing reasonably randomly orientated discontinuities at different spacings is illustrated in Fig. 15. For the condition shown in A, the cone will provide a reasonable measure of the large-scale strength for failures similar to that around a cone. However, in case C, the cone will provide a measure of the strength of the intact clay between the discontinuities which in stiff clays is often several times that mobilized along discontinuities. This is reflected in the values of the cone factor N_k obtained when values of c_u back analysed from large plate tests are used (Marsland, 1979; Marsland and Quarterman, 1982). In stiff clays values of 30 or more are obtained when the cones are small compared to the discontinuity spacing as in C, and values of around 15 or less are obtained when the condition in A applies. In typical strength stiff clays free of macro discontinuities, the values of N_k are close to 15 (Marsland and Powell, 1985). The lower values sometimes obtained in clays containing closely spaced discontinuities, as in A, could be due to loose packing of the clay caused by

Fig. 6 Photomicrograph of fairly loosely packed till from a depth 14.0 m near C in Fig. 1; photograph width represents approximately 0.35 mm

processes such as freeze–thaw. In practice the situation is never as simple as that illustrated in Fig. 15, since the discontinuities are seldom randomly orientated and there is often a hierarchy of discontinuities at difference scales, at different orientations, and when there are significantly different surface properties. For example, highly polished slickensided surfaces are often larger than the discontinuities found in samples which are often less polished. An indication of the different influences of scale and direction of shearing in the various tests used offshore can be obtained by reference to Fig. 16. It is fairly obvious that the smaller tests will measure the strength of the clays between the macro discontinuities, even when they are fairly closely spaced. This will generally differ significantly from the larger-scale strengths which take into account the discontinuities. Possible exceptions are when the clay between the macro discontinuities is also highly fractured at the macroscale as in Fig. 14. Even here some differences are likely to be due to the probable different nature of the smaller discontinuities and directional effects. The reduction in strength due to the presence of dis-

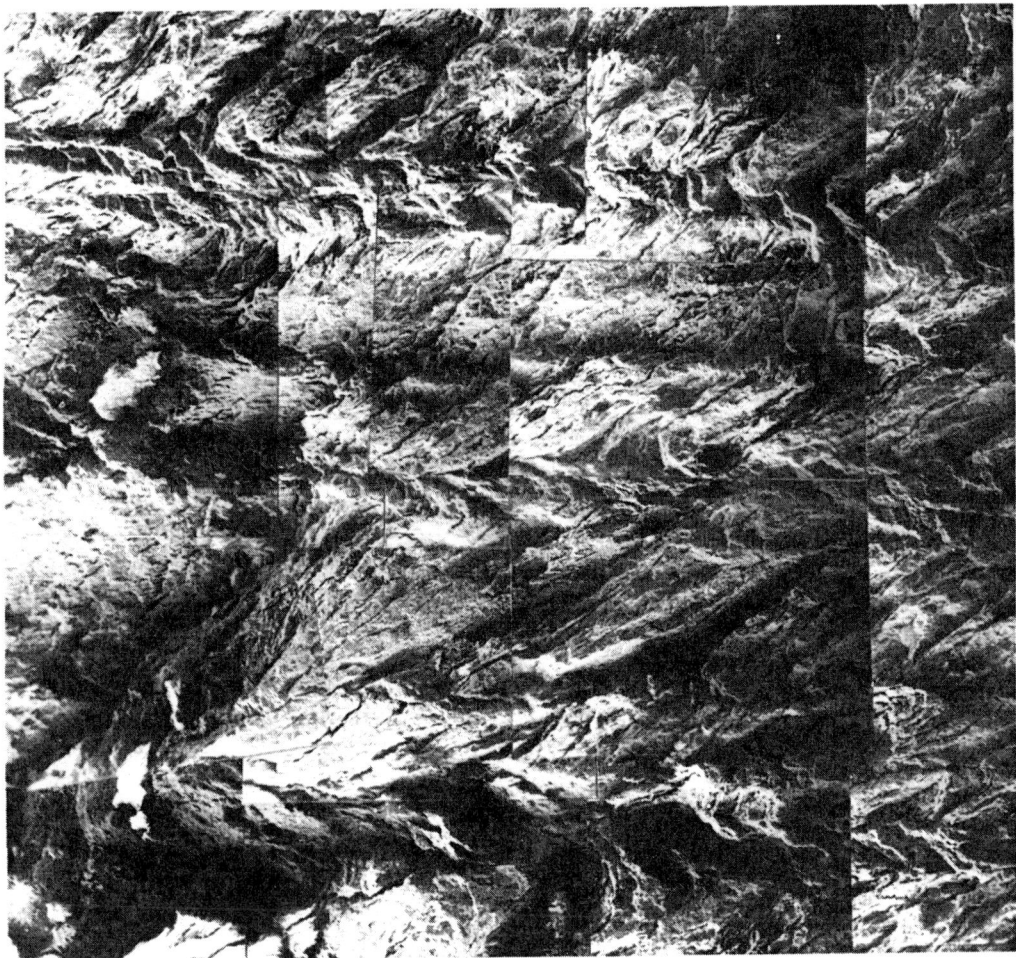

Fig. 7 Photomicrograph showing microfolding and shears in a sample from a depth of 4 m near C in
Fig. 1; photograph width represents approximately 4 mm

continuities depends on the degree of over-consolidation and the properties of the soil within or along the discontinuity (McGown, 1979; Lupini *et al.*, 1981).

The scale of the fabric features is even important in tests which stress a relatively large volume of soil when the stress levels applied by the tests vary within the stress field. This occurs in the pressuremeter test where the highest stresses occur adjacent to the expanding probe and decrease with the radial distance from the centre of the probe. The area within the annulus between the probe surface (broken circle) and the outer dashed line shown in Fig. 16 is subjected to ι average stress which is about 70% of the stresses at the surface of the probe. Thus strengths evaluated from pressuremeter tests in stiff clays may be significantly higher than the *in situ* strengths if the spacing between vertical discontinuities is comparable or greater than the radius of the probe. The pressuremeter will also give strengths approaching the intact strength of a stiff clay if the discontinuities are predominantly horizontal.

The degree of disturbance during sampling and specimen preparation, or during insertion of *in situ* test equipment, can also depend on the nature and scale of the soil fabric and the stiffness of the soil between the discontinuities.

Fig. 8 Highly polished shear discontinuity cutting across sample from depth of 90 m near E in Fig. 1; sample width 72 mm

Fig. 10 Lenticular fabric in sample from depth of 33 m near E in Fig. 1; sample width 72 mm

Fig. 9 Well-developed platy-lenticular structure in sample from a depth of 66 m near C in Fig. 1; sample width 72 mm

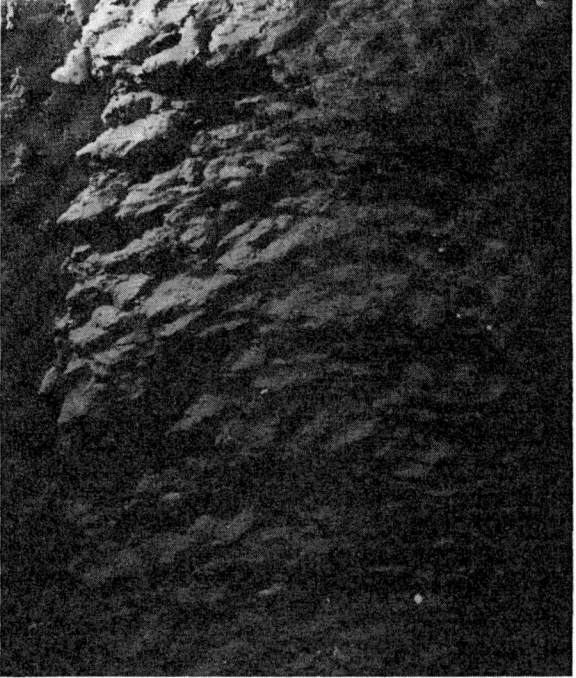

Fig. 11 Finer, more distorted lenticular fabric in sample from depth of 50 m near E in Fig. 1; sample width 72 mm

Fig. 12 Macroped fabric developed by a single artificial freeze–thaw cycle in a sample of previously intact uniform glacial lake clay from Winnipeg; photograph width represents 70 mm

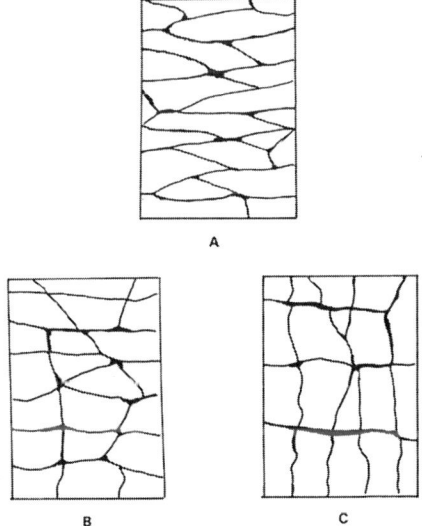

Fig. 13 Fabric features caused by freezing and thawing (after Van Vliet and Laughor, 1981): (a) Foliated or platy (0.2–3.0 mm thick and 0.5–10.0 mm long), (b) Platy or lenticular subangular blocks (3.0–15 mm thick and 8–50 mm long), (c) ' ngular blocky to prismatic (dimensions 10–50 mm)

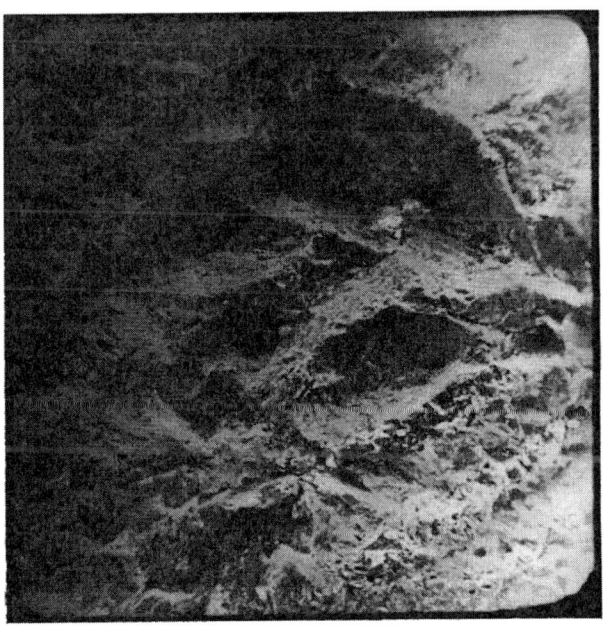

Fig. 14 Photomicrograph of microped fabric within macroped features from a depth of about 18.4 m at a location near E in Fig. 1; photograph width represents approximately 1 mm

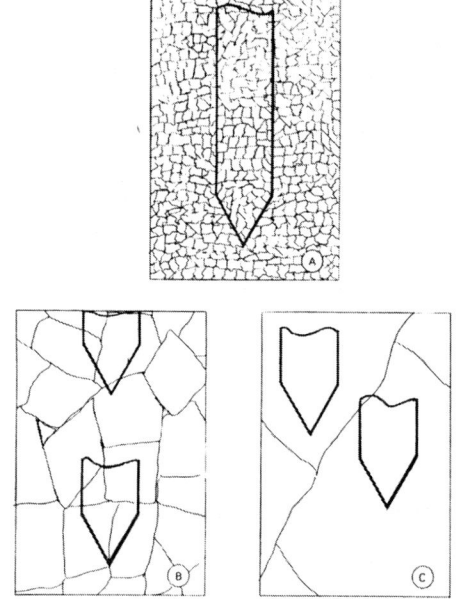

Fig. 15 Relative size of field cone and typical sized macro discontinuities: (A) closely spaced, (B) spacing approximately equal to cone diameter, (C) widely spaced

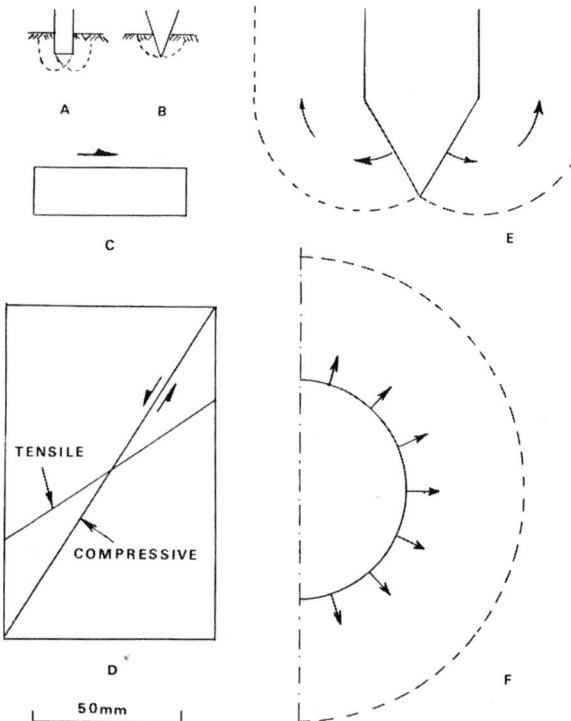

Fig. 16 Comparison of soil volume mobilized in different types of tests used offshore

Post-depositional processes which significantly modify the depositional fabric must also be considered when evaluating the *in situ* horizontal stresses required for the reconsolidation of samples or the interpretation of *in situ* tests. Both desiccation and freeze–thaw processes 'consolidate' the soil between discontinuities and result in the measurement of high OCRs and intact strengths which are unrelated to past loading–unloading events.

EFFECTS OF SOIL DISTURBANCE AND TEST PROCEDURES

The results of both laboratory and *in situ* tests can be affected to differing degrees by the disturbance which occurs prior to testing. Disturbance of samples for laboratory testing can occur as a result of physical disturbance, moisture transfer, and stress relief during drilling, sampling, extrusion, and specimen preparation. Physical disturbance during drilling and sampling operations was a major problem in early site investigations, but these have been largely overcome by improvements in compensation for heave during drilling and by pushing in thin-walled tubes from devices resting or anchored to the seabed (Marsland and Windle, 1982). Increased use of fixed piston samplers could result in higher quality samples being obtained in soft deposits. They would be equally beneficial in stiff soils since wet, soft material near the base of the borehole can be displaced and prevented from being incorporated in the sampling tubes. This could significantly reduce the amount of softening due to moisture transfer. The release of dissolved gas from the pore fluid on removal of high total stresses (water load plus saturated soil overburden) can result in severe physical disturbance which can only be prevented by keeping the soil pressurized at all times. This does not appear to have been a major problem in clays obtained from North Sea sites, but evidence of partial saturation has been reported (Schjetne and Brylawski, 1979).

Even when the pore water is free from dissolved gases, pore suctions within samples gradually break down when the reduction in effective stress exceeds a threshold value, which differs for each soil (Marsland and Windle, 1982; Kirkpatrick and Khan, 1984). The presence of macro- and microdiscontinuities can reduce this threshold and speed up the reduction in the all-round effective stresses locked up in the sample. Shear stresses induced in stiff soils during sampling and extrusion can cause dilation and partial cavitation of parts of the samples. As a consequence the effective stresses maintained within a sample after stress release decrease with increasing sample disturbance. Since there is increasing evidence that undrained shear strengths are dependent on the initial mean effective stress within the sample (Marsland and Powell, 1985), any decrease in the mean effective stress can lead to the measurement of low shear strengths. Whether or not this is the case depends not only on the degree of breakdown of pore suctions during sampling and storage but also on the increases in effective stresses which occur during applicaion of confining pressures on the now partially saturated soil. When stiff soils which have suffered small but significant shear strains during sampling are subjected to soil pressures significantly greater than the effective overburden, the mean effective stresses which develop in the sample may be substantially higher than the mean effective *in situ* stresses. As a consequence the use of confining pressures equal to the total *in situ* overburden (including water load) can lead to serious overestimation of strengths in some stiff clays. Not only can cavitation within the pore water of the clays result in the measurement of incorrect strengths, it can also lead to incorrect estimates of *in situ* stresses.

The assessment of the likely effects of sample disturbance on the parameters measured in laboratory tests is one of the most difficult and neglected aspects in the overall evaluation of laboratory data. For all geotechnical and fabric measurements it is vital to use the best available sampling techniques, but even these result in some sample disturbance. Direct assessment of the degree of disturbance of offshore samples taken by the best available techniques by comparing the results obtained with those from 'truly' undisturbed samples, such as carefully taken block samples or reliable large-scale tests, is not possible.

The first requirement in assessing the likely degree of disturbance is to have available a precise account, including the times, of all the significant stages of drilling and sampling or insertion, together with details of operational conditions and specific problems which may have arisen. For example, the possible reasons for poor or nil recovery and for obtaining heavily disturbed samples may be very relevant since many geotechnical problems arise due to the missing or poor recovery of strata which are critical in a particular foundation problem.

Indications of the degree of sampling disturbance can be obtained by examining radiographs of undisturbed samples, by visual inspection of split samples and by observing thin slices as they dry. Careful measurements of the moisture content gradients, variations in the microfabric and small-scale strength (using small penetrometers or fall cones) in different parts of a sample can provide more positive evidence.

Several investigators (for example, Broms, 1980) have suggested that the comparison of the mean all-round effective stress (σ'_m) on a sample with that which would act on an 'ideal' sample from the same location may provide a more quantitative measure of disturbance.

However, in practice problems arise in the determination of both the 'ideal' and the measured values of σ'_m. As a first approximation it is adequate to assume that stiff clays behave as elastic solids when

$$\sigma'_m \text{ (ideal)} = \frac{\sigma'_v + 2\sigma'_h}{3}$$

The measured values depend on the extent to which cavitation has occurred within the pore fluid. It is necessary to measure the equilibrium pore-water pressures under cell pressures of σ'_v, $2\sigma'_v$ and $4\sigma'_v$. Provided no cavitation has occurred, the values of σ'_m measured in a triaxial cell should be independent of the cell pressure. If σ'_m increases as the cell pressure increases, this provides evidence of possible stress release 'disturbance'. In such cases the cell pressure should be reduced to σ'_v and the corresponding value of σ'_m used in the relationship:

Degree of disturbance =

$$\frac{\sigma'_m \text{(ideal)} - \sigma'_m \text{(measured)}}{\sigma'_m \text{(ideal)}}$$

Values obtained from this expression should only be used as an indication of the degree of disturbance since low values of σ'_m (measured) can also be obtained on samples from localized weak zones. Low initial mean effective stresses and strengths measured on unconsolidated samples need to be very carefully assessed before they are neglected, since there is a possibility that these may be closer to the real *in situ* values than higher results obtained in other parts of the profile.

Although the above procedure provides a measure of the degree of sample disturbance, the effects on engineering parameters can best be determined by comparing the results obtained by different sampling and testing methods with those obtained from measurements of the foundation performance or the results of large *in situ* tests. Although significant progress has been made in comparative studies at onshore sites typical of some North Sea soils (Marsland, 1977, 1979; Marsland *et al.*, 1982; Marsland and Powell, 1985), there is a need for more basic studies both at these and other sites containing soils similar to those from other sections of the North Sea soil spectrum.

Reconsolidation under anisotropic effective stresses equal to the *in situ* values combined with sufficiently large back pressures to resaturate the specimens has been used to overcome some of the effects of

sampling disturbance (Andresen *et al.*, 1979). This approach is very reasonable in relatively insensitive soils, which have high *in situ* densities, but it can lead to the masking of real low *in situ* values, which may be present in looser soils. It also requires reasonable assessment of *in situ* effective horizontal stresses, which may be subject to considerable errors. Reconsolidation of clays with a very open microfabric can result in measured strengths considerably above the true *in situ* values. An extreme example of this is the Champlain Clay (Eden, 1970) where reconsolidation of good-quality thin-walled fixed-piston samples resulted in strengths significantly greater than those obtained from block samples.

The use of parameters from tests on specimens consolidated to pressures well above the *in situ* stresses and normalized with respect to the consolidation stress are advocated by some engineers (Young *et al.*, 1983). Although this approach can provide useful background data, the application of the results depends on obtaining reasonable estimates of overconsolidation ratios, which involve at least as much uncertainty as the measurements of undrained strengths. Such estimates are particularly difficult in soils that have been subjected to a sequence of post-depositional changes such as those which have occurred in North Sea soils. Even when applying this approach to soils exhibiting normalized behaviour, it is still necessary to use the best quality samples. Unfortunately in some soils consolidation to stresses well above the preconsolidation load can modify the soil fabric and lead to results that differ considerably from those at lower stresses.

The geotechnical properties of some 'fresh' undisturbed well-mixed tills, such as those at the Cowden, can be approximately reproduced in tests on remoulded reconstituted soils. Reasonable agreement is, however, only possible in a limited number of soils where the soil fabric can be artificially reproduced (Gens and Hight, 1979; Marsland and Powell, 1985).

The effects of the rate of shearing and

other time-dependent factors which arise in different types of tests require much more consideration. Recent studies on the effects of the rate of testing in a wide range of tests carried out by the BRE (Marsland and Powell, 1985) show that they are much more complex than is often realized. They not only vary with the type of test but also with the composition and fabric of the soil. In soils containing an extensive network of macrofabric discontinuities, increases in strength due to drainage during the tests can more than counteract reductions due to viscous rate effects, which occur as the rate of shear is decreased. Little attention has been given to the effects of time delays during sampling and *in situ* testing operations, even though these can have a considerable effect on the results. The breakdown of internal pore suctions in samples during storage (the stress relief effect) is another important time-dependent factor which varies with the stress levels, the type of soil and the soil fabric (Marsland and Powell, 1979; Marsland and Windle, 1982; Kirkpatrick and Khan, 1984).

Varying degrees of disturbance caused during installation of *in situ* tests, such as the pressuremeter test, can both affect the results and lead to difficulties in interpretation (Powell *et al.*, 1983; Powell and Uglow, 1985).

INTERPRETATION OF TEST DATA

Much more attention has recently been given to the interpretation of test data. Some aspects were considered by the author in a paper to the last SUT Conference (Marsland, 1979), and others formed the subject of the 24th Rankine Lecture (Wroth, 1984). However, the central problem of how to derive the most appropriate large-scale parameters from test data still remains. One way of improving interpretation is to establish better empirical relationships between test data and large-scale behaviour. Many existing correlations are poor because they are based on too wide a soil spectrum and often incorporate test data which, by todays

standards, would be considered inadequate. The author has attempted to overcome some of these problems by using results obtained from carefully executed large *in situ* plate tests made on land test-bed sites to evaluate results obtained from good-quality 'standard' tests such as can be made offshore. Although many of the results have been published (Marsland, 1971a,b,c, 1975; Marsland and Powell, 1979, 1985; Marsland *et al.*, 1982), there is still a need for greater intergration with detailed soil fabric studies at both land-based test-bed sites and offshore sites. Available comparisons have been used in the evaluation of test data from a recent North Sea site where more adequate details of the soil fabric were obtained (Marsland *et al.*, in literature).

Direct evaluation of relevant large-scale parameters from the result of one particular type and scale of tests is difficult in all but 'ideal' soils. This is particularly the case when the scatter of the results at a given depth down a profile is large. Averaging of results is only justified in soils containing well-distributed pockets (not layers) of relatively uniform soils containing no significant macrofabric features. In stiff clay containing randomly oriented discontinuities the large-scale strength is often close to or even below the lower bound values given by good-quality tests. Consistent results showing small scatter do not necessarily justify confidence since they may be as much a function of the test as of the large-scale properties of the ground. More confidence is possible when there is a reasonable measure of agreement between several different types and scales of tests, bearing in mind the different effects of stress paths and anisotrophy. Fabric profiles such as that given in Fig. 17, used together with detailed descriptions such as those illustrated in Fig. 18 and photographs of typical fabric features, are necessary to build up a more complete picture of the ground. They are also useful in explaining many of the anomalies and differences in test results which may occur.

The broad link between studies of soil

Depth (m)	Zone	Obser-vations Macro	Obser-vations Micro	Description of Principal Fabric Features	Fig Nos	Fabric Type
	I	●	●	Clay Till micro cracks No large discontinuities	12(a)(b)	
	II	●	●	Laminations of silt and clay Trans-sample discontinuities	13(a) 13(b)(c)	BC, EF
10	III (ı)	●	●	Near horizontal primary macro features still evident	14(a)	A → C EF
		●	●	Numerous small scale macro fissures and micro cracks	15	C, EF
		●		Trans-sample discontinuities	14(b)	
20	III (ıı)	●	●	Primary deposition features completely disrupted numerous small scale macro peds Some transample discontinuities Highly developed micro ped fabric	16(b) 16(a)	D, EF
30	III (ıı)	●		Primary depositional features completely disrupted Highly developed small scale macro ped fabric but larger and more interlocked than in III (ıı) Some trans-sample discontinuities	17	CD, E (few F)
40	IV (ı)	●		No large transample discontinuities observed except horizontal partings often covered with particles of sand Intact shells in upper part of zone Broken shells in lower part		BC
50	IV (ıı)	●		Small scale macro ped fabric well developed Some major discontinuities		C D E → F

Fig. 17 Fabric profile based on observations at location near E in Fig. 1. Note: figure numbers given in column 6 refer to figures in Marsland *et al*. (in literature)

fabric geological processes and geotechnical properties has already been established. It is possible to build up a reasonably comprehensive record of the past geological processes by combining detailed fabric studies with data obtained from investigations of fossil flora and fauna, estimates of overconsolidation, radiocarbon dating, stratigraphical variations from geophysical measurements, profiles obtained by static cone tests and piezocone tests, and borehole logging techniques. A good overall model of the 'fossil' geological processes, combined with observations of present-day depositional processes in similar environments, provides an indispensable background against which to interpret the soil data and its relevance to the large-scale situation.

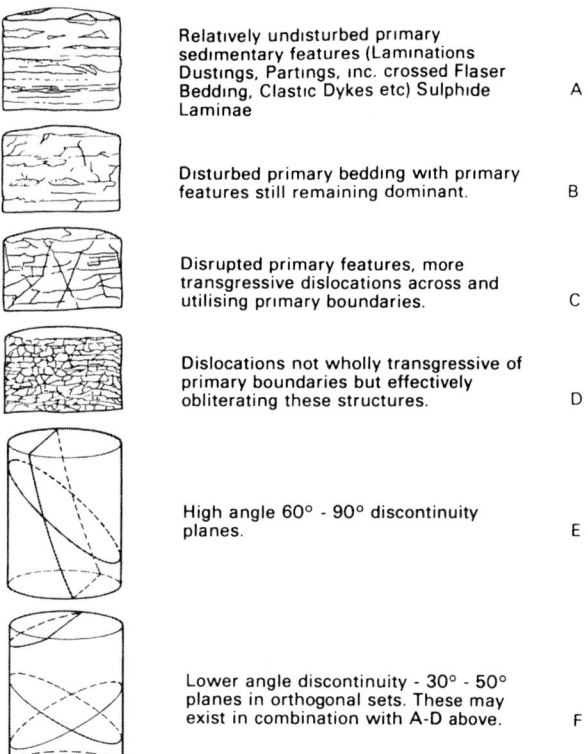

Relatively undisturbed primary sedimentary features (Laminations Dustings, Partings, inc. crossed Flaser Bedding, Clastic Dykes etc) Sulphide Laminae	A
Disturbed primary bedding with primary features still remaining dominant.	B
Disrupted primary features, more transgressive dislocations across and utilising primary boundaries.	C
Dislocations not wholly transgressive of primary boundaries but effectively obliterating these structures.	D
High angle 60° - 90° discontinuity planes.	E
Lower angle discontinuity - 30° - 50° planes in orthogonal sets. These may exist in combination with A-D above.	F

Fig. 18 Typical macrofabric features observed at location near E in Fig. 1. Marsland *et al*. (in literature)

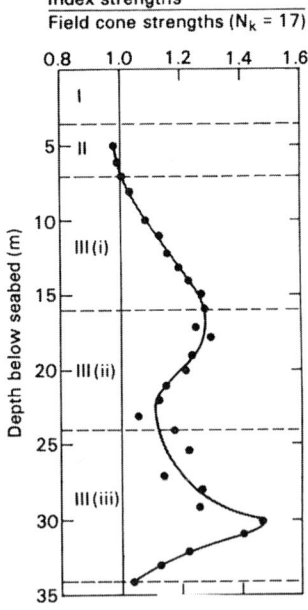

Fig. 19 Example of plot of ratios of two different scales of test down profile at location near E in Fig. 1: Marsland *et al*. (in literature)

Plots of the ratio of results from different types of tests when correlated with fabric types, such as in the example in Fig. 19, can provide additional clues to the value of data from particular tests and could even indicate the presence of fabric features not already observed.

CONCLUSIONS

Detailed knowledge of 'fossil' geological processes is essential for the adequate understanding and evaluation of geotechnical data. Detailed studies of soil fabric provide an important link between the geological processes and geotechnical behaviour. Fabric features at all scales have an important effect on both measured and large-scale geotechnical parameters. The effects of soil fabric on different types and scales of tests vary. Reliance should not be placed on the results of one or two different types of test. It is important to compare the results obtained from a range of tests and to relate them to the soil fabric, stress histories and applied stress paths. In this respect the Norwegian Geotechnical Institute's approach (Andresen *et al.*, 1979), which uses a combination of triaxial compression, triaxial extension and simple shear, has much to commend it. However, a more critical examination needs to be made of estimations of OCR and *in situ* stresses by taking into account all past geological processes. More consideration also needs to be given to the soil fabric and the representative nature of the test specimens. The inclusion of more comprehensive profiles of tests on unconsolidated specimens could provide additional checks and control. Measurements of porewater pressure responses prior to and during shearing should be made in as many of the unconsolidated specimens as possible (Marsland, 1977). Additional control could be obtained by simple measurements of the mean effective stresses within specimens soon after sampling. Such measurements, together with appropriate observation of soil fabric in different parts of the specimens, could help in the evaluation of the possible effects of disturbance during sampling and insertion of *in situ* test equipment. In order to explain the varying ratios between results from different types and scales of tests and between tests and large-scale behaviour, it is necessary to consider the changes in fabric features down the soil profile as well as differences between the stress paths relevant in each case. More correlations between large-scale behaviour and high-quality test data in soils with well-documented fabrics are urgently required. The interdependence of data from all the relevant sources and the importance of continual interaction throughout an investigation of specialists in the various fields cannot be overestimated.

ACKNOWLEDGEMENTS

The work described in this paper forms part of a research programme of the Building Research Establishment and is published by permission of the Director. The author wishes to thank Dr E. Derbyshire and Mr M. A. Love of Keele University for their close co-operation in many aspects of the work.

REFERENCES

1. Aarseth, I. and Sejup, H. P. (Eds) 1984. Quaternary stratigraphy of the North Sea. Abstract Volume, *Symp. Univ. Bergen*, Dec. 1984, Bergen, Norway.
2. Andersen, A., Berre, T., Kleven, A. and Lunne, T. 1979. Procedures to obtain soil parameters for foundation engineering. *Mar. Technol.* 3(3), 201–266.
3. Behre, K. E., Menke, B. and Streif, M. 1979. The Quaternary geological development of the German part of the N. Sea. In *The Quaternary History of the North Sea* (Eds E. Oele, R. T. E. Schuttenhelm and A. J. Wiggers). Acta Univ. Uppsala, Symp. Univ. Uppsala Annum Quingentesimum Celebrantis, Vol. 2. Uppsala, Sweden, pp. 85–113.

4. Bishop, A. W. and Little, A. L. 1967. The influence of size and orientation of sample on the apparent strength of the London Clay at Maldon, Essex. In *Proc. Geotechnical Conference on the Shear Strength Properties of Natural Soils and Rocks*, Oslo, 1967, Vol. 1, pp. 89–96.

5. Boulton, G. S. 1970. On the deposition of subglacial and melt-out tills at the margin of certain Svalband glaciers. *J. Glacoil.* **9**, 231–245.

6. Boulton, G. S. 1972. Modern Arctic glaciers as depositional models for former ice sheets. *J. Geol. Soc., London* **128**, 361–393.

7. Boulton, G. S. and Deynoux, M. 1981. Sedimentation in glacial environments and the identification of tills and tillites in ancient sedimentary sequences. *Precam. Res.* **15**, 397–422.

8. Boulton, G. S. and Paul, M. A. 1976. The influence of genetic processes on some geotechnical properties of glacial tills. *Q. J. Eng. Geol.* **9**, 159–194.

9. Boulton, G. S., Jones, A. S., Clayton, K. M. and Kenning, M. J. 1977. A British ice-sheet model and pattern of glacial erosion and deposition in Britain. In *British Quaternary Studies: Recent Advances* (Ed. F. W. Shotton). Clarendon Press, Oxford (551.79).

10. Broms, B. B. 1980. Soil sampling in Europe. *J. Geotech. Engng Div. Proc. Am. Civ. Eng.* **106**(GT1), 65–98.

11. Chamberlain, E. W. and Gow, A. J. 1978. Effects of freezing and thawing on the permeability and structure of soils. In *Proc. Int. Symposium on Ground Freezing*, Ruhr University, Bochum, pp. 31–34.

12. Derbyshire, E. and Love, M. A. 1985. Microshears in diamicts from the North Sea Basin. In *Scanning Electron Microscope in Geology* (Ed. W. B. Whalley). Geobooks, Norwich.

13. Derbyshire, E., Love, M. A. and Edge, M. J. 1985. Fabrics of probable segregated ground ice origin in some sediment cores from the North Sea Basin. In *Soils and Quaternary Landscape Evolution* (Ed. J. Boardman). John Wiley, Chap. 11.

14. Eden, W. J. 1970. Sampler trials in overconsolidated sensitive clay. *Proc. Symp. on Sampling of Soil and Rock*, Toronto, Canada. STP 483, American Society for Testing and Materials, pp. 132–142.

15. Gens and Hight. 1979. *Proc. 7th European Conference on Soil Mechanics and Foundation Engineering*, Brighton, September 1979, Vol. 4, pp. 57–65.

16. Jardine, W. G. 1979. The Western (United Kingdom) Shore of the North Sea in Late Pleistocene and Holocene times. In *The Quaternary History of the North Sea* (Eds E. Oele, R. T. E. Schuttenhelm and A. J. Wiggers). Acta Univ. Uppsala, Symp. Univ. Uppsala Annum Quingentesimum Celebrantis, Vol. 2. Uppsala, Sweden, pp. 159–174.

17. Jelgermsa, S. 1979. Sea level changes in the N. Sea Basin. In *The Quaternary History of the North Sea* (Eds E. Oele, R. T. E. Schuttenhelm and A. J. Wiggers). Acta Univ. Uppsala, Symposium Univ. Uppsala Annum Quingentesimum Celebrantis, Vol. 2. Uppsala, Sweden, pp. 233–248.

18. Johnston, G. H. (Ed.) 1981. *Permafrost Engineering Design and Construction*. John Wiley, pp. 52–53.

19. Kirkpatrick, W. M. and Khan, A. J. 1984. The reaction of clays to sampling stress relief. *Geotechnique* **34**(1), 29–42.

20. Liljequist, G. H. 1974. Notes on meteorological conditions in connection with continental land-ices. In *Pleistocene Gelogiska Foreningens i Stockholm Forhandlingar*, Vol. 96, pp. 293–298.

21. Løken, T. 1976. Geology of superficial sediments in the Northern North Sea. In *Proc. 1st Int. Conf. on Behaviour of Offshore Structures*, Trondheim, August 1976, pp. 501–515.

22. Love, M. A. and Derbyshire, E. 1985. Micro fabric of glacial soils and its quantitation measurements. In *Proc. Int. Conf. Construction of Glacial Tills and Boulder Clays*, Edinburgh, March 1985, pp. 129–133.

23. Lupini, J. F., Skinner, A. E. and Vaughan, P. R. 1981. The drained residual strength of cohesive soils. *Geotechnique* **31**(2), 181–213.

24. Mackay, R. J. 1974. Reticulate ice veins in permafrost Northern Canada. *Can. Geotech. J.* **II**, 230–237.

25. Marsland, A. 1971a. The shear strength of stiff fissured clays. In *Proc. Roscoe Memorial Symp*, Engineering Department, University of Cambridge, 29–32 March 1971. T. J. Foulis, Henley-on-Thames, pp. 59–68.

26. Marsland, A. 1971b. Large *in-situ* tests to measure the properties of stiff fissured clays. In *Proc. First Australian—New Zea-*

land Conf. on Geomechanics, Melbourne 1971, Vol. 1, pp. 180–189.

27. Marsland, A. 1971c. Laboratory and *in-situ* measurements of the deformation moduli of London Clay. In *Proc. Symp. on Interaction of Structure and Foundation*, Birmingham, July 1971. Midland Soil Mechanics and Foundation Engineering Society, pp. 7–17.

28. Marsland, A. 1975. *In-situ* and laboratory test on glacial clays at Redcar. In *Proc. Symp. on Behaviour of Glacial Materials*, Birmingham, April 1975. Midland Soil Mechanics and Foundation Engineering Society, pp. 164–180.

29. Marsland, A. 1977. The evaluation of the engineering design parameters for glacial clays. *Q. J. Eng. Geol.* 10(1), 1–26.

30. Marsland, A. 1979. The interpretation of *in-situ* tests in glacial clays. In *Proc. Int. Conf. on Offshore Site Investigation*, London, March 1979. Society for Underwater Technology, pp. 218–228.

31. Marsland, A. and Butler, M. E. 1967. Strength measurements on stiff fissured Barton Clay from Fawley Hampshire. In *Proc. Geotechnical Conference on the Shear Strength Properties of Natural Soils and Rocks*, Olso, 1967, Vol. 1, pp. 139–146.

32. Marsland, A. and Powell, J. J. M. 1979. Evaluating the large scale properties of glacial clays for foundation design. In *Proc. 2nd Conf. on Behaviour of Offshore Structures*, Vol. 1, pp. 193–214.

33. Marsland, A. and Powell, J. J. M. 1985. Field and laboratory investigations of the clay tills at the Building Research Establishment test site at Cowden, Holderness. In *Proc. Int. Conf. on Construction in Glacial Tills and Boulder Clays*, Edinburgh, March 1985, pp. 147–168.

34. Marsland, A. and Quaterman, R. S. T. 1982. Factors affecting the measurement and interpretation of quasi static penetration tests in clays. In *Proc. 2nd European Symp. on Penetration Testing*, Amsterdam, May 1982 (Eds A. Verruijt, F. L. Beringen and E. H. de Leeuw). Published by A. A. Balkema, Vol. II, pp. 697–702.

35. Marsland, A. and Windle, D. 1982. Developments in offshore site investigation. *Proc. Oceanology International Conference*, Brighton, March 1982, Vol. 1, Paper 0182.7.

36. Marsland, A., Love, M. A., Loken, T. and Lunne, T. (in literature). The use of fabric studies in evaluating the engineering properties of clays in a North Sea Site investigation. (Presented for publication.)

37. Marsland, A., Prince, A. and Love, M. A. 1982. The role of soil fabric studies in the evaluation of the engineering parameters of offshore deposits. In *Proc. 3rd International Conference on the Behaviour of Offshore Structures*. MIT, Boston, Vol. 1, pp. 181–202.

38. McGown, A. 1979. Discussion on paper by Gens and Hight, 1979. *Proc. 7th European Conference on Soil Mechanics and Foundation Engineering*, Brighton, September 1979, Vol. 1, pp. 131–133.

39. McGown, A. and Radwan, A. M. 1975. The presence and influence of fissures in the boulder clays of west central Scotland. *Can. Geotech. J.* 12, 24–97.

40. Oele, E. and Schuttenhelm, R. T. E. 1979. Development of the North Sea after the Saalian glaciation. In *The Quaternary History of the N. Sea* (Eds E. Oele, R. T. E. Schuttenhelm and A. J. Wiggers). Acta Univ. Uppsala, Symposium Univ. Uppsala Annum Quingentesimum Celebrantis, Vol. 2, Uppsala, Sweden, pp. 191–215.

41. Powell, R. D. 1984. Glaciomarine processes and inductive lithofacies modelling of ice shelf and tide water glacier sediments based on Quaternary examples. *Mar. Geol.* 57, 1–52.

42. Powell, J. J. M. and Uglo, I. M. 1985. A comparison of Menard, self-boring and push-in pressuremeter tests in a stiff clay till. Paper 13, this volume.

43. Powell, J. J. M., Marsland, A. and Al Khafazi, A. N. 1983. Pressuremter Testing of Glacial Tills. In *Proc. International Symposium on In-Situ Testing*, Paris, Vol. 2, pp. 373–378.

44. Redding, J. 1976. Glacial genesis of North Sea soils. In *Offshore Soil Mechanics* (Eds P. George and D. Wood). Engineering Department, University of Cambridge.

45. Rowe, P. W. 1972. The relevance of soil fabric to site investigation. *Geotechnique* 22, 193–300.

46. Schjetne, K. and Brylawski, E. D. 1979. Offshore soil sampling in the North Sea. In *Proc. Int. Symp. on Soil Sampling*, Singapore, 1979. The Sub-Committee on Soil Sampling, Int. Soc. Soil Mech. and Found. Engng, pp. 139–156.

47. Schjetne, K., Anderson, K. H., Lauritssen, R. and Hansteen, O. E. 1979. Foundation Engineering for Offshore Gravity Structures. *Mar. Geotech.* 3(4).

48. Van Vliet, N. and Langohr, R. 1981. Correlation between fragipans and permafrost — with special reference to Weichsel silty deposits in Belgium and Northern France. *Cantena* 8, 137–154.

49. Ward, W. H., Marsland, A. and Samuels, S. G. 1965. Properties of the London clay at the Ashford Common Shaft: *in-situ* and undrained strength tests. *Geotechnique* 15, 321–344.

50. Wroth, C. P. 1984. The interpretation of *in-situ* soil tests. *Geotechnique* 34(4), 449–489.

51. Young, A. G., Quiros, G. W. and Ehlers, C. J. 1983. The effects of offshore sampling and testing on undrained soil shear strength. *Proc. 15th Annual Offshore Technology Conference*, Houston, Texas. Paper OTC 4465.

16

Evaluation of Geotechnical Parameters from Triaxial Tests on Offshore Clay

D. W. Hight, Geotechnical Consulting Group, UK,
A. Gens, Universidad de Cataluña, Barcelona, Spain,
and R. J. Jardine, Imperial College of Science and Technology, London, UK

The pattern of undrained behaviour for young sedimented clays of low and medium plasticity is described. This pattern is used to assess the potential effects of different forms of sampling on the stress–strain–strength properties measured in laboratory tests on typical offshore clays. Methods of determining *in situ* behaviour from laboratory tests on retrieved soil samples are considered.

INTRODUCTION

The undrained stress–strain–strength properties of young sedimented clays, as measured in the triaxial apparatus, are described in this paper. These properties are used to examine some of the potential effects of sampling and to consider how the behaviour observed on unconsolidated undrained tests on samples retrieved from the seabed relates to *in situ* behaviour.

Results from studies on the following three reconstituted soils will be presented:

(a) Lower Cromer Till (LCT);
(b) London Clay (L); and
(c) a glacial clay from a site in the Northern North Sea (R).

Gradings of these clays are shown in Fig. 1, together with details of their plasticity. In a review of clay sediment types in the Northern North Sea (Hight, 1983), four groupings emerged on the basis of soil grading and plasticity. Grading envelopes for these four groups are also shown in Fig. 1, from which it can be seen that the soils under study provide a reasonable coverage of typical Northern North Sea clays.

For each clay, blocks of reconstituted soil have been formed by K_0 consolidation from slurry, using a 225 mm diameter oedometer. For LCT the soil blocks were generally swelled to an overconsolidation ratio (OCR) of 4 ($\sigma'_{v,max}$ = 200 kPa) and for L and R the soil blocks were swelled to an OCR of 2 ($\sigma'_{v,max}$ = 400 kPa). Specimens cut from these blocks have been reconsolidated in the hydraulic triaxial cell and taken along the virgin

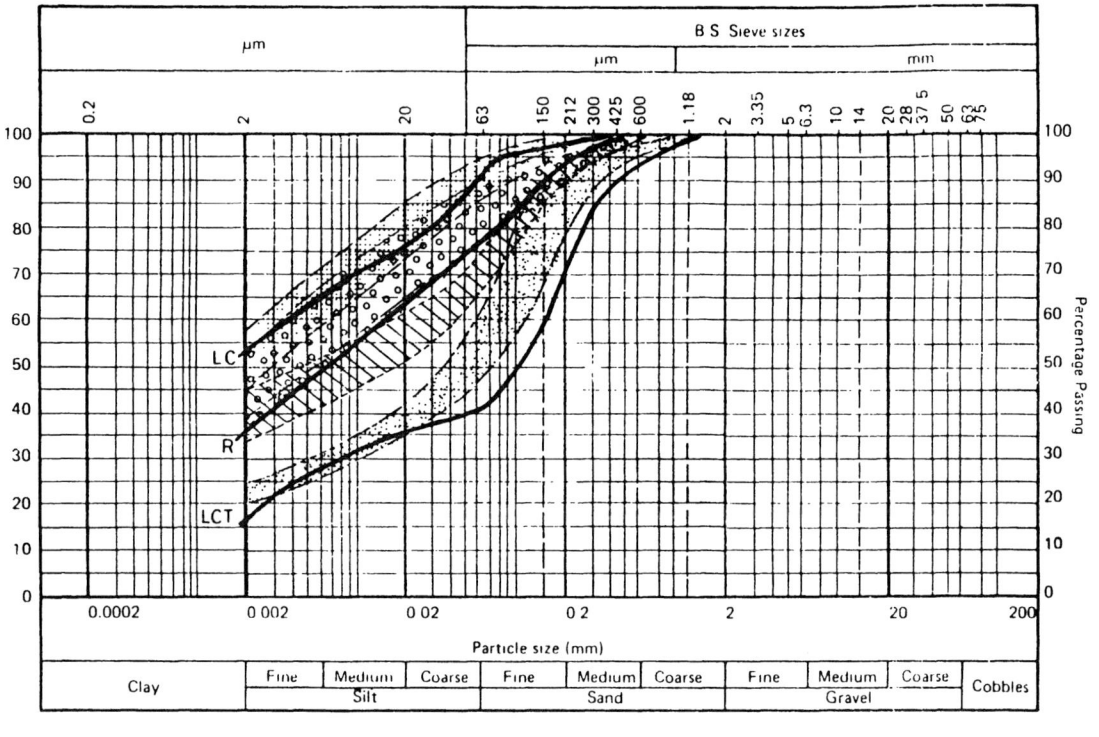

Symbol	Clay	LL (%)	PL (%)	PI (%)
LCT	Lower Cromer Till	25	12	13
R	North Sea Clay	32	15	17
L	London Clay	75	28	47
	North Sea Clay Groups			
	A	26 – 30	12 – 14	13 – 15
	B	35 – 40	14 – 17	18 – 23
	C	42 – 48	17 – 20	25 – 28
	D	50 – 55	18 – 22	31 – 34

Fig. 1   Gradings and index properties of the clays

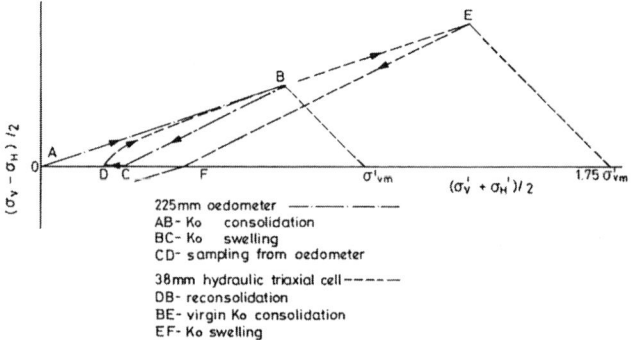

Fig. 2   Stress paths prior to undrained shear

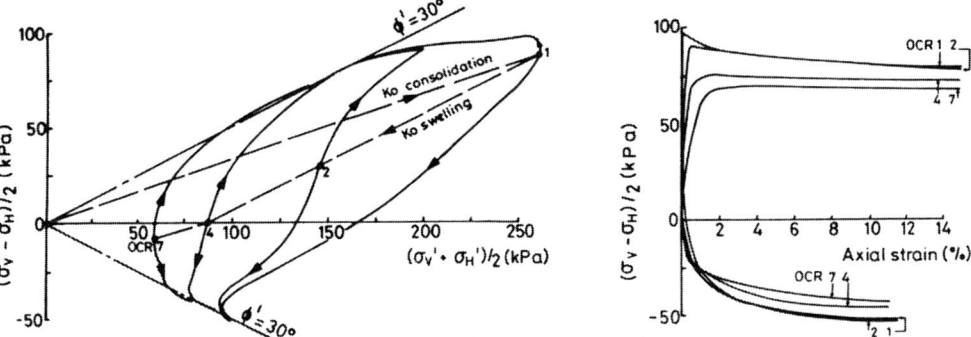

'ig. 3   Effective stress paths and stress–strain curves for undrained shear in Lower Cromer Till

K_0 consolidation line to a vertical effective stress at least 1.75 times the maximum value applied in the oedometer (Fig. 2); in this way the effects of sampling from the blocks have been removed (Gens, 1982). Specimens have been sheared undrained in triaxial compression and extension from a normally consolidated state, or from a variety of overconsolidated states produced by swelling back under K_0 conditions.

The relevance of studies on reconstituted soil to the *in situ* behaviour of sedimented clays has been demonstrated by Gens and Hight (1979) for the case of low-plasticity clays. Their data suggest that for these soils the resedimented material provides a good model for *in situ* behaviour, even if the deposit contains unfilled fissures and other inhomogeneities. With more plastic soils, which are aged or fissured and which may develop residual fabric during shear, complete agreement between the behaviour of soil *in situ* and of that reconstituted is unlikely. An important distinction also has to be made between the behaviour of sedimented clays, such as glaciomarine tills, which are deposited at a high liquidity index, and the behaviour of clays deposited at a low liquidity index, such as lodgement tills; the latter require a simple extension to the framework for sedimented clays that will be described herein.

The behaviour observed in the K_0-consolidated tests on reconstituted soil will be assumed, therefore, to represent the *in situ* behaviour of a wide range of sedimented North Sea clays.

PATTERN OF UNDRAINED BEHAVIOUR

The basic pattern for the undrained behaviour of sedimented clays can be illustrated by way of the effective stress paths and stress–strain curves shown in Fig. 3 for Lower Cromer Till. The key features to note are:

(a) In triaxial compression, the normally consolidated soil exhibits pronounced undrained brittleness (or strain softening); this brittleness reduces with increasing OCR and is absent in triaxial extension.

(b) Although ultimate* strengths can be consistently related to water content, in the manner described by Schofield and Wroth (1968), peak triaxial compression strengths cannot (Fig. 4).

(c) There is anisotropy of ultimate strength and stiffness, both being lower in triaxial extension.

(d) Strains to mobilize strength are small, except for heavily overconsolidated samples.

(e) The undrained stress paths for triaxial compression and extension of normally consolidated soil appear to act as a boundary to the paths for lightly overconsolidated soil. Indeed, when normalized using the methods suggested by Burland (1967), these two normalized

---

* Ultimate strength is taken here to be the strength measured at large strains (20% in compression and 10% in extension) in the triaxial apparatus.

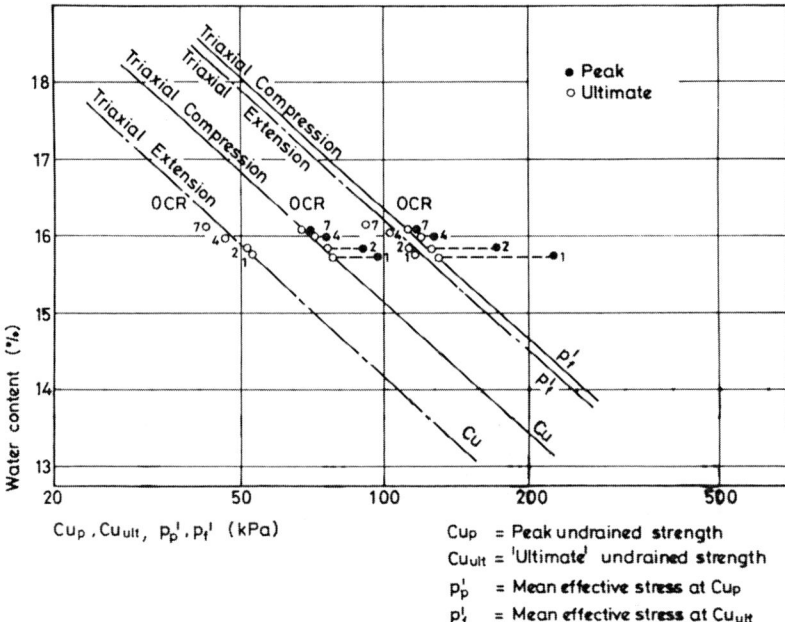

Fig. 4 Undrained strength and water content relationship for Lower Cromer Till

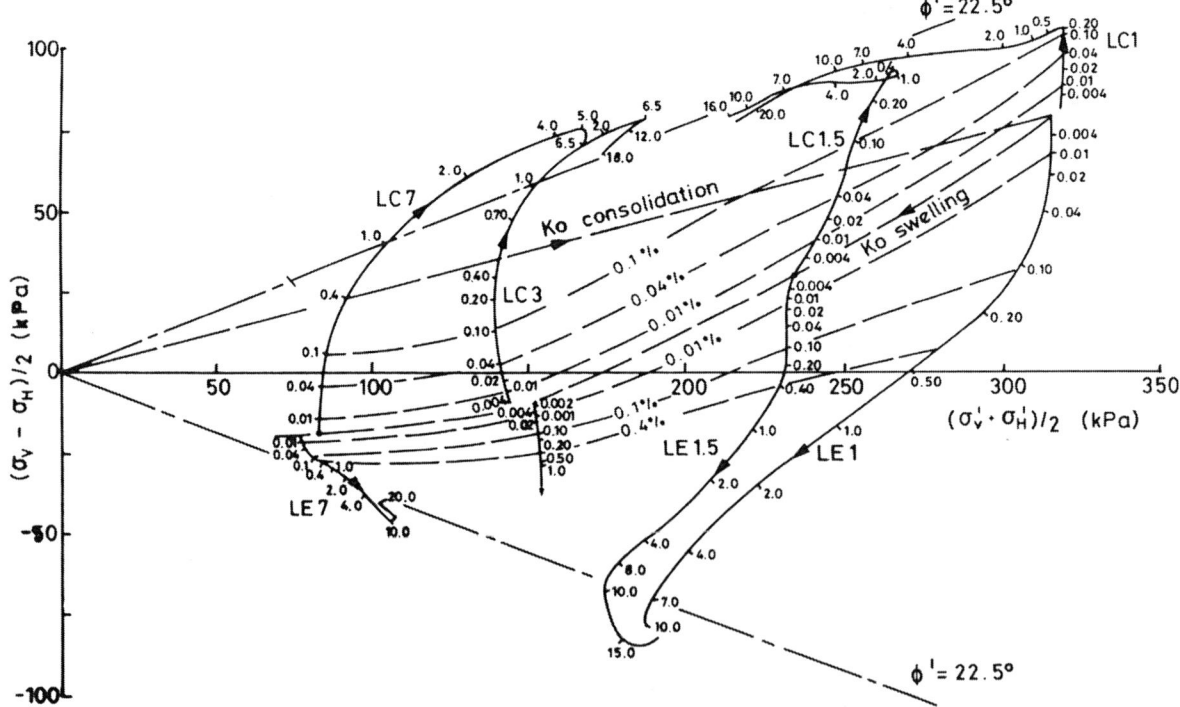

Fig. 5 Pattern of undrained behaviour for reconstituted London Clay (axial strains indicated in %)

stress paths form part of a bounding surface to the undrained behaviour of the
K_0-consolidated soil.

learly, the overall behaviour is governed
Clearly, the overall behaviour is governed
by the pre-shearing position on the swelling
line, EF in Fig. 2, and by direction of shearing. The changes in behaviour with OCR can
be considered to result from changes in
stress state and water content. Normaliza-
tion to eliminate the effects of water content
changes reveals that it is the effect of the
stress state that dominates. This has an
important bearing on sampling since, as will
be seen, one of its effects is to change the
effective stress state.

This same pattern of behaviour has been
found to apply to other young sedimented
clays of higher plasticity. Figures 5 and 6
show the effective stress paths from a similar series of triaxial compression and extension tests on reconstituted specimens of
London Clay and the North Sea clay. In the
experimental work on these two clays, the

stress–strain properties have been explored
in more detail, particularly over the strain
range 0.001 to 1%, using internally mounted
displacement-measuring devices based on
electrolevels (Burland and Symes, 1982).
The internally measured axial strains are
shown plotted along each stress path in Figs
5 and 6, and contours of equal axial strain
have been sketched.

The internal measurements of strain have
revealed an initially very stiff and non-linear
response inside the 0.1% major principal
strain contour. This response appears to be
typical of a wide range of soils (Jardine *et
al.*, 1984) and has an extremely important
bearing on their *in situ* behaviour and on the
interpretation of *in situ* tests carried out in
them (Jardine *et al.*, 1985). The 0.1% major
principal strain contour will be taken as the
boundary to the small-strain region or zone.

These features of the stress–strain properties can best be visualized by plotting
stress–strain data on a semi-logarithmic
scale; data for the North Sea clay are shown

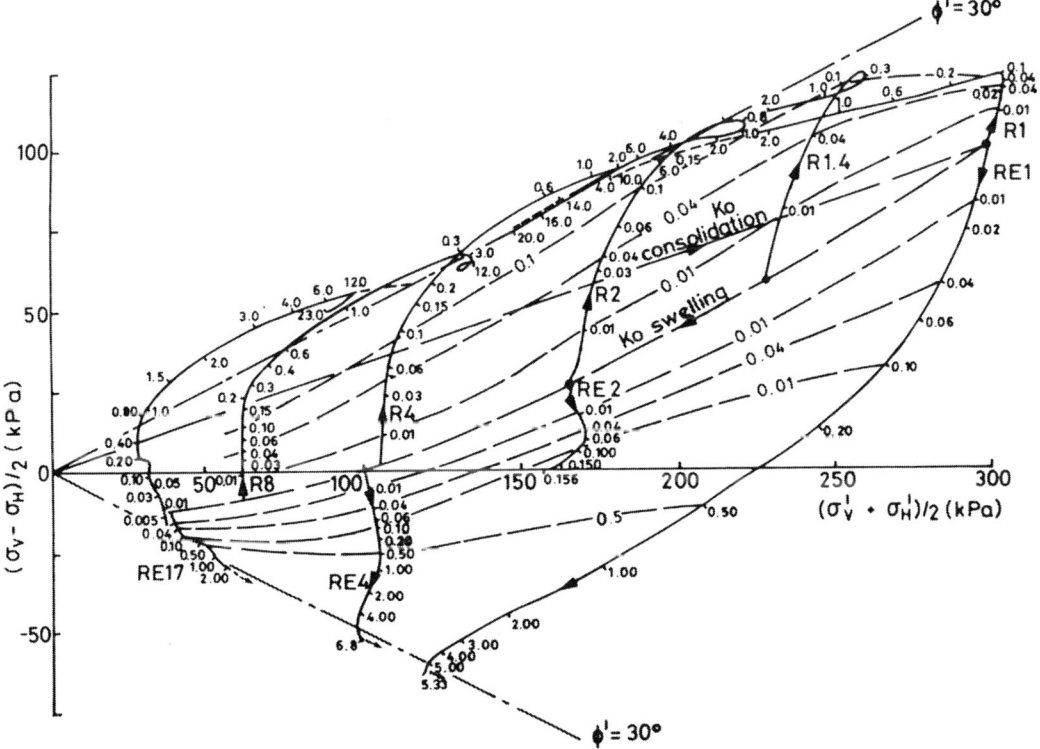

6   Pattern of undrained behaviour for reconstituted North Sea clay (axial strains indicated in %)

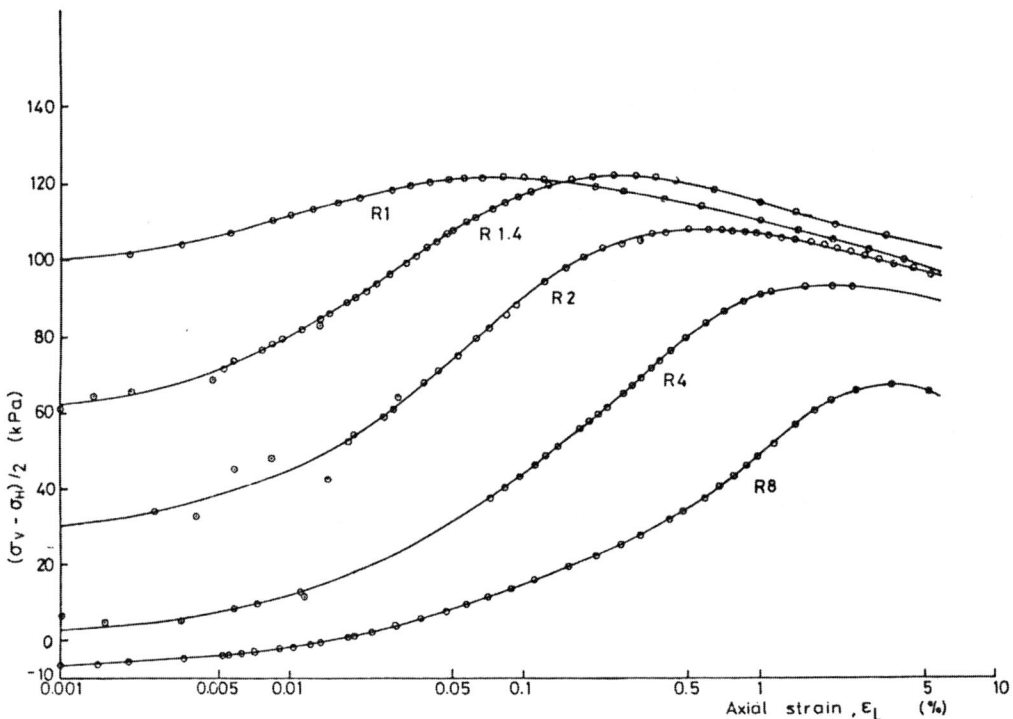

Fig. 7 Stress–strain details for reconstituted North Sea clay

TABLE 1
Stiffness indices

	$\dfrac{(E_u)_{0.01\%}}{p'_0}$	$L = \dfrac{E_{u(0.1)\%}}{E_{u(0.001)\%}}$	OCR	$\dfrac{p'_0}{(kPa)}$
'In situ'				
R1	830	0.185	1	267
R1.4	2180	0.270	1.4	206
R2	2270	0.353	2.05	158
R4	2130	0.386	3.73	106
R8	1740	0.407	7.4	65
After 'perfect sampling'				
PS1	1371	0.404	1*	193
PS8	852	0.58	7.4*	62
After 'block sampling'				
IS1	1661	0.435	2*	55
IS2	1436	0.42	2*	72
After 'tube sampling'				
I1 (UU)	1080	0.333	1.1[†]	474
I2 (reconsolidated)	1460	0.187	1.1[†]	508
I3 (UU)	2030	0.340	>50[†]	46

*Before sampling
[†]Estimated

plotted this way in Fig. 7. With non-linear materials it is impossible to define stiffness without reference to the level of strain or stress at which it is evaluated. A convenient description is provided by the normal secant modulus, E_u, determined over fixed strain increments and plotted against strain, again on a logarithmic scale. Jardine *et al.* (1984) present data for the North Sea clay this way, with E_u normalized by triaxial compression strength, C_u. Here, we normalize E_u at any particular strain by E_u at 0.01% strain and display this normalized ratio, $L(\varepsilon)$, against log strain (Fig. 8). This plot, together with one showing the variation of E_u at 0.01% strain with OCR (Fig. 9), provides a complete description of the stress–strain properties. In Fig. 9, E_u at 0.01% strain has been normalized by p'_0, the initial mean effective stress prior to shear; it could equally well have been normalized by vertical effective stress, σ'_v, or by C_u. Jardine *et al.* (1984) propose the use of two indices to characterize these stress–strain properties: $(E_u)_{0.01\%}/p'_0$ and L, the value of $L(\varepsilon)$ at an axial strain of 0.1%. The former provides a measure of the size of the small strain region and the latter is an indicator of

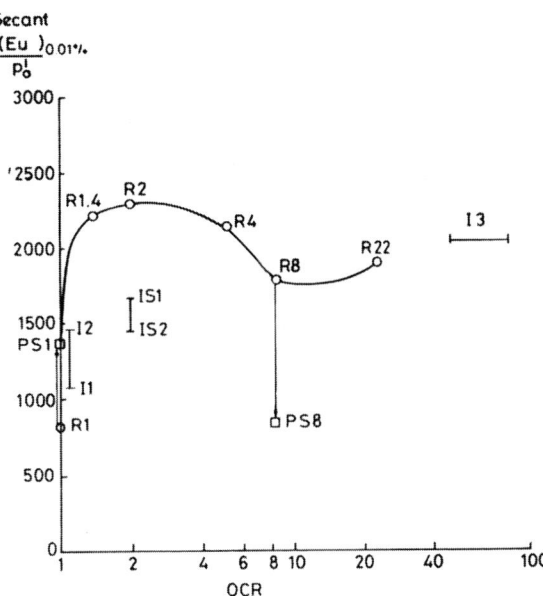

Fig. 9 $(E_u)_{0.01\%}/p'_0$ versus OCR before and after sampling

non-linearity, $L = 1$ indicating linear behaviour. These two indices will be used to examine the effects of sampling on stress–strain properties (refer to Table 1).

The features to note in Figs 5 to 9 regarding the stress–strain properties of these

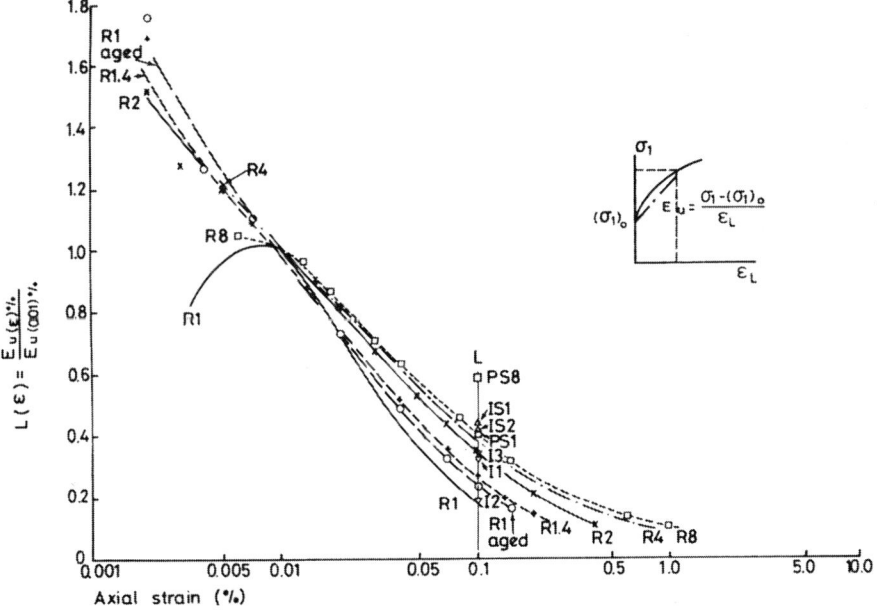

Fig. 8 Variation of secant stiffness with axial strain

sedimented clays are:

(a) the size of the small-strain region is dependent on OCR and p'_0;
(b) for each OCR, the initial stiffness varies with strain, although the degree of non-linearity reduces with increasing OCR;
(c) yielding, evident from a sharp deviation in the effective stress path, occurs at axial strains less than 0.2%;
(d) major dilation/contraction and, therefore, major distortion to the inital soil fabric begins at strains of 1 or 2%.

This general pattern of behaviour for sedimented clays, especially the location of the bounding surface and the $\varepsilon_1 = 0.1\%$ contour, is modified by both rate of shear and ageing. These considerations are beyond the scope of this paper.

SAMPLING EFFECTS

Traditionally the effects of sampling have been considered by dealing separately with:

(a) *perfect sampling*, in which the *in situ* shear stress is removed and undrained shear to failure starts from an isotropic stress, and
(b) *imperfect sampling*, in which some arbitrary stress path is assumed to be applied before undrained shearing to failure.

We will retain this distinction and subdivide

imperfect sampling into *block* and *tube sampling*. Although it is recognized that block sampling is not feasible offshore, its treatment allows the effects of sampling to be introduced in stages. No account is taken here of the effects of advancing the borehole.

PERFECT SAMPLING

The advantage of considering *perfect* sampling is that it is a well-defined path which illustrates certain features of behaviour that can then be anticipated in *imperfect* samples. It is well known that, even for the hypothetical case of perfect sampling there can be a change in strength and stress–strain behaviour from that for the *in situ* soil (e.g. Skempton and Sowa, 1963). This is illustrated for the case of the North Sea clay in Fig. 10; perfect sampling is shown from two *in situ* conditions:

(1) normally consolidated, perfect sampling path AB and reloading path BC (test PS1); and
(2) overconsolidated soil (OCR = 7.4), perfect sampling path DE and reloading path EF (test PS8).

The path for reloading in test PS1 has moved inside the bounding surface provided by the stress paths for triaxial compression and extension on the normally consolidated soil. The path intercepts the bounding sur-

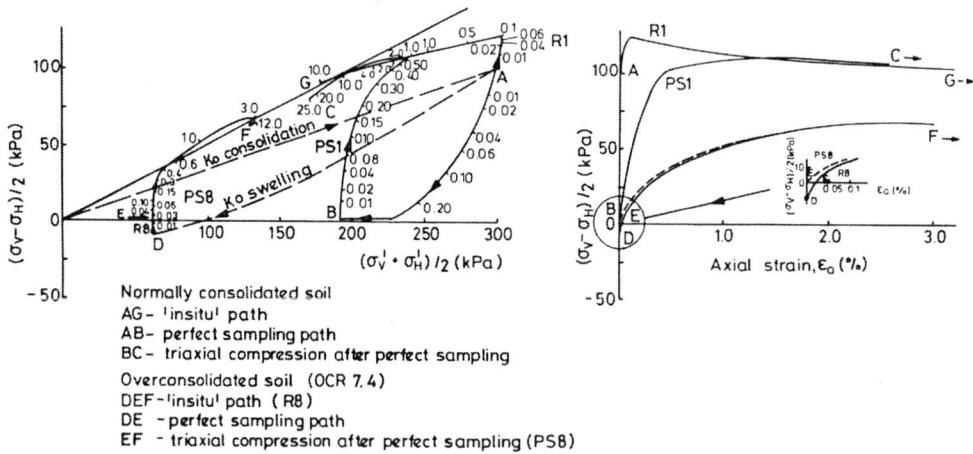

Normally consolidated soil
AG – 'insitu' path
AB – perfect sampling path
BC – triaxial compression after perfect sampling

Overconsolidated soil (OCR 7.4)
DEF – 'insitu' path (R8)
DE – perfect sampling path
EF – triaxial compression after perfect sampling (PS8)

Fig. 10   Perfect sampling of normally consolidated and heavily overconsolidated soil

face at a lower point than the path for the *in situ* soil, and, as a result, its peak undrained strength and undrained brittleness are reduced by perfect sampling. Since the water content has not been changed, the ultimate strength is unaffected and this parameter provides a lower bound to strength after sampling.

The changes in the stress–strain properties may be viewed in terms of the two indices $(E_a)_{0.01\%}/p'_0$ and L. The former, based on p'_0 prior to shear, increases from 830 to 1371 (cf. R1 and PS1 in Fig. 9 and Table 1); the latter increases from 0.185 to 0.404 (cf. R1 and PS1 in Fig. 8 and Table 1).

For the case of the overconsolidated soil, in which $K_0 > 1$, the perfect sampling path is identical to the initial section of the triaxial compression path from *in situ* conditions. There is, therefore, no change in strength. However, the stress–strain properties of the perfect sample have as their origin point E, as opposed to point D for the *in situ* soil. Part of the small-strain region has already been traversed along DE. Consequently, both indices of stiffness and linearity are modified; in this case $(E_u)_{0.01\%}/p'_{00}$ has been reduced and L increased (see Figs 8 and 9 and Table 1, and compare PS8 and R8).

The contrasting changes in stress–strain characteristics after perfect sampling of specimens with OCRs of 1 and 7.4 reveal an additional important aspect of sampling, namely the direction of its final stress path relative to that followed in the subsequent triaxial tests. This can be illustrated in terms of the small-strain zone which, based on the data for PS1, appears to be carried around stress space. In Fig. 11 this small-strain zone is sketched for specimens which are:

(a) normally consolidated (OCR = 1);
(b) perfectly sampled from a normally consolidated state (PS1); and
(c) heavily overconsolidated (R8/PS8).

These zones have been identified on the basis of the data shown in Figs 6 and 10 and on additional tests. Processes such as perfect sampling can modify the size and shape of the zones and also alter their position in relation to the state of stress prior to shear.

The following points should be noted:

(a) When the direction of the stress path reverses (as for PS1) the full small-strain zone is traversed; this can result in an increase in $(E_u)_{0.01\%}$ even if the overall size of the zone has diminished.
(b) When the direction of the stress path in triaxial compression is a continuation of that followed during sampling (PS8) some or all of the small-strain zone is missed.
(c) When the small-strain zone approaches the bounding surface it becomes distorted and the stress–strain characteristics are modified accordingly.

In reality, sampling involves a release of total stresses and a delay between unloading and subsequent reimposition of total stresses and shearing to failure. Both delays and the reimposition of stress are likely to modify the location of the small-strain zone. Evidence suggests that the distance to the nearest point on the zone's boundary is likely to increase with time.

On the basis of the simple stress paths involved in *perfect* sampling, we can anticipate that, as a result of *imperfect* sampling, there will be changes in strength and stress–strain response in subsequent undrained loading. Furthermore, the size and form of these changes will depend on the initial *in situ* stress state and stress history, and on the directions of straining involved in the sampling process.

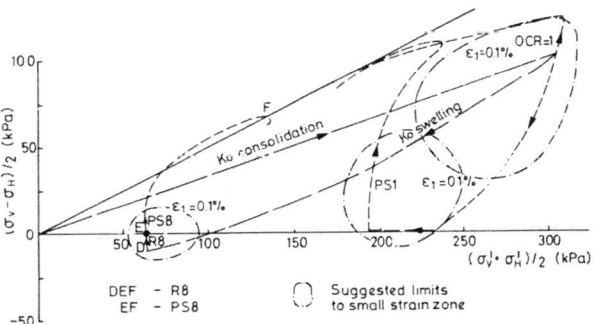

Fig. 11 Small-strain zones

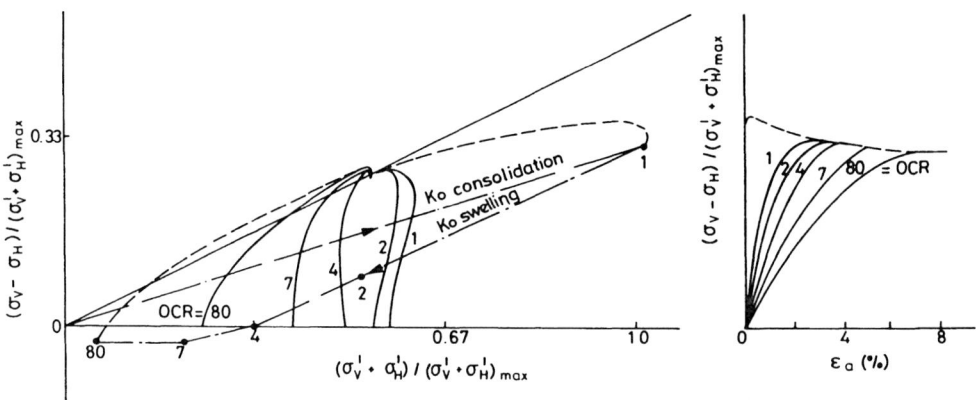

Fig. 12 Unconsolidated undrained triaxial compression of samples cut from blocks of reconstituted Lower Cromer Till

BLOCK SAMPLING

Releasing the blocks of reconstituted soil from the large oedometer and trimming specimens from the blocks can be considered to represent a number of the features of block sampling. The results of unconsolidated undrained triaxial compression tests run on specimens of Lower Cromer Till prepared in this way are presented in Fig. 12. The specimens were cut from blocks having different stress histories (OCRs of 1, 2, 4, 7 and 80). It can be seen that the effect of block sampling largely obliterates the important effects of stress history on *in situ*

behaviour. Specimens tend towards a similar initial effective stress and, as a consequence, show similar behaviour. As could have been anticipated from the results of perfect sampling, peak strengths and undrained brittleness are reduced in the normally and lightly overconsolidated soil.

The effect of 'block sampling' on the detailed stress–strain characteristics can be assessed from the results of two similar UU tests on specimens of the North Sea clay cut from reconstituted soil blocks (Fig. 13). The values of $(E_u)_{0.01\%}/p'_0$ are 1661 and 1436; the values of L are 0.435 and 0.42. These are plotted in Figs 8 and 9 on the basis of the

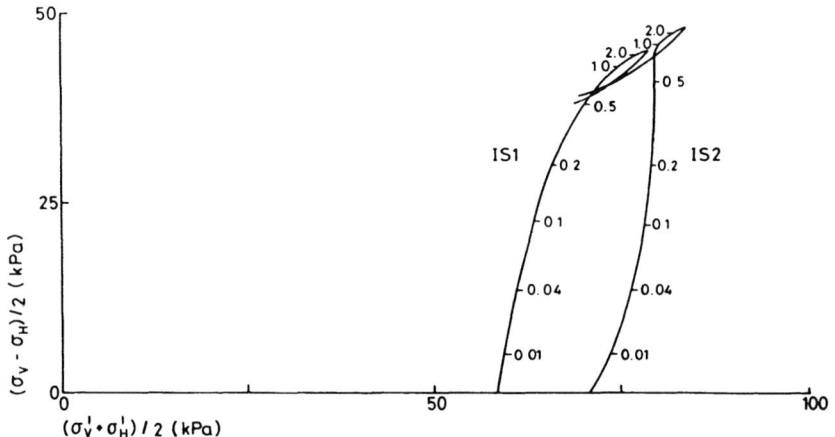

Fig. 13 Unconsolidated undrained triaxial compression of samples cut from blocks of reconstituted North Sea clay (axial strains indicated in %)

OCR before sampling (OCR = 2). It is evident that both the initial stiffness and the degree of non-linearity have been reduced.

TUBE SAMPLING

In contrast to perfect sampling, the stress or strain paths involved in tube sampling and subsequent extrusion are complex. We will simplify the problem and consider the conditions at the *periphery* of a tube sample and along its *centreline*.

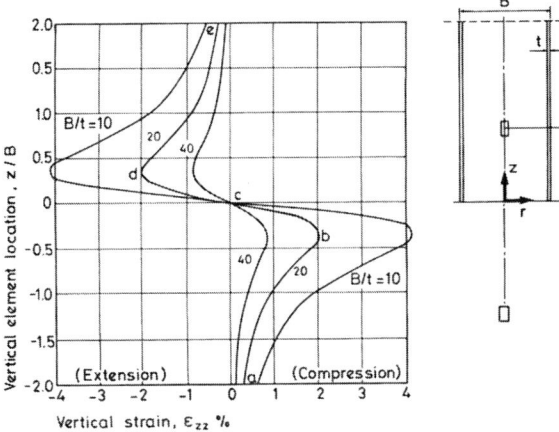

Fig. 14 Strain paths for an element on the centreline of a tube sampler (after Baligh, 1984)

The levels of distortion which occur as soil enters a sampler have been most elegantly elucidated by application of the Strain Path Method (Baligh, 1984). Figure 14 has been taken from Baligh's work and illustrates the sequence of strains experienced by an element of soil on the centreline of samplers of different geometry. The strains involve triaxial compression followed by triaxial extension. Their effect can be assessed from the pattern of behaviour displayed in Fig. 3 and from the results of an extensive study of these soils under cyclic triaxial compression and extension loading. In Figs 15 and 16 the estimated stress paths corresponding to the strain paths in Fig. 14 are shown for normally and heavily overconsolidated soil. In the former, there is a major reduction in effective stress, in the latter an increase in effective stress.

For certain geometries of sampler, the strain levels predicted for an element on the centreline in Fig. 14 exceed those quoted above as leading to gross distortion of the initial fabric. There must, in these cases, be uncertainty as to whether the true undrained characteristics of the *in situ* soil can ever be measured on unconsolidated retrieved samples.

The distortion which occurs around the *periphery* of a tube sample is often apparent

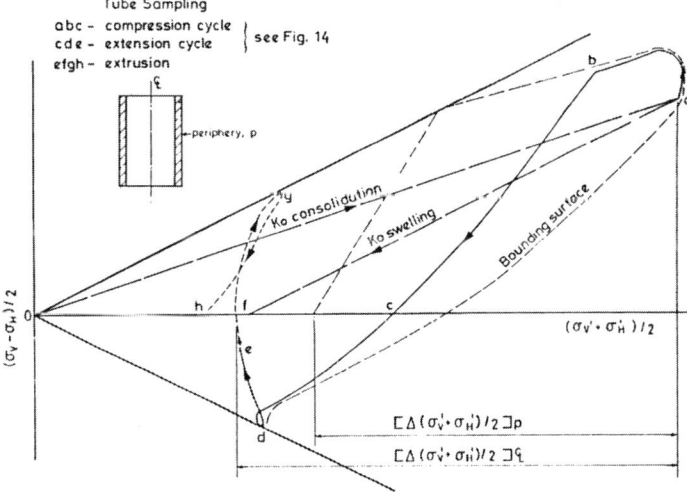

Fig. 15 Stress paths for tube sampling of normally consolidated soil

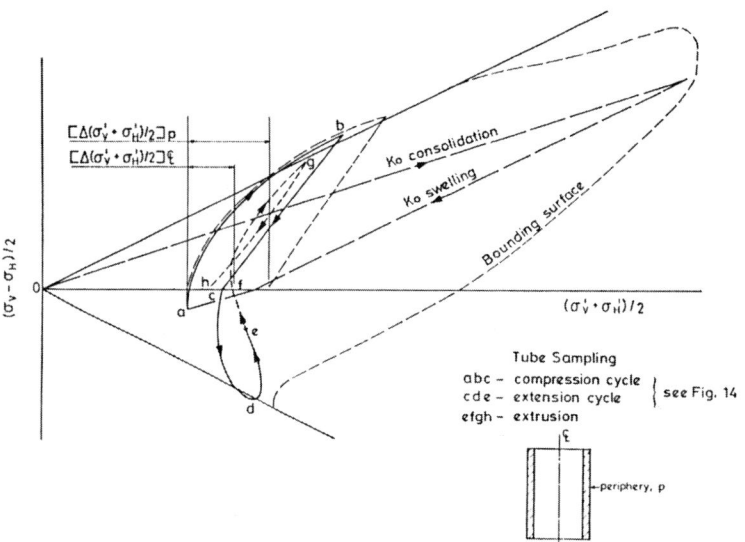

Fig. 16   Stress paths for tube sampling of heavily overconsolidated soil

when such a sample is split to expose its fabric. Obviously the strain paths followed in this outer zone are inadequately modelled in triaxial tests but we can, on the basis of Fig. 3, reasonably anticipate that:

(a) soil in an initially normally consolidated or lightly overconsolidated state will develop positive pore pressure increments (Fig. 15);
(b) soil in a heavily overconsolidated state will develop negative pore pressure increments (Fig. 16).

Since shear is actually taking place against a steel interface, residual fabric will be developed and the thickness of the distorted zone will be affected by surface roughness, soil plasticity and rate of displacement.

There will be equilibration of the pore-pressure differences between the centreline and periphery of the sample. At equilibrium there will have been an overall reduction in pore pressure in the heavily overconsolidated soil and an overall increase in the normally consolidated soil. Correspondingly, there will have been redistribution of water content.

Extrusion involves additional distortion, somewhat analagous to triaxial compression, and final removal of the total stresses on the sample. Its path is indicated arbitrarily by efgh in Figs 15 and 16. Reductions in effective stress which would result from the pore-pressure parameter B being less than unity are neglected.

From the findings on perfect sampling, the direction of the extrusion path relative to that followed in subsequent triaxial testing is of major importance in determining the initial stress–strain properties that are observed. If the final phase of the process is to unload from compression, then the soil's stiffest response would be shown in tests that recompress the sample.

The response that could be anticipated in normally consolidated and overconsolidated soil after tube sampling and extrusion is shown by path IS in Fig. 17. This response is compared to that for the *in situ* soil, and for the *in situ* soil after perfect sampling (PS) and block sampling (BS). For completeness, the response of soil subject to gross distortion (G) is also shown. Even after extrusion, further changes in effective stress state can occur as a result of: changes in temperatures; gas exsolution; drying; sample trimming and installation. Provided, however, that these changes are not excessive, the basic effects of sampling will not be altered.

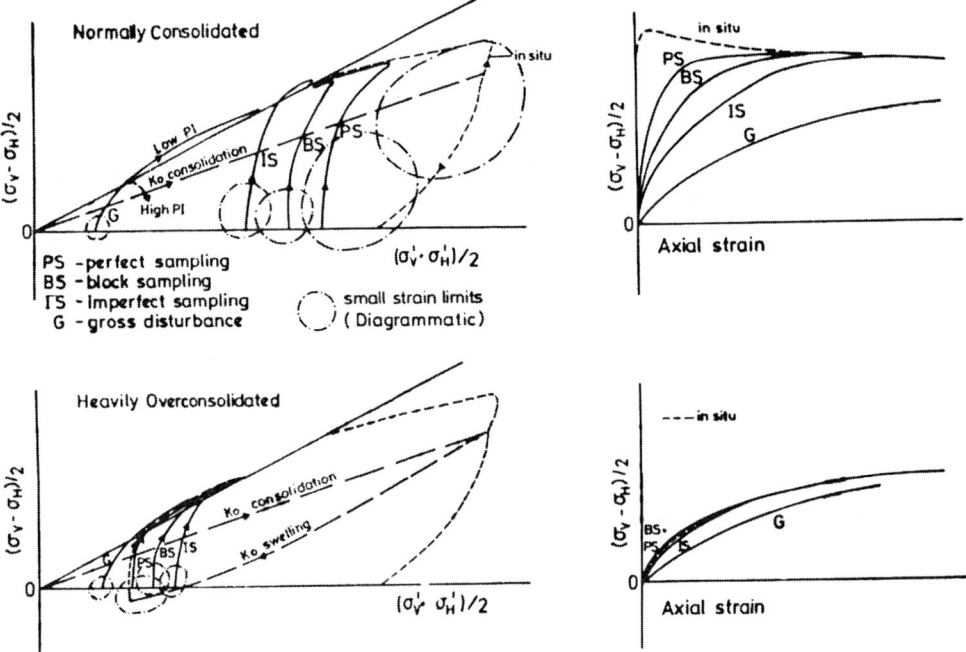

Fig. 17 Summary of sampling effects on normally consolidated and overconsolidated offshore clays

BEHAVIOUR OF TUBE SAMPLES FROM THE NORTH SEA

The results of three triaxial compression tests on tube samples taken from the seabed in the Northern North Sea are shown in Fig. 18. Two of the tests (I1 and I2) were run on samples taken from the same depth; on the basis of their ultimate strength and depth, these were estimated to have an OCR *in situ* of 1.1. The third test (I3) was run on a sample with an estimated OCR greater than 50. These particular tests have been selected from a more extensive programme described by Jardine (1985) and are used only to illustrate the essential features.

The initial stress states measured after setting up samples I1, I2 and I3 reflected the predicted effects of sampling. Samples I1 and I2 displayed initial effective stresses, p', below those estimated *in situ*, but I3 showed a large overall increase in p'. Sample I1 was tested without consolidation, I3 was swelled back to its estimated *in situ* p' and I2 was anisotropically consolidated to

the supposed ground stresses, following the stress path shown in Fig. 18.

The response to undrained shear is summarized in Fig. 18 and Tables 1 and 2. Regarding these results, it is clear that none of the three intact tests gives a satisfactory model for *in situ* behaviour. As expected for normally consolidated soil, both I1 and I2 give higher values of $(E_u/p'_0)_{0.01}$ than the corresponding test on reconstituted soil, but only I2 reproduces its strong non-linearity. Both samples show behaviour after 0.1% strain that is quite different from that expected *in situ*.

The comparison between the response of the heavily overconsolidated sample and the *in situ* soil is more encouraging. There are differences in initial stress path direction, and the stiffnesses around 0.1% strain, but reasonable agreement is found, providing that the initial state of stress is suitably adjusted before shearing.

The tests on intact samples reported by Jardine (1985) at intermediate OCRs show mixed behaviour; the initial stiffnesses are

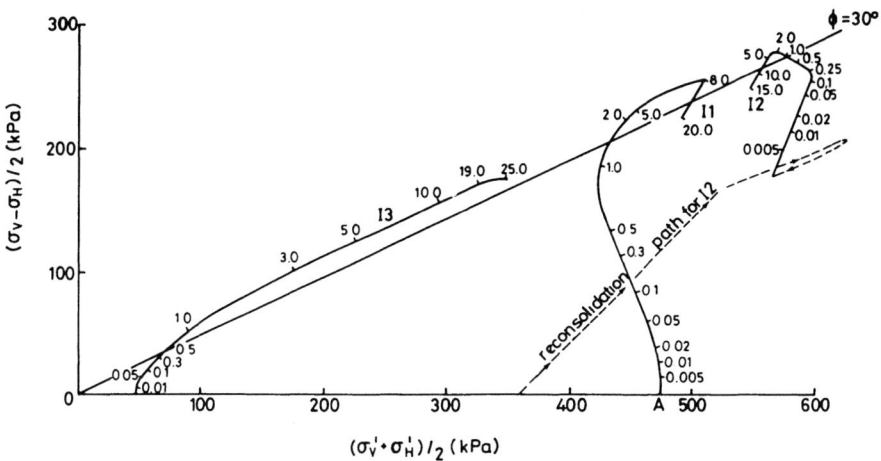

Fig. 18 Stress paths for triaxial compression tests on tube samples from the Northern North Sea (axial strains indicated in %)

TABLE 2
Comparison of 'in situ' behaviour with that measured on tube samples of North Sea clay

Stage of test	Test I1	Test I2	Test I3
Initial condition	p' far lower than estimated *in situ* conditions	Reconsolidated from lower p' than estimated *in situ*	p' far higher than estimated *in situ* conditions. Sample swelled back to 46 kPa
Compression to $\varepsilon_a = 0.01\%$	Stiffness similar to *in situ*, sample generates more pore pressure	Higher stiffness than *in situ*, similar pore pressure generation	Stiffness similar to that expected *in situ*, but less pore pressure generated
Compression to $\varepsilon_a = 0.1\%$	Higher stiffness than *in situ*, more pore pressure generated	Behaviour similar to *in situ*	Stiffness lower than expected *in situ*, less pore pressure generated
Compression to $\varepsilon_a = 1.0\%$	*In situ* sample fails, strain softens and produces large pore pressures. I1 stress path climbs almost vertically	I2 shows yield and slow strain hardening, with some pore-pressure generation	*In situ* sample yields and behaviour is similar to I3
Compression to $\varepsilon_a = 10.0\%$	I1 shows a slow strain hardening with sample dilating. *In situ* behaviour shows continuous strain softening and pore pressure generation	I2 shows strain softening but with less pore-pressure generation than *in situ*	I3 shows strain hardening in a similar way to *in situ* behaviour
Peak strength	Less than *in situ*	Greater than *in situ*	Approximately equal to *in situ*
Axial strain at failure	I1 8%, *in situ* 0.1%	I2 2.5%, *in situ* 0.1%	I3 25%, *in situ* 10–15%

generally slightly lower than those expected *in situ*, but the large-strain behaviour becomes progressively more realistic with increasing OCR.

SUMMARY OF SAMPLING EFFECTS

Sampling modifies the effective stress state in an element of soil so that the behaviour measured in unconsolidated undrained triaxial tests does not match that of an element *in situ*, subject to the same stress path. Important differences in stress–strain characteristics, peak strength and undrained brittleness can occur. In these young sedimented clays, peak strengths for triaxial compression may be reduced by as much as 30%; the ultimate strength is not affected, provided that water content changes are avoided. Stress–strain properties are modified by the changes in size and location of the small-strain zone which accompany changes in effective stress state. The potential effects of different forms of sampling on normally consolidated and heavily overconsolidated soil are summarized in Fig. 17.

RECONSTRUCTING *IN SITU* BEHAVIOUR FROM TESTS ON TUBE SAMPLES

Provided that the sampling process does not involve excessive strains, the behaviour measured in unconsolidated undrained triaxial tests fits into the overall pattern of behaviour for the soil. However, the important effects of stress history are largely obliterated by the changes in effective stress state which accompany sampling (Fig. 12). Unconsolidated undrained triaxial tests are not, therefore, a good indicator of stress history or of detailed *in situ* behaviour. Furthermore, measurements of initial effective stress on relatively small tube samples retrieved from the seabed are unlikely to provide a reliable means of estimating the *in situ* stress state.

Nevertheless, UU tests can be used to determine ultimate strength, although efforts must be made to minimize the opportunities for uncontrolled water content change.

The use of the measured strengths in the ratio C_u/σ'_v allows an estimate to be made of the *in situ* degree of overconsolidaiton. Such apparent OCRs give some guidance in evaluating the *in situ* horizontal stresses, but there is a clear requirement for the independent measurement of this stress *in situ*.

For samples estimated as having a high OCR, isotropic swelling to the *in situ* mean effective stress has been found to provide a reasonable model of *in situ* behaviour. This approach was proposed by Gens and Hight (1979) on the basic of work with the Cowden Till. Reimposition of anisotropic stresses, if these are known, will improve the fit for stress–strain properties.

For soils estimated as being normally or lightly overconsolidated engineers face a dilemma. Reconsolidation to estimated *in situ* stresses does not completely recover *in situ* behaviour (refer to test I2 and Table 2). Reconsolidation beyond the *in situ* stresses following the SHANSEP approach (Ladd and Foott, 1974) recovers the features observed in young clays, but eliminates the potential effects of ageing and cementing. We therefore advocate an overlapping approach which relies on:

(a) UU tests to identify special features of behaviour, such as bonding;
(b) tests run on samples anisotropically consolidated to estimated *in situ* stresses which, together with the results of UU tests, enable the small and intermediate strain response to be bracketed; and
(c) either the interpretation of the data and the evaluation of design parameters within the framework described in this paper, or the performance of a suite of SHANSEP tests on identical samples.

With regard to interpretation, recognition should be given to the existence of:

(a) a bounding surface within which undrained stress paths during sampling and subsequent shear must be contained; and

(b) a zone bounded by a principal strain contour of 0.1% in which response is stiff and non-linear; the zone is translated and altered by stress paths moving inside and along the bounding surface, and the size and position of the zone are dictated by the location of the stress state within the bounding surface and by the recent direction of the stress path.

In this approach, realistic assessment of the behaviour of sampled soil and its relation to *in situ* behaviour requires careful measurement of:

(a) initial effective stress states;
(b) full effective stress paths during shear; and
(c) detailed stress–strain characteristics and their relation to effective stresses.

Techniques for routinely making these measurements with the required level of accuracy are available. Hight (1982) describes a simple piezometer probe which can achieve measurement of the full effective stress path within reasonable testing times. Jardine *et al.* (1984) describe the application of the internal electrolevel displacement gauges to routine triaxial testing. For anisotropic reconsolidation, the hydraulic triaxial cell (Bishop and Wesley, 1975) is available and its operation by computer is well established (Hight, 1983).

ACKNOWLEDGEMENTS

This work has been funded by the Marine Technology Directorate of SERC. Dr A. Fourie conducted some of the tests on London Clay and Mr P. Smith assisted in testing the North Sea clay. Samples for work on the North Sea clay were kindly provided by Mr W. J. Rigden of BP International.

REFERENCES

1. Baligh, M. M. 1984. The strain path method in geotechnical engineering. MIT Report No. R84–01.
2. Bishop, A. W. and Wesley, L. D. 1975. A hydraulic triaxial apparatus for controlled stress path testing. *Geotechnique* 25(4), 657–670.
3. Burland, J. B. 1967. Deformation of soft clay. PhD thesis, University of Cambridge.
4. Burland, J. B. and Symes, M. 1982. A simple axial displacement gauge for use in the triaxial apparatus. *Geotechnique* 32(1), 62–65.
5. Gens, A. 1982. Stress–strain and strength characteristics of a low plasticity clay. PhD thesis, University of London.
6. Gens, A. and Hight, D. W. 1979. The laboratory measurement of design parameters for a glacial till. In *Proc. 7th European Conf. on Soil Mech. and Fdn. Engng*, Brighton, Vol. 2, pp. 57–65.
7. Hight, D. W. 1982. A simple piezometer probe for the routine measurement of pore pressure in triaxial tests on saturated soils. *Geotechnique* 32(4), 396–401.
8. Hight, D. W. 1983. Laboratory investigations of sea bed clays. PhD thesis, University of London.
9. Jardine, R. J. 1985. Investigations of pile-soil behaviour with special reference to the foundations of offshore structures. PhD thesis, University of London. (In preparation).
10. Jardine, R. J., Symes, M. J. and Burland, J. B. 1984. The measurement of soil stiffness in the triaxial apparatus. *Geotechnique* 34(3), 323–340.
11. Jardine, R. J., Potts, D. M., Fourie, A. B. and Burland, J. B. 1985. Studies of the influence of non-linear stress–strain characteristics in soil-structure interaction. Submitted to *Geotechnique*.
12. Ladd, C. C. and Foott, R. 1974. New design procedure for stability of soft clays. ASCE, 100, GT7, 763–786.
13. Schofield, A. N. and Wroth, C. P. 1968. *Critical State Soil Mechanics*. McGraw-Hill, New York.
14. Skempton, A. W. and Sowa, V. A. 1963. The behaviour of saturated clays during sampling and testing. *Geotechnique* 13(4), 269–290.

17

Design Parameters for Offshore Sands: Use of *In Situ* Tests

T. Lunne, S. Lacasse, G. Aas and C. Madshus,
Norwegian Geotechnical Institute, Norway

ABSTRACT

Traditionally, North Sea sands have been tested by laboratory tests on reconstituted samples and by *in situ* cone penetration tests. A number of other *in situ* test devices are already available for offshore use or will most likely be available in the near future. These include the piezocone, the pressure-meter, the dilatometer, nuclear density and electrical resistivity probes, the seismic cone and the screw plate.

The present status of interpretation of the above tests in terms of the following soil design parameters is discussed: soil density, drained shear strength, deformation characteristics, initial shear modulus, and soil layering and soil identification. In addition, the need for obtaining samples and performing laboratory tests is stressed. The results from *in situ* tests give the basis for reconstituting specimens for laboratory testing soil density and consolidation stresses).

INTRODUCTION

The use of *in situ* tests is particularly important in sands where it is often impossible to take representative undisturbed samples.

Traditionally, *in situ* testing in connection with offshore soil investigations has predominantly consisted of cone penetration tests (CPTs), including frequently for the last three years measurement of pore-water pressure (piezocone). A few other *in situ* tests applicable in sandy soils have been used in selected cases, and other *in situ* tests that have proved particularly useful on land will undoubtedly be used offshore in the future. These include the use of pressure-meter, dilatometer, special sand-density methods, and tests measuring shear-wave velocity and screw plate.

Some of the interpretation methods have a purely theoretical basis, but the most commonly used interpretation methods are semi- or purely empirical. In the latter case, the methods are mostly based on research data from large calibration chamber tests and to some extent on actual field tests performed at uniform, well investigated test sites.

IN SITU TEST TYPES CONSIDERED

The test types selected for the present review had to obey three criteria:

- Test must be applicable in sand.
- Test must be operational for offshore use or likely to become operational offshore in the near future.

- Test must have been shown to given meaningful results in terms of soil design parameters in sandy soils.

Cone Penetration Test (CPT)

The cone penetration test measures cone-tip resistance, q_c, and sleeve friction, f_s. The newer devices (piezocones) also measure pore-water pressure (u). The standard cone has a 60° apex angle and cross-sectional area of 10 cm² (Fig. 1a). In the North Sea, 15 cm² cones are frequently used. For more details, see, for instance, de Ruiter (1982).

Special Density Probes

Special density probes directly or indirectly measure *in situ* density or porosity. They

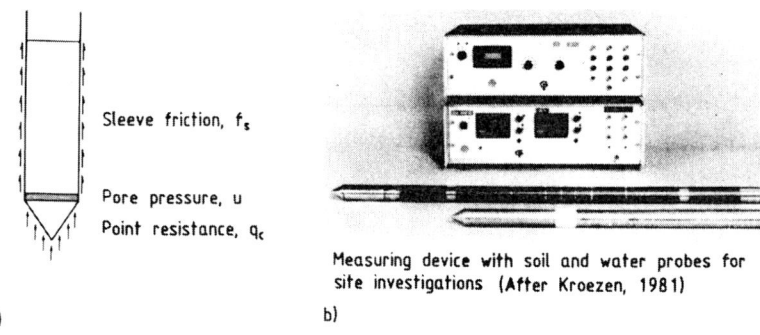

a)

Measuring device with soil and water probes for site investigations (After Kroezen, 1981)

b)

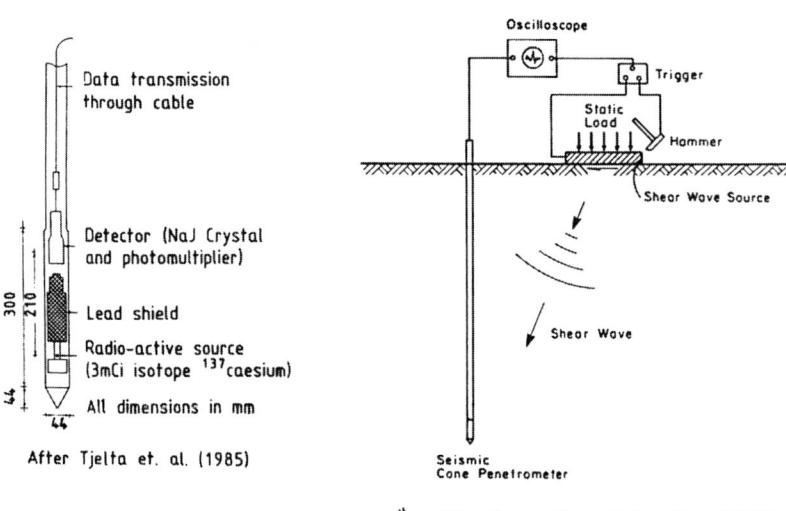

c)

d)   After Campanella and Robertson (1984)

Fig. 1   Sketches of *in situ* devices used in sand: (a) cone penetration test with pore-pressure measurement; (b) electrical resistivity probe; (c) nuclear density probe; (d) seismic cone;

have been developed by the Delft Soil Mechanics Laboratory.

The electrical resistivity probe measures cone resistance and sleeve friction in addition to the electrical resistivity of the soil volume (Fig. 1b). A separate probe measures the resistivity of the pore water. Readings are generally taken at 0.2 m intervals. The field measurements are correlated to laboratory calibrations on specimens reconstituted to different porosities (see Kroezen, 1981).

The nuclear density probe has a radioactive source and detector built into a 15 cm^2 cone (Fig. 1c). The bulk density of the soil is determined by comparing the energy level of the emitted photons from the source and the photons entering the detector after backscattering from the soil mass. The equipment is calibrated by lowering the probe into fluids with known densities between 10 and 22 kN/m^3. The nuclear density probe was recently used offshore (Tjelta *et al.*, 1985).

Seismic Cone
In the seismic cone a piezocone is combined with a set of miniature seismometers built into the cone (Fig. 1d). The device was developed at the University of British Col-

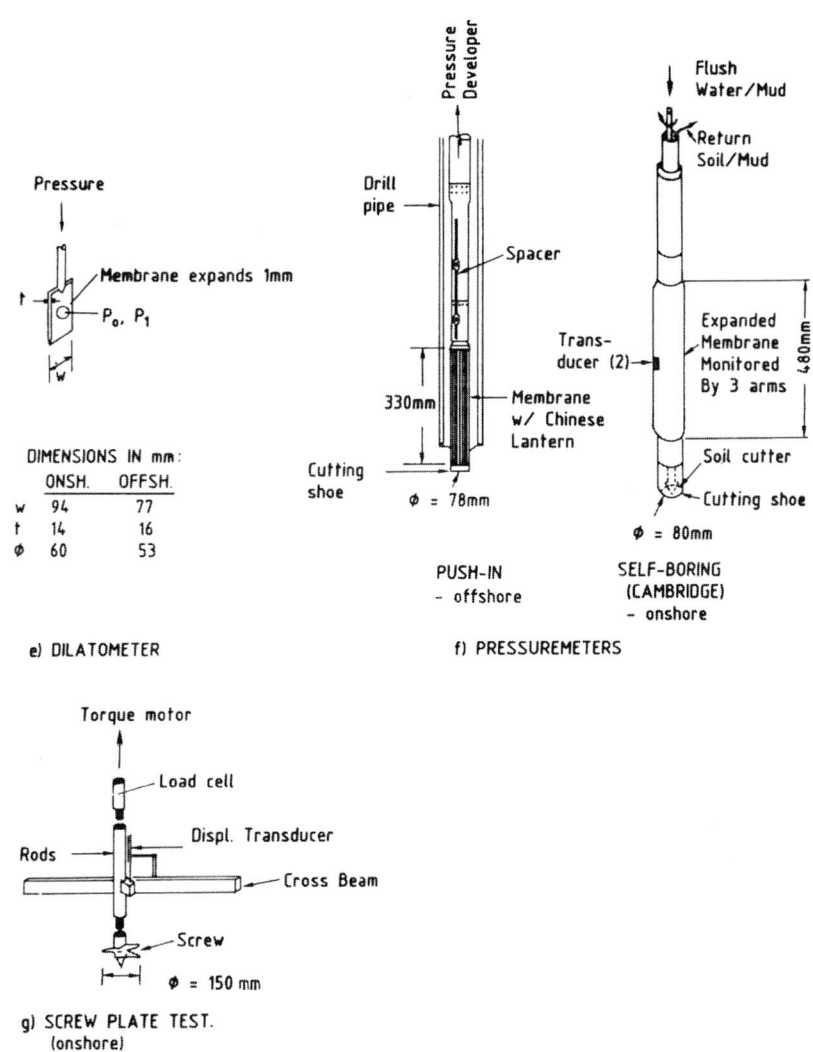

Fig. 1 (e) dilatometer; (f) pressuremeters; (g) screw-plate test

umbia (Campanella and Robertson, 1984). The cone resistance, sleeve friction and pore-pressure measurements are used to determine the soil layering at a site during penetration, and the downhole shear-wave velocity is measured at intervals (normally 1 m). A profile of the *in situ* dynamic shear-wave velocity, and initial modulus (i.e. at very small strain level) is therefore obtained. During on-land tests a sledgehammer is used to generate the shear wave. The wave generation system is presently being modified for use offshore in the Beaufort Sea (Campanella, 1985).

On land the crosshole technique is also used to measure shear-wave velocity. However, it is not practical to perform this test offshore except perhaps underneath already installed platforms.

Dilatometer Test (DMT)

The dilatometer test measures soil contact pressure and the 1 mm expansion pressure (Fig. 1e). The device, 94 mm wide and 14 mm thick for onshore use and 77 mm wide and 16 mm thick for offshore use, provides nearly continuous data with depth. The test is performed very rapidly, as the expansion test every 20 cm takes about 15 seconds. Empirical correlations enable one to predict *in situ* horizontal stress, stress history, constrained modulus and effective friction angle. The device is promising in sand where few reliable methods for obtaining K_0 are available, with the exception of the more sophisticated pressuremeter. Robertson and Campanella (1984) also use dilatometer results to predict the liquefaction potential of sands. The device has been used offshore in limited water depths (Burgess *et al.*, 1983), but a device for deep-water use is under development.

Pressuremeter Tests

Pressuremeter tests provide *in situ* stress–strain curves (Fig. 1f). *The push-in pressuremeter (PIP)* was developed mainly for offshore use (Reid *et al.*, 1982). The tests are usually carried out alternately with other tests in the same borehole. Tests in sands are fully drained. The friction angle

ϕ', dilation angle and drained shear modulus are derived from the field curve.

The self-boring pressuremeter (SBP) is also used offshore (Faÿ and Le Tirant, 1982). The device tunnels itself into the ground and should induce the least disturbance of all the *in situ* devices. The experience offshore with the SBP is so far limited. When successful, self-boring tests offer the best alternative for measuring the *in situ* horizontal stress in a sand deposit.

The Screw Plate

The screw plate (Fig. 1g) can perform load tests at depth more economically than plate load tests. The plate consists of a flat pitch auger screwed into the ground and loaded vertically (Janbu and Senneset, 1973; Berzins and Campanella, 1981). The settlement versus time is measured at selected depth intervals, but penetration is limited to about 15 or 20 m in a medium dense sand. The screw-plate test has been used only onshore up to now.

STRATIFICATION AND SOIL IDENTIFICATION

General

It is always important to establish a clear picture of the soil layering or stratigraphy at a site. *In situ* tests are extremely useful in this context, and they are a necessary supplement to laboratory tests on obtained samples.

Very frequently CPTs are performed prior to sampling and a careful interpretation of CPT results can yield optimum planning of the sampling programme. Further, since performing *in situ* tests is usually much cheaper than obtaining and testing samples, *in situ* test results are used to extrapolate the laboratory results and to interpolate between boreholes.

Cone Penetration Test

It is generally accepted that the cone penetration test with measurement of pore pres-

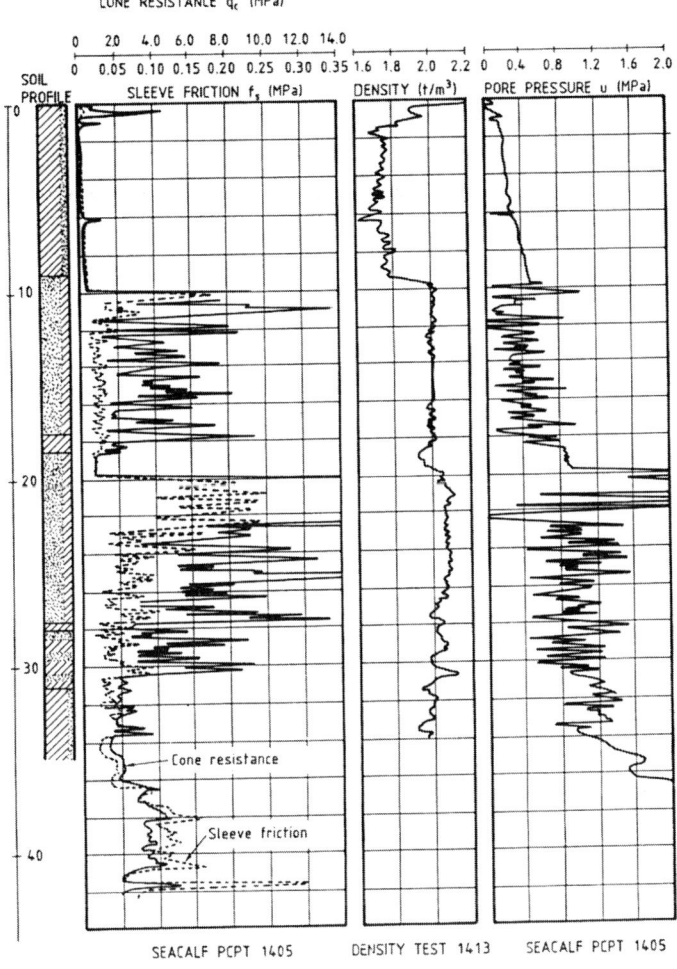

CONE RESISTANCE q_c (MPa)

SLEEVE FRICTION f_s (MPa)

DENSITY (t/m^3)

PORE PRESSURE u (MPa)

SOIL PROFILE

Cone resistance

Sleeve friction

SEACALF PCPT 1405   DENSITY TEST 1413   SEACALF PCPT 1405

Fig. 2 Piezocone and nuclear density test results, Gullfaks 'C' (after Tjelta *et al.*, 1985)

sure is the best and most efficient tool for logging stratigraphy at a site. A thin pore-pressure filter (either at the tip or immediately behind the cone tip) can sense layers only a few centimetres thick. However, Robertson and Campanella (1983) point out that, for interbedded deposits, the thinnest stiff layer for which the cone resistance measures full response is about 10–20 cone diameters. For the standard 10 cm² cone, the minimum stiff-layer thickness to ensure full or plateau response of cone resistance is therefore about 35–70 cm.

Figure 2 shows as an example the results of a piezocone profile performed at Gullfaks 'C' in the Norwegian sector of the North Sea (Tjelta *et al.*, 1985). The usefulness of the piezocone in determining soil stratigraphy is evident from this figure.

With reference to soil classification from CPT results, several charts relate the friction ratio to soil type. The most comprehensive work on this has been done by Douglas and Olsen (1981). Their chart (Fig. 3) correlates CPT data with other soil-type indices, such as those provided by the Unified Soil Classification System. The measured sleeve friction is usually far less accurate than the cone reading, and more liable to variability from one cone to another. Among other factors, the pore pressure induced during cone penetration affects the sleeve friction measurements (e.g. Campanella *et al.*, 1982).

For classification purposes, one should

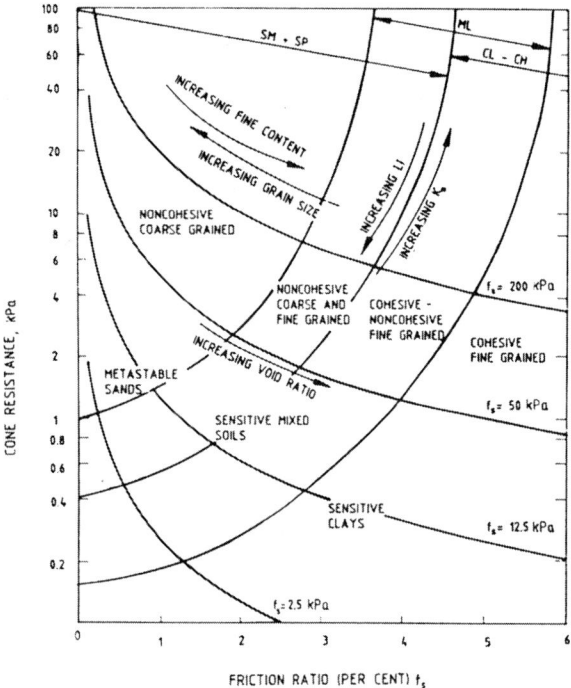

FRICTION RATIO (PER CENT) f_s

Fig. 3 Classification chart based on q_c and R_f for electrical friction cone (after Douglas and Olsen, 1981)

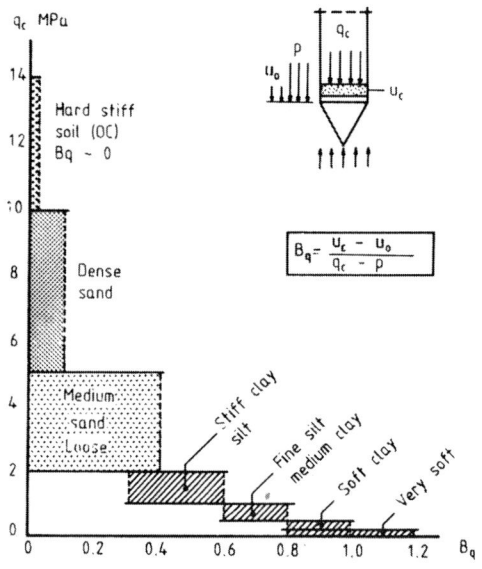

Fig. 4 Tentative classification chart based on q_c and B_q for piezocone (after Senneset and Janbu, 1984)

Dilatometer Tests

rather use the pore-pressure ratio, B_q, as defined by Senneset and Janbu (1984):

$$B_q = \frac{\Delta u}{q_T - p_o}$$

where Δu = excess pore pressure; q_T = cone resistance corrected for pore-pressure effects (e.g. see Aas *et al.*, 1984); p_o = total overburden pressure.

The classification chart based on q_T and B_q proposed by Senneset and Janbu is shown in Fig. 4. One should note that:

(a) the measured penetration pore pressure depends on the location of the porous filter (e.g. see Campanella *et al.*, 1985);
(b) the cone resistance should be properly corrected for pore-pressure effects.

At present, with the existing piezocone geometries, the authors recommend that pore pressure should be measured by or corrected for a filter location immediately behind the neck of the cone.

The dilatometer also provides a good picture of soil variability, as shown by the examples in Fig. 5a. Marchetti and Crapps (1981) provide a soil classification chart (Fig. 5b), which also enables an approximate determination of the soil density. The material index, I_D, is defined as the normalized difference between the 1 mm expansion pressure (P_1) and the contact pressure (P_0). In the future, the dilatometer may be fitted with a pore-pressure sensor which may improve the interpretability in terms of soil type (Boghrat, 1982; Campanella and Robertson, 1983).

SOIL DENSITY

General

The *in situ* density of a sand layer is normally expressed in terms of porosity or relative density, where the soil density is related to maximum and minimum laboratory densities.

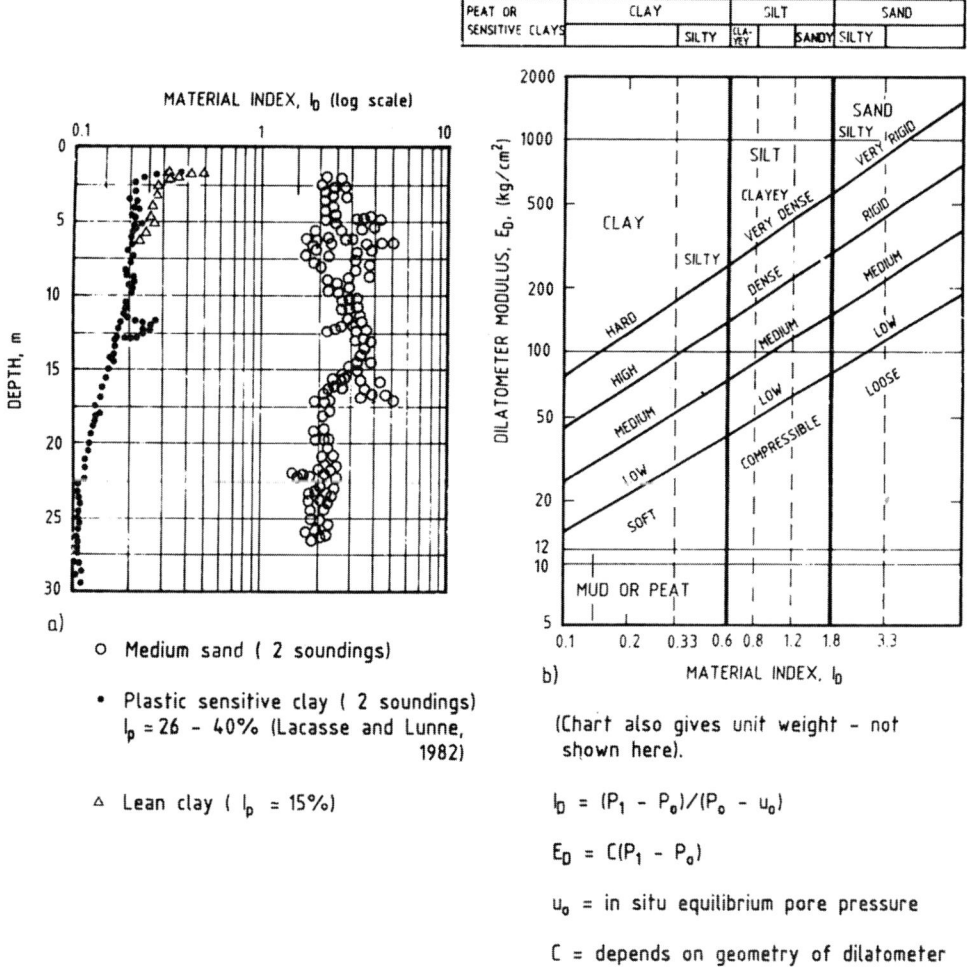

a)

○ Medium sand (2 soundings)

• Plastic sensitive clay (2 soundings)
$I_p = 26 - 40\%$ (Lacasse and Lunne, 1982)

△ Lean clay ($I_p = 15\%$)

b)

(Chart also gives unit weight - not shown here).

$$I_D = (P_1 - P_0)/(P_0 - u_0)$$

$$E_D = C(P_1 - P_0)$$

u_0 = in situ equilibrium pore pressure

C = depends on geometry of dilatometer

Fig. 5 Soil classification from dilatometer test: (a) examples of profiling with dilatometer; (b) soil classification chart

For years the relative density (D_R) has been used by engineers as an important parameter to describe sand deposits. For example, the *in situ* relative density is often evaluated from CPTs or standard penetration tests and used with empirical relationships to evaluate the angle of internal friction, the liquefaction potential, etc.

Recent trends try to bypass the intermediate determination of D_R and rather correlate the *in situ* data directly to the relevant engineering parameter. The most important reasons for this are that the maximum and minimum laboratory densities vary with the testing method used and that the relative density alone does not determine the engineering behaviour of a sand. Other very important factors include stress history, age, sand structure, grain size and shape, mineralogy and cementation.

It is essential to have a good estimate of the *in situ* soil density or relative density since running laboratory tests on sand requires recompacting the material to an *in situ* density. For these reasons, a reliable *in situ* method for measuring sand density is very important.

Electrical Resistivity Tests

The *in situ* porosity of a sand can be determined from electrical resistivity measure-

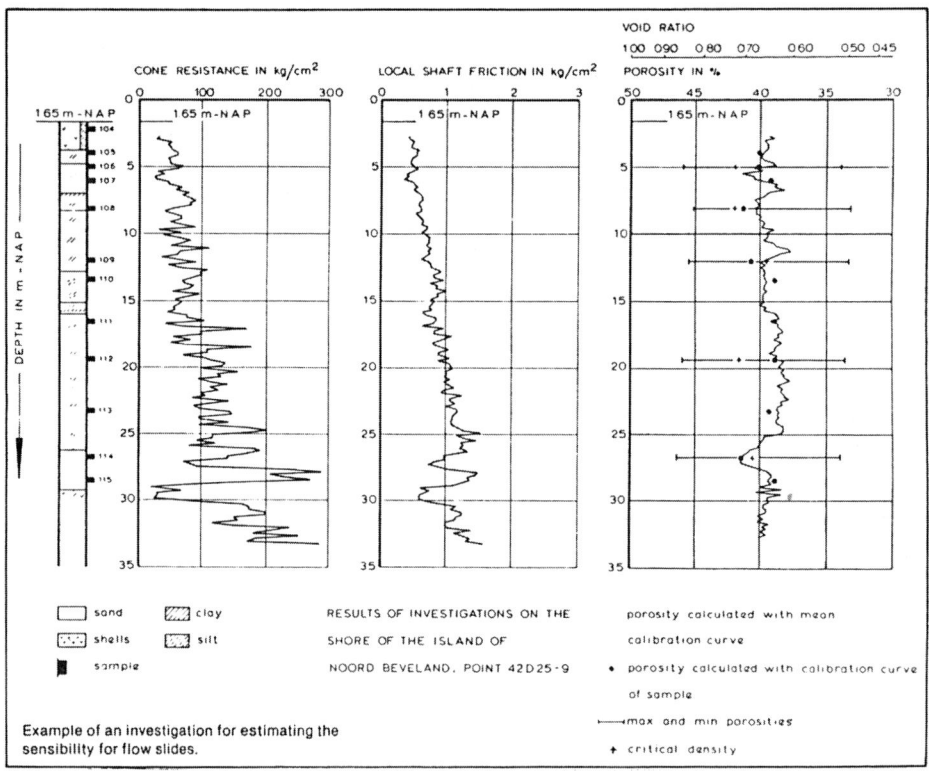

Fig. 6 Example result from electrical resistivity test (after Kroezen, 1981)

ments and laboratory calibrations. For this test, sand samples have to be taken throughout the sand profile. It is therefore routine also to determine maximum and minimum porosity, and to calculate the relative density, D_R. Figure 6 shows an example of measurements (Kroezen, 1981). The electrodes of the soil probe are 0.5 m apart, and it is expected that the bulk of the soil in which the average soil resistivity is measured is within 0.1 to 0.5 m from the probe. It is therefore thought that the disturbance of the soil caused by the penetrating 10 cm² cone in most cases does not influence the resistivity readings significantly, i.e. that the measured soil resistivity is representative for undisturbed soil. However, little data have been provided which can be used to check the accuracy of the electrical resistivity probe.

Nuclear Density Tests

This test gives directly the soil density of

the penetrated sand. Figures 2 and 7 show the results of a test performed in 220 m of water at the Gullfaks Field in the Norwegian Sector of the North Sea (Tjelta *et al.*, 1985). The detector is placed 210 mm from the radioactive source inside the 15 cm² cone, and the result of this test is perhaps more influenced by the disturbance around the penetrating cone than in the case of the electrical resistivity test. However, the test is less cumbersome to perform than the electrical resistivity test where two probes need to be used and soil samples are required for calibration purpose. The nuclear density test can also be used in cohesive soils. As for the electrical resistivity tests, the accuracy of the nuclear density test needs to be documented.

Cone Penetration Test

The most commonly used method for deriving relative density from CPT results is that given by Schmertmann (1978), who per-

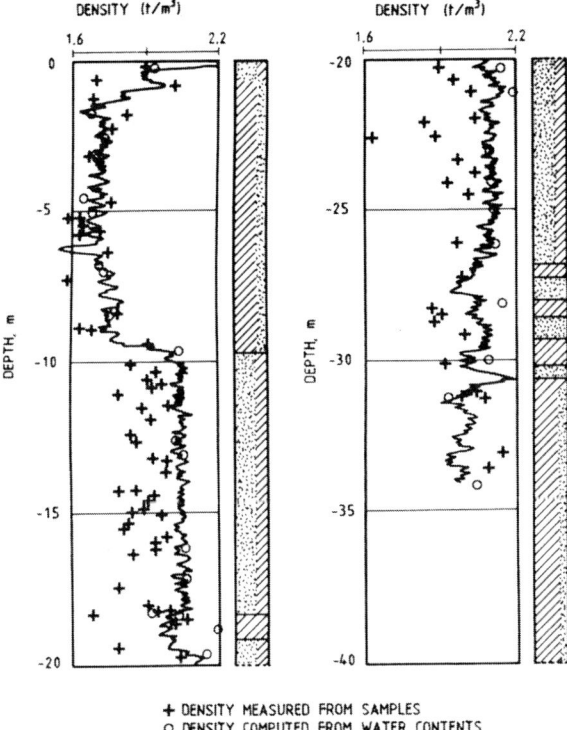

+ DENSITY MEASURED FROM SAMPLES
○ DENSITY COMPUTED FROM WATER CONTENTS
− DENSITY FROM IN-SITU DENSITY TEST

Fig. 7 Comparison of *in situ* nuclear density test and laboratory results from samples, Gullfaks 'C' (after Tjelta *et al.*, 1985)

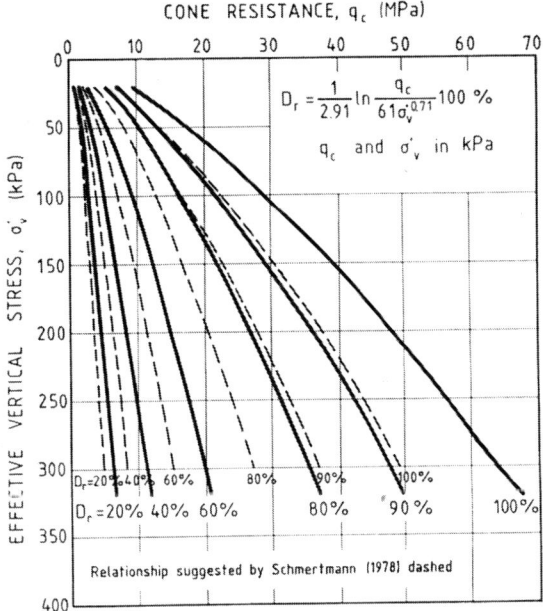

Fig. 8 Recommended relationship among σ'_v, q_c and D_R for NC fine–medium quartz sand (from Lunne and Christoffersen, 1983)

formed a statistical analysis of calibration chamber test results at the University of Florida. Schmertmann's correlation (Fig. 8) can be expressed by the following relationship:

$$D_R = \frac{1}{C_2} \ln \frac{q_c}{C_0 \, \sigma'_{vo} C_1} \qquad (1)$$

where σ'_{vo} = effective overburden pressure with q_c and σ'_{vo} in kPa. Table 1 lists

Schmertmann's empirical constants, C_0, C_1 and C_2.

Based on a review of recent calibration chamber results from NGI (Parkin *et al.*, 1980), Italy (Baldi *et al.*, 1982) and Australia (Chapman and Donald, 1981), as well as Schmertmann's results, Lunne and Christoffersen (1983) suggested a revised D_R, q_c, σ'_{vo} correlation, also shown in Fig. 8. This correlation was based on q_c-values corrected for sample size and boundary effects (Parkin and Lunne, 1982). The revised coefficients are given in Table 1.

On the basis of even more calibration chamber data, Jamiolkowski *et al.* (1985) recommend that D_R should be computed from the following formula:

$$D_R = -98 + 66 \lg \frac{(q_c)}{\sqrt{\sigma'_v}} \qquad (2)$$

where chamber-size effects (Parkin and Lunne, 1982) may be taken into account by dividing the *in situ* q_c value by a factor K_q as follows:

$$K_q = 1 + \frac{0.2 \, (D_R = 30)}{60} \qquad (3)$$

TABLE 1
Empirical constants for calculating relative density from cone resistance
(see equation 1)

Method	C_0	C_1	C_2
Schmertmann (1978)	46.54	0.71	2.91
Lunne and Christoffersen (1983)	61	0.71	2.91

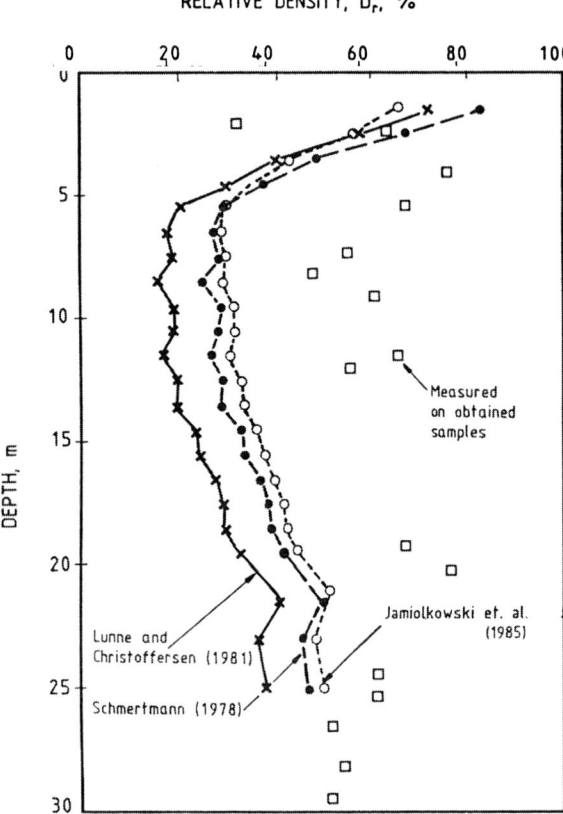

RELATIVE DENSITY, D_r, %

Fig. 9   Relative densities computed from CPT results in medium-to-coarse sand

As an example, the cone resistance (mean of 42 tests) for a normally consolidated medium sand has been interpreted using the methods listed above (Fig. 9). Generally, the absolute value D_R predicted by Schmertmann's correlation is 10% higher than that predicted by Lunne and Christ offersen's chart, while the D_R-values based on Jamiolkowski et al. (1985) lie slightly above Schmertmann's values.

Figure 9 also includes relative densities as derived from measured densities on 'undisturbed' samples. The CPT predicted lower densities than the ones based on the actual sand samples. Because the samples are likely to become densified during sampling and handling, the authors believe that the CPT values may be more representative of the in situ conditions.

The q_c, σ'_{vo}, D_R relationships referred to above are only valid for normally consoli-

dated, uniform, fine-to-medium, completely dry or saturated sand, consisting mainly of non-crushable grains (quartz). An equivalent overconsolidated cone resistance, $q_{c,co}$, may be found by the following formula suggested by Schmertmann (1978):

$$\frac{q_{c,oc}}{q_{c,nc}} = 1 + \lambda \frac{(K_{0,oc} - 1)}{K_{0,nc}} \qquad (4)$$

where

$$\frac{K_{0,oc}}{K_{0,nc}} = OCR \ \beta \qquad \text{(ratio of } K_0\text{-value for normally consolidated and overconsolidated sand)}$$

λ and β are experimental constants, and nc and oc denote normally and overconsolidated respectively (Schmertmann suggested $\lambda = 0.75$ and $\beta = 0.42$).

Baldi et al. (1985) tentatively suggested that β increases with D_R, i.e. $\beta \simeq 0.27 + 0.26 D_R$, and that π decreases with OCR, being about 0.50 for OCR = 2 and 0.25 for OCR = 15. The authors recommend the use of Baldi et al.'s findings when evaluating $q_{c,nc}$, and hence D_R, for overconsolidated sands. Unfortunately, a reliable assessment of OCR and/or K_0 of natural sand deposits is a very difficult task. However, as discussed in the next section, the dilatometer shows promise in this respect.

Dilatometer Test

Robertson and Campanella (1984) proposed a relationship between dilatometer parameters and relative density. Empirical data are available for D_R between 15 and 70%.

IN SITU HORIZONTAL STRESS

General

The determination of the horizontal stress and the coefficient of earth pressure at rest in sand is probably one of the most difficult tasks of in situ testing. The action of the measurement itself defeats the purpose of obtaining an at-rest condition. The more promising devices for such measurement are the self-boring pressuremeter test and the dilatometer tests, although both methods present drawbacks.

Pressuremeter Test

The self-boring pressuremeter can provide reliable measurements of the horizontal stress, as reported by Jamiolkovski *et al.* (1985), but only if the self-boring operation is carried out with the utmost care. Otherwise the starting pressure of the expansion test is invariably equal to the *in situ* pore pressure (e.g. Fahey and Randolph, 1984). The available experience is still very limited. The push-in pressuremeter *does not* provide the *in situ* stress in sand because of the disturbance caused by the insertion method.

Dilatometer Test

The dilatometer can provide an estimate of the variation of the horizontal stress with depth in a sand deposit, as shown by Fig. 10. For the same sand with two different relative densities, significant differences are seen in the material index, I_D, horizontal stress index, K_D, and the derived values of K_0 and OCR. The latter two soil characteristics are obtained from correlations which still have limited applications, and more evaluation work is needed in this area. Marchetti (1980) and Schmertmann (1982) present the interpretation procedures.

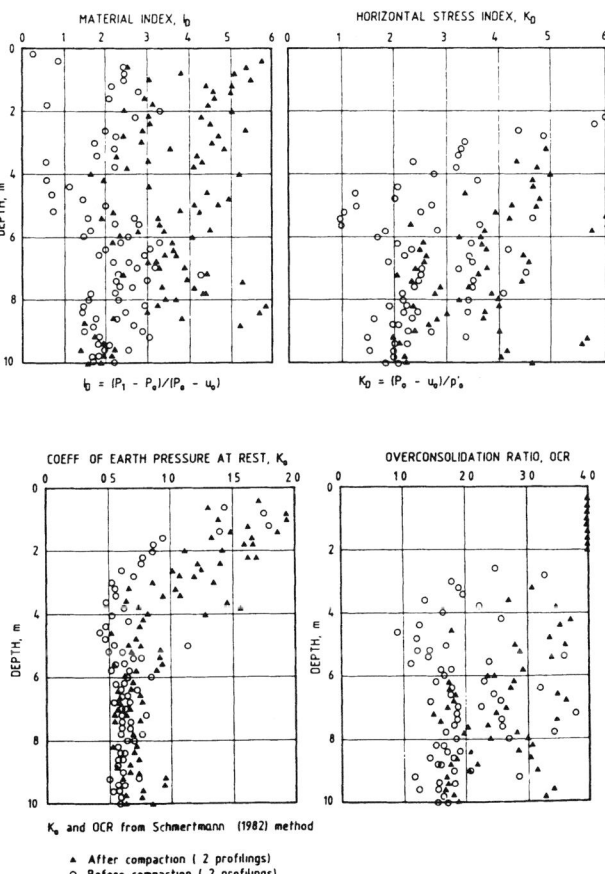

Fig. 10   Dilatometer parameters I_D and K_D and derived K_0 and OCR in medium-to-coarse sane (four profilings)

Marchetti's method is an empirical correlation based on two sites with loose sand. Schmertmann recommends using a bearing capacity method and the thrust needed to push the dilatometer to obtain a value of K_0 and a friction angle ϕ'. Many uncertainties reside in the thrust measurements. A more promising approach would be to extend Schmertmann's method by using cone penetration resistance directly instead of the thrust measured at the top of the rods. Research is underway on calibration chambers to improve the interpretation of K_0 from dilatometer tests (Jamiolkowski et al., 1985).

Cone Penetration Test

A special cone has been developed at the University of California (Mitchell, 1985) which can measure the horizontal stress acting on the sleeve friction. Potentially, this cone can determine the in situ horizontal stress. Both equipment and procedure are still under development.

DRAINED SHEAR STRENGTH

General

The drained shear strength of sand is needed for computing drained bearing capacity such as penetration of foundation skirts. The traditional approach in North Sea soil investigations has been to interpret the CPTs in order to estimate in situ density and to perform triaxial tests on sand reconstituted to the estimated density. The sand strength found from these tests has then been compared to the strength derived from the interpretation of the CPTs, as discussed below.

In the future it is thought that this procedure will also be extended to include the results of other in situ tests offshore.

Cone Penetration Test

Several theories or methods are available

for interpreting cone penetration tests in terms of the drained shear strength of sand (e.g. Mitchell and Lunne, 1978). Most of these methods are based on bearing capacity theory which assumes slip surface shear failure (classical bearing capacity) worked out for plane strain conditions and empirically modified for circular foundations. Other theories are based on cavity expansion theory. In addition, purely empirical correlations have been presented in the literature.

Lunne and Christoffersen (1983) reviewed available theories and evaluated them using calibration chamber data with parallel triaxial tests. They concentrated on evaluating the methods of Meyerhof (1961), Durgunoglu and Mitchell (1975), Janbu and Senneset (1975) and Schmertmann (1978). Lunne and Christoffersen (1983) ended up by making the following recommendations.

For normally consolidated or slightly overconsolidated sands:

(1) Use a modified Janbu and Senneset method. Since the plastic zone of stress (Fig. 11) is believed to increase with increasing friction angle (i.e. β decreases with increasing friction angle), the relationship in Fig. 12 is proposed (Lunne and Christoffersen, 1983).

$$N_q = \tan^2 (45° + \phi'/2) \exp (\lambda/3 + 4\phi') \tan \phi' \qquad (5)$$

This relationship is almost indentical to that presented by Robertson and Campanella (1983).

(2) Use the Durgunoglu and Mitchell method where the roughness steel/soil, δ/ϕ', and the lateral earth pressure coefficient, K_0, must be given in addition to q_c and p_o'. Use $K_0 = 1 - \sin \phi$ or K_0 as found by other in situ tests. Bellotti et al. (1983) evaluated the Durgunoglu and Mitchell method using calibration chamber and triaxial test results and concluded that the method was acceptable, provided a curved failure envelope was taken into account.

(3) Use the modified Schmertmann method.

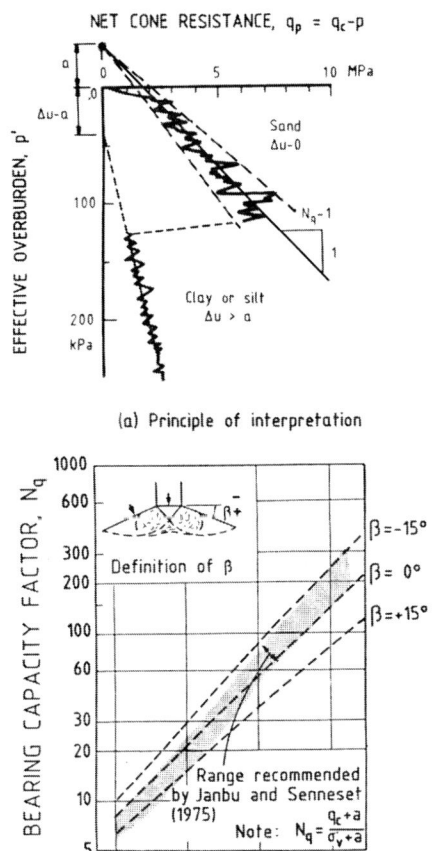

Fig. 11 Janbu and Senneset's interpretation method of drained strength in sand

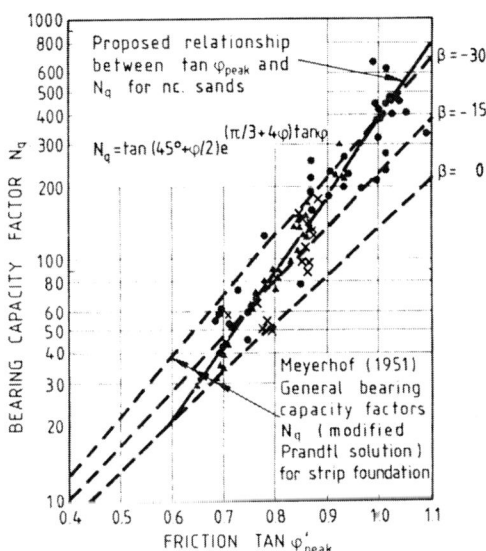

Fig. 12 Bearing capacity factor N_q vs $\tan\phi'$ peak from tests on NC sand (after Lunne and Christoffersen, 1983)

Find D_R from the revised q_c, σ'_v, D_R chart in Fig. 8 and carry out triaxial tests on sand specimens reconstituted to different D_R values. Find ϕ' by interpolation. If no triaxial tests all available, ϕ' may be found by the chart given in Fig. 13, as recommended by Schmertmann.

For overconsolidated sands, OCR and K_0 should be estimated by using the relationships recommended above under the heading 'Pressuremeter Test'.

For both NC and OC sands, the use of the more conservative ϕ'-value computed by the three methods for the problem at hand is recommended. For computing bearing capacity, a low ϕ' is considered conservaive, but for the calculation of skirt penetra-

tion or pile-driving resistance, a high ϕ' may be considered conservative. If the Janbu and Senneset method yields a value for attraction (cohesion), this must be taken into account when selecting the design value. For sands with low relative density, the structure of the sand may significantly influence the triaxial (and *in situ*) ϕ' value. In forming sand for tests in the laboratory a

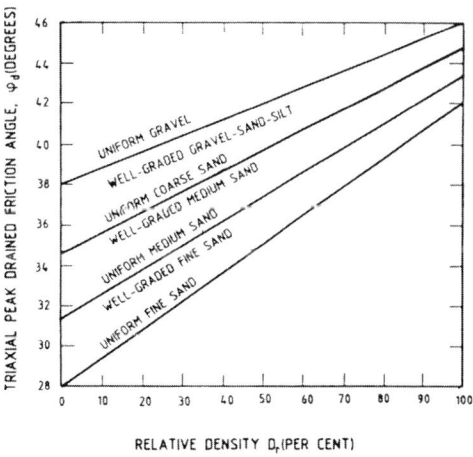

Fig. 13 Relationship between ϕ' and D_R suggested by Schmertmann (1978)

method should be used that will reproduce the *in situ* sand as closely as possible. The ϕ' value found using the above procedures will generally correspond to the peak value in a drained triaxial test on sand consolidated to *in situ* stresses.

It should be noted that bearing capacity theories based on the assumption that soil is incompressible cannot adequately describe the physical behaviour of a sand subjected to a penetrating cone. Especially at large depths, failure through compression and grain crushing will probably occur. It is thus likely that the expansion cavity theory can more realistically model soil behaviour during cone penetration, since it takes soil compressibility into account (e.g. Vesic, 1973). However, the method requires a knowledge of soil parameters which are as yet very difficult to determine. Thus a semi-empirical approach as outlined above may still be required, although upgrading of the method is warranted as more data become available.

Dilatometer Test

As mentioned earlier, two methods exist for obtaining an effective friction angle in sands from dilatometer tests. Figure 14 gives examples of friction angles obtained in a medium sand before and after compaction. The two methods provide slightly different answers. However, it is not clear whether the dilatometer friction angle should be correlated with the friction from a drained or undrained triaxial test. The dilatometer penetrates very quickly, and one should consider whether the bearing capacity failure takes place under undrained conditions. Future developments with measurements of pore pressure in connection with the dilatometer test will reduce or eliminate this uncertainty.

Pressuremeter Test

The pressuremeter test results may be interpreted, as illustrated in Fig. 15, to determine the effective friction angle ϕ'. The method shown, proposed by Hughes *et*

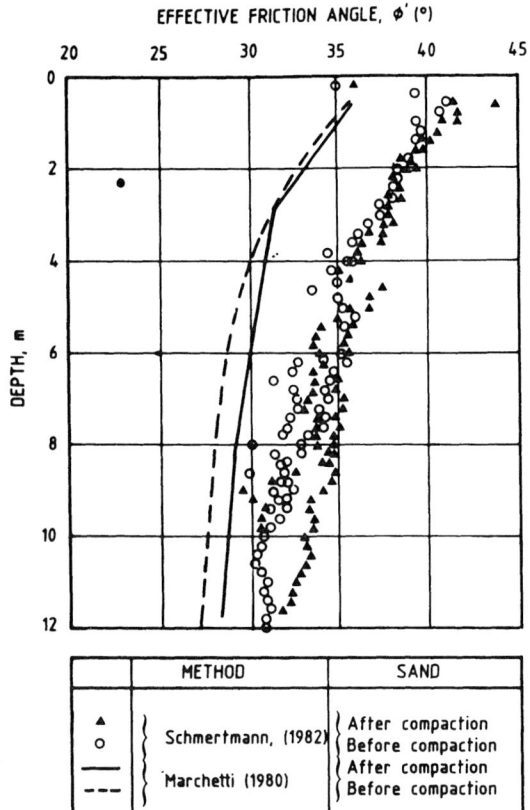

Fig. 14   Effective friction angle predicted from dilatometer test

al. (1977), is the one most commonly used today. The field expansion curve is plotted on log-log scales, and the linear portion of the curve gives a slope related to the friction angle (Fig. 15b). The inital strain can also be adjusted if necessary. The friction and dilation angles are found by assuming a ϕ'_{cv} value in Fig. 15c. Figure 16 compares the expansion curves of push-in and self-boring pressuremeter tests in sand from the same depth. Neither of the tests predicted the *in situ* horizontal stress well. The two tests result in different effective friction and dilation angles. It is interesting to note that the two expansion curves are very similar, and that the PIP test appears to have had a translation of about 100 kPa along the stress axis.

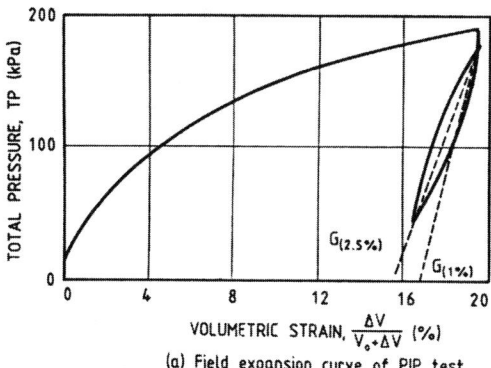

(a) Field expansion curve of PIP test

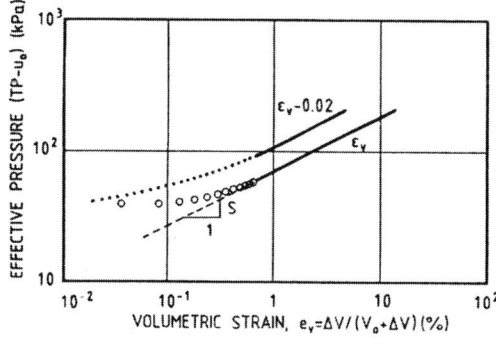

(b) PIP expansion curve on log-log plot

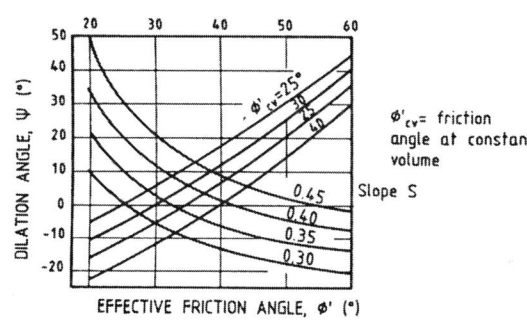

(c) Calculation of friction angle ϕ'

Fig. 15 Interpretation of pressuremeter test in sand: (a) field expansion curve of PIP test; (b) PIP expansion curve on log-log plot; (c) calculation of friction angle ϕ'

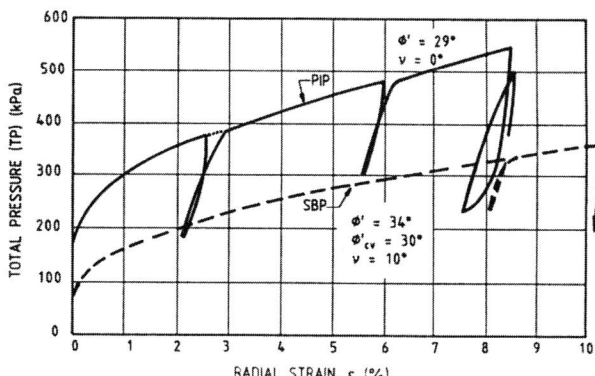

Fig. 16 Comparison of typical push-in and self-boring pressuremeter tests in medium dense sand

(1) Constrained modulus, M, valid for one-dimensional cases corresponding to oedometer testing.
(2) Young's modulus, E, which is equivalent to the modulus from triaxial tests where normally the secant modulus at 50 and 25% of failure stress is given.
(3) The initial shear modulus, G_{max}, measured at a very low strain level; it is used in the calculation of dynamic soil behaviour (see 'Initial Shear Modulus' below).

Constrained Modulus, M

Cone Penetration Test

Correlations between constrained modulus and cone resistance are most reliably determined from calibration chamber tests where M is measured directly on a large sample prior to penetrating the cone (e.g. Parkin et al., 1980; Baldi et al., 1982). Lunne and Christoffersen (1983) carried out an extensive review of calibration chamber test results, as shown in Fig. 17 where the initial tangent modulus M_0 for normally consolidated sands has been plotted against cone resistance q_c. Recent results from tests at the University of Southampton using the Norwegian Technical Institute's chamber (Last et al., 1985) have been included. Lunne and Christoffersen recommended that M_0 in normally consolidated

DEFORMATION CHARACTERISTICS

General

Depending on the problem under consideration, it may be necessary to evaluate several different deformation moduli. Three different moduli are covered below.

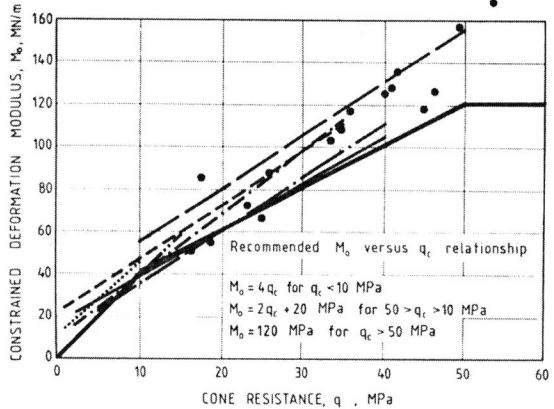

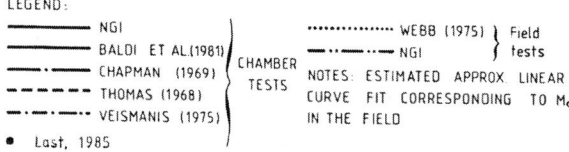

LEGEND:

───── NGI		············· WEBB (1975) ⎫ Field
───── BALDI ET AL.(1981) ⎱ CHAMBER		─··─··─ NGI ⎰ tests
─── CHAPMAN (1969) ⎰ TESTS		NOTES: ESTIMATED APPROX. LINEAR
───── THOMAS (1968)		CURVE FIT CORRESPONDING TO M_o
─·─·─ VEISMANIS (1975)		IN THE FIELD
● Last, 1985		

Fig. 17 Summary of relationships between constrained modulus M_0 and cone resistance q_c for NC sands (based on Lunne and Christoffersen, 1983

sands should be estimated as follows:

$M_0 = 4\,q_c$ for $q_c < 10$ MPa
$M_0 = 2\,q_c + 20$ MPa for 10 MPa $< q_c$ < 50 MPa
$M_0 = 120$ MPa for $q_c > 50$ MPa

For overconsolidated sands with OCR > 2 they proposed:

$M_0 = 5\,q_c 1$ for $q_c < 50$ MPa
$M_0 = 250\,q_c$ for $q_c > 50$ MPa

The constrained modulus applicable for the stress range σ'_{vo} to σ'_{vo} to $\Delta\sigma'_v$ can be estimated as

$$M = M_0 = \left(\frac{\sigma'_{vo} + \Delta\sigma'_v/2}{\sigma'_{vo}}\right)^{\frac{1}{2}}$$

The relationships given above are meant to be used as a guide for a first estimate of settlements. As pointed out by Jamiolkowski *et al.* (1985), both M/q_c and E/q_c depend on several factors including stress and strain history, stress level, relative density of sand and mineralogical composi-

tion, which is related to crushability of sand grains.

However, despite these limitations, engineers need correlations like those given above between M and q_c because there is a lack of technically and economically valid alternatives. For future applications, the dilatometer may become an attractive alternative.

Dilatometer Test

Marchetti (1980) provides empirical correlations for the calculation of the vertical constrained modulus from the horizontal stress index K_D, the material index i_D, and the dilatometer modulus E_D. The correlations are known to work well in Norwegian clays (Lacasse and Lunne, 1982) but have been little used in sand. Figure 18 gives examples of the constrained modulus obtained from dilatometer tests on a medium coarse sand before and after compaction and compares them with modulus values obtained via Janbu's (1970) relationship for normally consolidated sand:

$$M = m(p_o' p_a)^{1/2}$$

where M = constrained modulus; m = modulus number = 150–250 for medium sand; p_o' = effective overburden stress; p_a = reference pressure = 100 kPa.

The comparison of the modulus based on Janbu's relationship and the modulus based on dilatometer tests on the normally consolidated sand before compaction is relatively good.

Screw-plate Test

In general, the first load increment of the screw plate is p'_o, and the load is then increased in steps to 500 or 1000 kPa. The deformation pattern is shown in Fig. 19a. The sand volume compresses vertically, and a small shear deformation also takes place horizontally. This shear deformation is believed to have little significance at low stresses, but becomes significant at high stresses. One should calculate the modulus

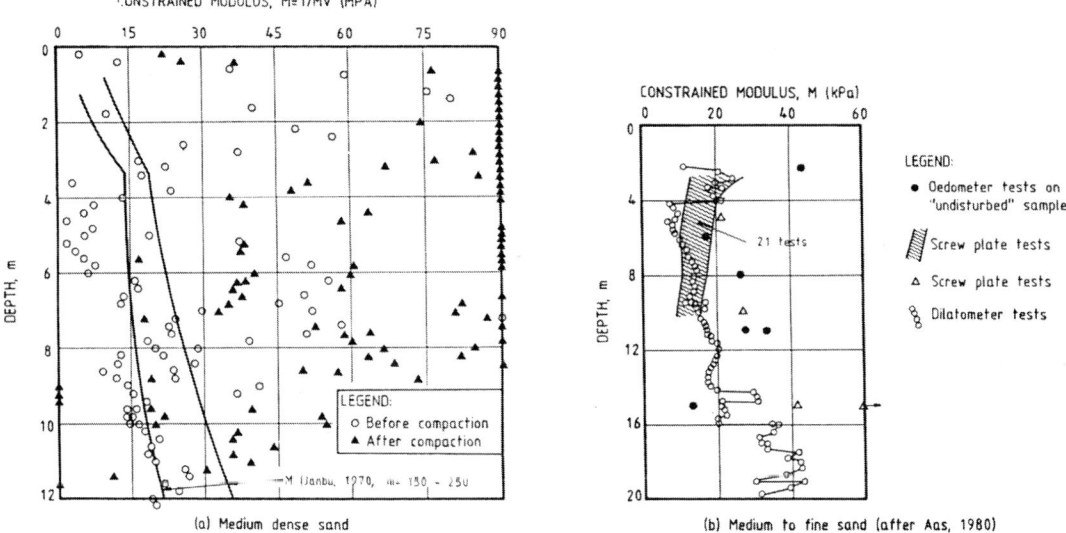

Fig. 18 Constrained modulus from dilatometer test: (a) medium coarse sand; (b) medium-to-fine sand

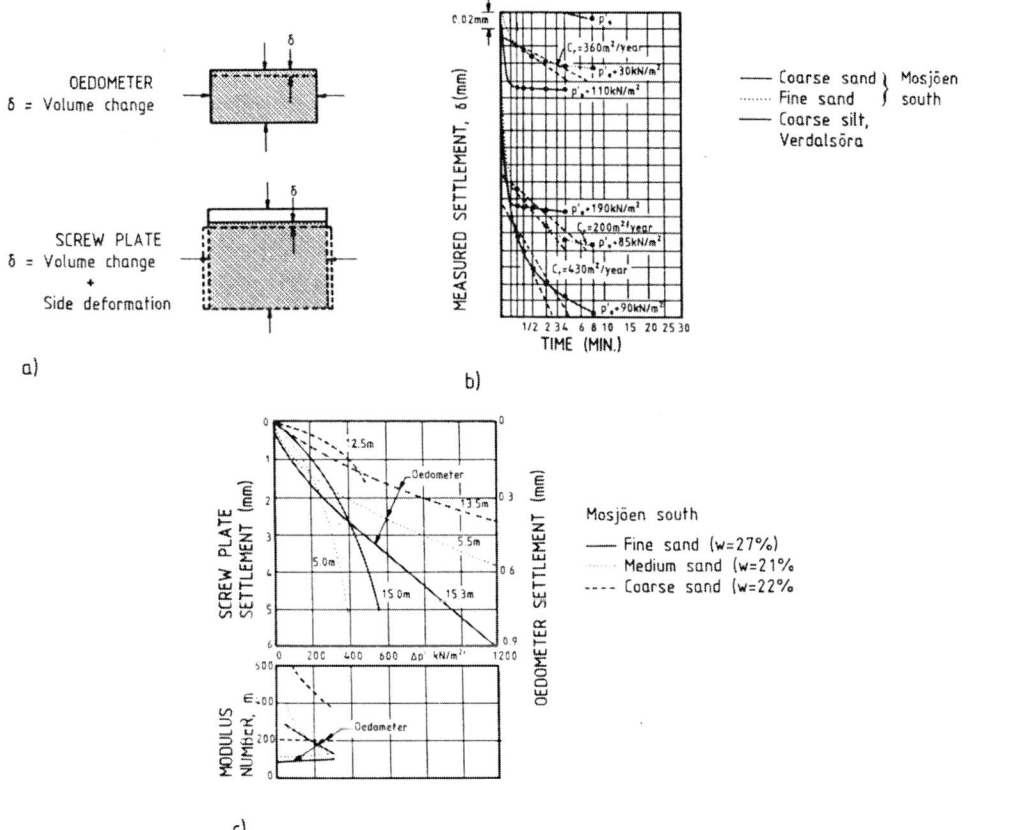

Fig. 19 Results of screw-plate tests in different sand/silt types (Aas, 1977): (a) deformation pattern in oedometer and screw-plate tests; (b) time-settlement curves in 3 soil types; (c) compression curve from oedometer and screw-plate tests in three sand types

based on screw-plate tests for the stress range p'_o to $2p'_o$. The screw-plate test provides a time–settlement curve which can be used to calculate the coefficient of consolidation, as shown in Fig. 19b. Figure 19c compares screw-plate and oedometer compressibility curves on three sand types. The screw plate tends to give a higher modulus number m $(m = dM/d\sigma'_v)$ than the oedometer tests (Aas, 1977).

Screw-plate tests were run in a loose sand in Drammen and compared to the results of oedometer tests on 'undisturbed' samples.

One dilatometer test was also carried out. Figure 18 compares the constrained modulus based on oedometer, screw-plate and dilatometer tests. The agreement of the screw-plate and dilatometer tests is very good.

Aas (1981) compared the modulus numbers backfigured from settlement measurements with the modulus numbers obtained from screw-plate and oedometer tests at six sites where silt and fine, medium and coarse sands were found (Fig. 20). The modulus numbers are presented as a function of

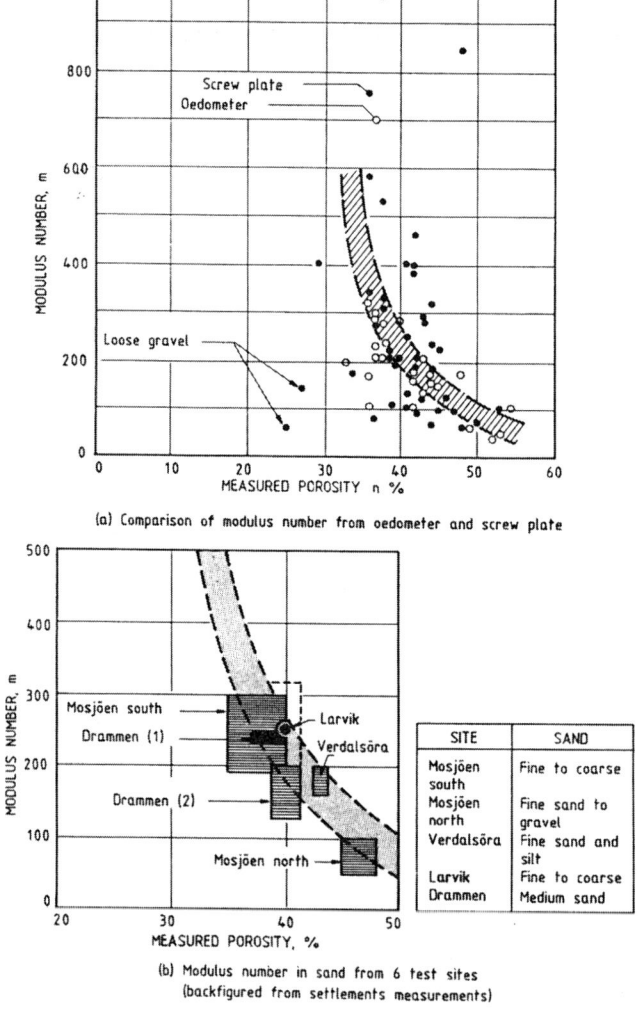

(a) Comparison of modulus number from oedometer and screw plate

(b) Modulus number in sand from 6 test sites
(backfigured from settlements measurements)

Fig. 20 Modulus numbers from six test sites, based on screw plate and oedometer tests (Aas, 1981): (a) comparison of modulus numbers from oedometer and screw plate; (b) modulus number in sand from 6 test sites

porosity. The porosity was calculated from the water content, the unit weight and the specific gravity of solids measured on undisturbed samples from the same depth as the screw-plate tests. There is some scatter in the data, but they show a definite trend. The agreement between the screw-plate and the backfigured data is very good.

Young's Modulus, E

Cone Penetrations Tests

For cases which are not one-dimensional, Young's modulus, E, is more appropriate than the constrained modulus, M. Young's modulus depends on several factors, including stress level, stress and strain history, relative density and sand type.

Robertson and Campanella (1983) reviewed available calibration-chamber test results in which cone resistance had been compared to drained secant modulus from parallel triaxial tests prepared in the same manner as the large calibration samples and to the same densities. Robertson and Campanella (1983) argued that E_{25} (secant modulus at 25% failure stress) is most appropriate since the overall safety factor against bearing capacity failure is usually around 4 for foundations on sand. For computing bearing capacity of piles, E_{50} may be more adequate. Calibration-chamber test results give values of E_{25}/q_c varying between 1.5 and 3.0 which are in good agreement with the recommended value of 2 given by Schmertmann (1978) for the computation of settlements of shallow foundations on sand. Schmertmann (1978) has since amended his earlier suggestion by recommending two values, 2.5 and 3.0, to allow for the variation of shape factors for square and strip footings. Robertson and Campanella (1983) recommend the relationship given in Fig. 21 for the computation E_{25} or E_{50}.

Results from calibration-chamber tests suggest higher values of E_{25}/q_c for overconsolidated sands (i.e. $6 \leqslant \alpha \leqslant 18$). However, the application of these larger factors to overconsolidated sands should be used with

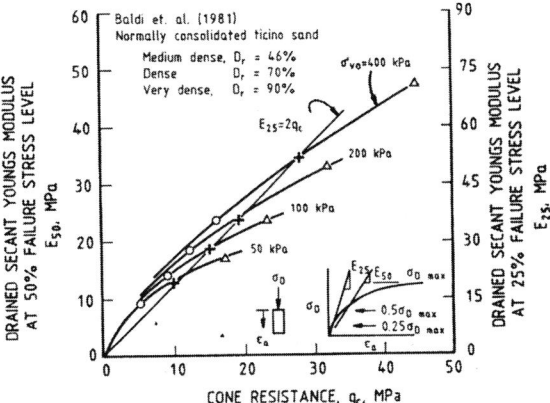

Fig. 21 Secant Young's modulus values for uncemented n.c. quartz sands (after Robertson and Campanella, 1983)

caution, since the increase depends on the degree of overconsolidation and density (Baldi et al., 1982).

When using the relationship given in Fig. 21 it is important to be aware of the limitations of this approach, which are the same as for the M/q_c relationship presented earlier.

Pressuremeter Test

Figure 16 compared the expansion curves of push-in and self-boring pressuremeter tests. The unload–reload loops shown are used to compute shear modulus. As shown by this particular test, the modulus values are very similar. In practice, however, this drained modulus is little used in geotechnical calculations. Campanella et al. (1985) describe a new cone pressuremeter which may be used to determine modulus and soil strength. The equipment is still being developed. It is also intended for offshore use.

INITIAL SHEAR MODULUS

Cone Penetration Test

Shear modulus at any strain level can as a first approximation be estimated from the initial tangent modulus (G_{max}) using the reduction curves given by Seed and Idriss

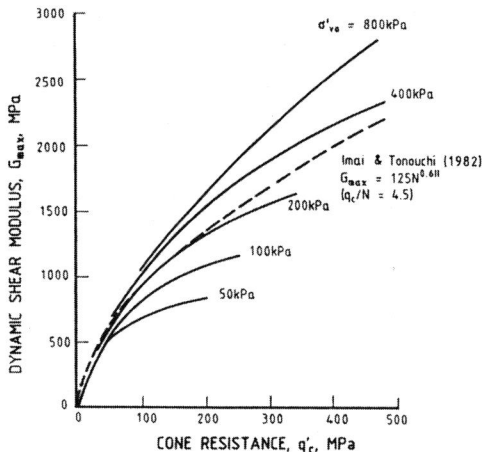

Fig. 22 Dynamic shear modulus values for uncemented n.c. quartz sands (after Robertson and Campanella, 1983)

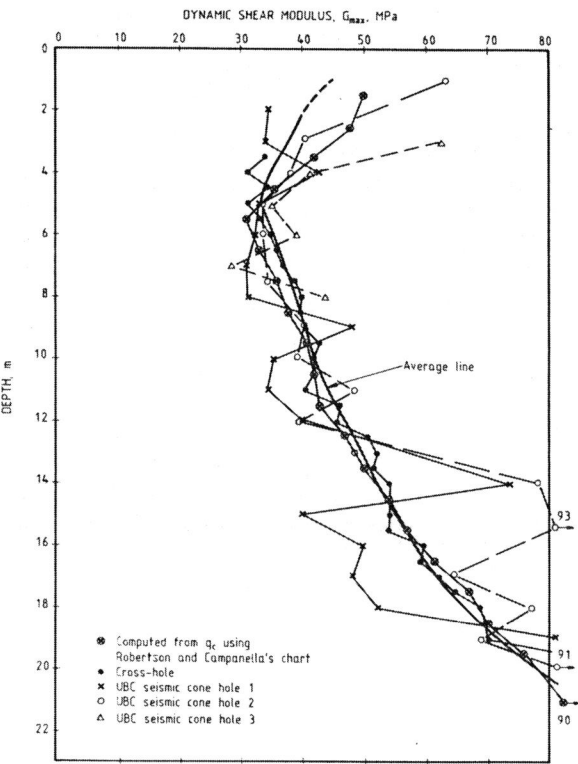

Fig. 23 Dynamic shear modulus values for Drammen sand

(1970), if no specific laboratory test results are available.

Robertson and Campanella (1983) presented the correlation chart in Fig. 22 which relates dynamic shear modulus, G_{max}, cone resistance and vertical effective stress. Robertson and Campanella based their chart on laboratory correlations between dynamic shear modulus and relative density (Seed and Idriss, 1970) and the relationship between D_R, q_c and σ'_{vo} recommended by Baldi *et al.* (1982).

Figure 23 compares G_{max} as measured by the cross-hole method and as computed from CPT results using the relationships in Fig. 22. It may be observed that there is a very good fit between the two sets of results, and this indicates that Robertson and Campanella's chart works well for the Drammen sand. As for the q_c, σ'_{vo}, D_R relationships discussed earlier under 'Soil Density', the chart in Fig. 22 is, strictly speaking, valid only for normally consolidated, fine-to-medium, uniform, non-cemented sands.

Seismic Cone

The seismic cone has been used in Drammen sand (Eidsoen *et al.*, 1985), and Fig. 23 shows the G_{max} values computed from three tests plotted together with G_{max} computed

from CPT results and measured by the cross-hole method. There appears to be more variation in the seismic cone measurements, but on average the results are close to those obtained by the two other methods. One reason for the larger variation in the seismic cone test results may be that this method is more sensitive to local layering than the cross-hole method.

The seismic cone appears to give reasonable and representative initial shear modulus values for the soils in which it has been tested out to date (see also Campanella and Robertson, 1984).

The chart in Fig. 22 also seems to give good results in clean, normally consolidated sand. However, its reliability should be assessed by comparative studies in the calibration chamber or field studies in well-known sand deposits. An attempt should also be made to develop such charts for overconsolidated sands.

IMPORTANCE OF TAKING SAMPLES AND PERFORMING LABORATORY TESTS

The selection of final design parameters should not be based on the results of *in situ* tests alone. A thorough laboratory investigation of the samples obtained is most important. The soil boring(s) should be carried out close to a CPT or to other *in situ* tests. Normally it is not possible to obtain undisturbed sand samples, and strength or deformation tests have to be carried out on recompacted sand samples.

It is advisable to compact samples to various densities and to obtain a relationship between the friction angle and the relative density. This should be compared with the one given by Schmertmann (1978).

Several researchers (e.g. Mitchell, 1976) have shown that the engineering properties of sand vary with sand fabric. This should be considered when deciding what method to use for recompacting the samples in the laboratory.

For a complete description of the sand, it is recommended that the following classification tests should be carried out on disturbed samples: grain-size distribution, specific gravity of solid particles, angularity of sand grains, maximum and minimum porosity, and mineralogy. From the sand description, the application of empirical correlations such as the q_c, σ'_v against D_R chart (Fig. 8) can be evaluated.

In some cases, it is possible to take samples in sand that can be transported undisturbed to the laboratory for testing. When evaluating results of tests on 'undisturbed' samples, the following aspects should be consider:

- loose samples tend to become denser during sampling, transport and handling;
- very dense samples tend to increase in volume and become less dense during sampling, transport and handling.

For special situations such as those encountered in offshore structures, no *in situ* test can yet determine the cyclic behaviour of a sand. It is therefore necessary to perform cyclic triaxial and simple shear tests *in situ* and resonant column tests on samples in the laboratory. On large projects, with uniform sand layers and a large quantity of sand available, tests may be performed in large calibration chambers. However, the effects of sample size and chamber boundary conditions must be considered (e.g. Parkin and Lunne, 1982).

SUMMARY AND CONCLUSIONS

1. *The cone penetration test*, especially when combined with measurement of pore pressures, is at present the most useful single *in situ* testing device. It should be included in all offshore soil investigations of any importance. The piezocone gives an excellent picture of soil layering and can be used to identify the penetrated soil with reasonable accuracy. Reasonably good correlations exist between piezocone results and soil density, drained shear strength, deformation characteristics and dynamic shear modulus, at least for normally consolidated sands. Stress history or *in situ* horizontal stress cannot be determined with the commercial devices at present available. Research to develop equipment and methods for improving this situation is in progress.

2. *The electrical resistivity and the nuclear density probes* should definitely be considered when it is particularly important to assess the *in situ* density of soils. Electrical resistivity tests work mainly in sands whereas nuclear density tests may be used in most soil types. However, both methods need to be checked in the laboratory and in the field against known densities before their full potential can be evaluated.

3. *The dilatometer* has as yet been little used offshore, but it has gained wide popularity onshore because of its practicality and because empirical correlations exist that can be used to evaluate soil type, soil density, *in situ* horizontal stress, stress history, shear strength and constrained modulus. Because

of its potential for determining *in situ* horizontal stress and overconsolidation ratios, it is expected that the dilatometer will be a powerful tool in offshore soil investigations when coupled with CPT results. The future development of pore-pressure readings in the dilatometer is also likely to make the tool even more useful.

4. *The push-in pressuremeter* is available for routine use offshore and may be used to determine internal friction and dilation angles of sand as well as shear modulus backfigured from the unload–reload loop. *The self-boring pressuremeter* tests, if carefully performed, may be used to determine horizontal stress *in situ*, even in sand deposits. The self-boring pressuremeter is, however, very complicated to use offshore, although equipment already exists which enables the test to be carried out as a special operation.

5. *The screw plate* is not yet available for offshore use but may be available for future marine soil investigations. Research and practice on land has shown the usefulness of this device, which provides very good estimates of the settlement parameters.

6. *The seismic cone* is at present being modified for use in shallow water, and it is hoped that it will be available for investigations offshore in the near future. Onshore research has shown this tool to be excellent for directly measuring down-hole shear-wave velocity, which may be used for determining the initial shear modulus.

Even if in future *in situ* testing is used more extensively, this will not eliminate the need to take good samples and test them in the laboratory. Final soil design parameters should be evaluated from both *in situ* and laboratory test results.

ACKNOWLEDGEMENTS

The authors wish to thank Messrs D. Gillespie, T. Eidsmoen and B. Thune, who participated in the field testing programme; Professor Jamiolkowski, who provided valuable pressuremeter test data; and K. Høeg, who reviewed the paper.

REFERENCES

1. Aas, G. 1977. Sammenstilling av styrkeparametre og spenningsmoduler for sand bestemt ved in situ målinger og laboratoriekorsøk. *Geoteknikkdagen Proc.*, pp. 32.1–32.20
2. Aas, G. 1981. Setninger av bygg på sand. Bestemmelse av jordartsparametre for bruk ved setningsberegninger. NGI Report 52409-8, 10 February 1981.
3. Aas, G., Lacasse, A., Lunne, T. and Madshus, C. 1984. *In situ* testing: new developments. In *Proc. Nordiska Geoteknikermøtet*, Linkøping, 1984, Vol. 2, pp. 705–716.
4. Baldi, G., Bellotti, R., Ghionna, V., Jamiolowski, M. and Pasqualini, E. 1982. Design parameters for sands from CPT. In *Proceedings of European Symposium on Penetration Testing*, Amsterdam, 1982. Balkema, Rotterdam, Vol. 2, pp. 425–432.
5. Baldi, G., Belotti, R., Ghionna, V., Jamilkowski, M. and Pasqualini, E. 1985. Penetration resistance and liquefaction of sands. Paper to *11th International Conference on Soil Mechanics and Foundation Engineering*, San Francisco, August 1985.
6. Bellotti, R., Ghionna, V., Jamiolkowski, M., Manassero, M. E. and Pasqualini, E. 1983. Evaluation of sand strength from CPT. Presented at *International Symposium on Soil and Rock Investigations by In Situ Testing*, Paris, 18–20 May 1983.
7. Berzins, W. E. and Campanella, R. G. 1981. Development of the screw plate test for *in situ* determination of soil parameters. Soil Mech. Series No. 48, Department of Civil Engineering, University of British Columbia, May 1981.
8. Boghrat, A. 1982. The design and construction of a piezoblade and an evaluation of the Marchetti dilatometer in some Florida soils. PhD thesis, University of Florida.
9. Burgess, N. C., Hughes, J. M. O., Innes, R. and Gladowe, J. 1983. Site investigation and *in situ* testing techniques in Arctic seabed sediments. OTC, Houston, May 1983.
10. Douglas, B. J. and Olsen, R. S. 1981. Soil classification using electric cone penetrometer. In *Cone Penetration Testing and Experience*, Proceedings of a session at the ASCE National Convention, St Louis, Missouri, 1981. American Society of Civil Engineers, pp. 209–227.

11. Durgunoglu, H. T. and Mitchell, J. K. 1975. Static penetration resistance of soils, I–II. In *Proceedings of Conference on In Situ Measurement of Soil Properties*, Raleigh, North Carolina, 1975. American Society of Civil Engineers, Vol. 1, pp. 151–189.

12. Eidsmoen, T., Gillespie, D., Lunne, T. and Campanella, R. G. 1985. Tests with UBC seismic cone at three Norwegian research sites. NGI and UBC joint report, 59040-1, 20 December 1984.

13. Fahey, M. and Ranolph, M. F. 1984. Effect of disturbance on parameters derived from self-boring pressuremeter tests in sand. *Geotechnique* 34(1), 81–97.

14. Faÿ, J. B. and Le Tirant, P. 1982. Offshore self-boring pressuremeter for deep water. *Symposium on the Pressuremeter and its Marine Application*, Paris. Editions Technip. Paris, pp. 305–324.

15. Campanella, R. G., Gillespie, D. and Robertson, P. K. 1982. Pore pressures during cone penetration testing. In *Proceedings of the 2nd European Symposium on Penetration Testing*, Amsterdam, Vol. 2, pp. 507–512.

16. Campanella, R. G. and Robertson, P. K. 1983. Flat plate dilatometer testing: Research and development. Soil Mech. Series No. 68, Department of Civil Engineering, University of British Columbia, January 1983.

17. Campanella, R. G. and Robertson, P. K. 1984. A seismic cone penetrometer to measure engineering properties of soil. Paper presented at the Society of Exploration Geophysicists' 54th Annual Meeting, Atlanta.

18. Campanella, R. G. 1985. Personal communication.

19. Campanella, R. G., Robertson, P. K., Gillespie, D. G. and Greig, J. 1985. Recent developments in *in situ* testing of soils. Paper to *11th International Conference on Soil Mechanics and Foundation Engineering*, San Francisco, August 1985.

20. Chapman, G. A. 1979. The interpretation of friction cone penetrometer tests in sand. PhD thesis, Department of Civil Engineering, Monash University, Australia.

21. Hughes, J. M. O., Wroth, C. P. and Windle, D. 1977. Pressuremeter tests in sands. *Geotechnique* 27(4), 455–477.

1. Jamiolkowski, M., Ladd, C. C., Germaine, J. T. and Lancelotta, R. 1985. New developments in field and laboratory testing of soils. Paper to *11th International Conference on Soil Mechanics and Foundation Engineering*, San Francisco, August 1985.

22. Janbu, N. C. 1970. Grunnlag i Geoteknikk. Tapir, 426 p.

23. Janbu, N. and Senneset, K. 1973. Field compressometer — principles and applications. In *Proceedings of 8th International Conference on Soil Mechanics and Foundation Engineering*, Moscow, Vol. 1.1, pp. 191–198.

24. Janbu, N. and Senneset, K. 1975. Effective stress interpretation of *in situ* static penetration tests. In *Proceedings of European Symposium on Penetration Testing*, Stockholm, 1974, Vol. 2.2, pp. 181–193.

25. Kroezen, M. 1981. Measurement of *in situ* density in sandy/silty soils. *Can. Geotech. Soc. Newslett.* 18(4), 13–15.

26. Lacasse, S. and Lunne, T. 1982. Penetration tests in two Norwegian clays. In *Proceedings of 2nd European Symposium on Penetration Testing*, Amsterdam, 1982. Balkema, Rotterdam, Vol. 2, pp. 661–669.

27. Last, N. 1985. *Proceedings from Seminar on Cone Penetration Testing in the Laboratory*, Southampton University, November 1984.

28. Lunne, T. and Christoffersen, H. P. 1983. Interpretation of cone penetrometer data for offshore sands. In *Proceedings of 15th Offshore Tech. Conf.*, Houston, Texas, 1983, Vol. 1, pp. 181–192.

29. Marchetti, S. 1980. *In situ* tests by flat dilatometer. ASCE, JGED, V. 106, No. GT3, pp. 299–321.

30. Marchetti, S. and Crapps, D. K. 1981. *Flat Dilatometer Manual*. Gainesville, Florida.

31. Meyerhof, G. G. 1961. The ultimate bearing capacity of wedge-shaped foundations. In *Proceedings of 5th International Conference on Soil Mechanics and Foundation Engineering*, Paris, Vol. 2, pp. 105–109.

32. Mitchell, J. K. 1976. *Fundamentals of Soil Behaviour*. John Wiley, New York.

33. Mitchell, J. K. and Lunne, T. A. 1978. Cone resistance as measure of sand strength. *Proc. Am. Soc. Civ. Engrs* 104 (GT7), 995–1012.

34. Mitchell, J. K. 1985. Personal communication.

35. Parkin, A., Holden, J., Aamot, K., Last, N. and Lunne, T. 1980. Laboratory investigations of CPT's in sand. Report 52108-9, Norwegian Geotechnical Institute.

36. Parkin, A. K. and Lunne, T. 1982. Boundary effects in the laboratory calibration of a cone penetrometer for sand. In *Proceedings of 2nd European Symposium on Penetration Testing*, Amsterdam, 1982. Balkema, Rotterdam, Vol. 2, pp. 761–768.

37. Reid, W. M., St John, H. D., Fyffe, S. and Rigden, W. J. 1982. The push-in pressuremeter. *37th Symposium on the Pressuremeter and its Marine Applications*. Editions Technip, Paris, pp. 247–261.

38. Robertson, P. K. and Campanella, R. G. 1983. Interpretation of cone penetration tests. Part I: Sand. *Can. Geotech. J.* 20(4), 718–733.

39. Robertson, P. K. and Campanella, R. G. 1984. Liquefaction potential of sands using the DMT. Technical note, submitted to ASCE, JGED, April 1984.

40. Ruiter, J. de 1982. The static cone penetration test state-of-the-art report. In *Proceedings of 2nd European Symposium on Penetration Testing*, Amsterdam, Vol. 2, pp. 389–405.

41. Schmertmann, J. H. 1978. Guidelines for cone penetration test: performance and design. Report TS-78-209, Department of Transportation, Federal Highway Administration Offices of Research and Development, Washington, DC.

42. Schmertmann, J. S. 1982. A method for determining the friction angle in sands from the Marchetti dilatometer test. In *Proceedings of 2nd European Symposium on Penetration Testing*, Amsterdam, Vol. 2, pp. 853–861.

43. Senneset, K. and Janbu, N. 1984. Shear strength parameters obtained from static cone penetration tests. *ASTM Symposium*, San Diego, 1984.

44. Tjelta, T. I., Tieges, A. W. W., Smits, F. P., Geise, J. M. and Lunne, T. 1985. *In situ* density measurements by nuclear backscatter for an offshore soil investigation. *Offshore Technology Conference Proceedings*, May 1985, Paper 4917. (In press.)

45. Vesic, A. S. 1973. Analysis of ultimate loads of shallow foundations. *Proc. Am. Soc. Civ. Engrs* 99 (SM1), 45–73.

18

The Applications of Centrifugal Modelling to the Design of Jack-up Rig Foundations

W. H. Craig, University of Manchester, UK,
and M. D. Higham, McClelland Ltd, UK

INTRODUCTION

In recent years the use of physical modelling aboard large centrifuges has been considerably extended in many areas of geotechnical research. In offshore and coastal engineering, where loadings on structures are often complex and non-static, such models have been widely used and are believed to have given considerable qualitative insights into the modes of behaviour of a number of different structural forms on both uniform and stratified soil deposits, as well as yielding quantitative data for performance predictions which add to the confidence of designers, operators, insurers and regulatory agencies. Perhaps surprisingly, the mobile jack-up structure has only recently been the subject of model studies of this type. Such studies as have been carried out are very preliminary and the present paper aims to add to these and consider possible areas of useful future extension.

JACK-UP PROBLEMS

Since their introduction in the early 1950s, the number of jack-up mobile drilling platforms available for petroleum exploration and production purposes has increased to its present level of more than 400. Most of these jack-up rigs are supported by three or four independently jacked legs with individual tank-type footings (spud-cans), although earlier models sometimes had more legs. The alternative jack-up concept is that of the mat-supported rig. Two excellent review papers, one by McClelland *et al.* (1983) and the other by Young *et al.* (1984), cover most aspects of the operation of all such rigs. While there are wide variations in design specifications, individual spud-can foundations are commonly of the order of 100–150 m² carrying loads up to 30 MN and working at bearing stresses of 200–300 kPa. Such footings may be operating at depths up to 30 or 40 m below the seabed, and in such instances huge volumes of soil are subjected to gross shearing at the time of installation. In contrast, the mat-supported rigs may have bearing areas an order of magnitude larger and working stresses an order lower, being particularly suitable for work on very soft cohesive seabeds where working depths are typically 2–3 m.

For both types of rig the normal practice is to pre-load the foundations to the maximum design load on first arrival on site. The design load includes the fairly reliable known all-up weight of the rig itself, together with an assessment of environmental loads which may be known to a considerably lower degree of precision. Thus, unlike permanent structures which may never be subjected to the design loads at any time in their working life, each mobile structure is proof loaded to the design load at each location. A high proportion of jack-up rig accidents occur during this proof testing stage of spud-can foundations, when punch-through failures occur beneath individual footings as they penetrate through a stiff stratum into a softer one. While most authors in the literature indicate that analytical methods are adequate to assess the likely performance of a single footing under static vertical loading, regardless of the soil statification, provided the soil profile and properties are known, the practice of installing rigs on sites without adequate pre-investigation continues and this may largely account for the high incidence of failure at the pre-load stage. In short, site investigation is as crucial for a mobile structure as for a fixed structure.

Young et al. (1984) list footing-supported rig failures which have resulted in major economic losses in recent years. In addition to these there are no doubt a greater number of failures which were not included because the consequences were less severe — the table may in reality only offer up the tip of the iceberg. Five problem areas are identified:

(1) Punch-through during pre-load
(2) Excessive penetration during storm loading.

(3) Footing instability caused by scour undermining footings.
(4) Mass movement of the sea floor.
(5) Jack-down and extraction of footings, where pull-out loads exceed available forces.

Category 5 is generally associated with deeply embedded footings, while category 3 is more likely at shallow depths in cohesionless soils.

For mat-supported rigs bearing failure at the pre-load stage is less likely to be catastrophic, but this form of structure is, by virtue of its low bearing pressures and shallow penetration, particularly vulnerable to failure under environmental loading with sliding displacements being difficult to predict.

PUNCH-THROUGH FAILURES

Most cases of punch-through occur in stratified soil profiles when there is a relatively thin layer of sand or stiff clay overlying a weaker layer. When installing a rig on this type of soil profile instability may occur after the footing has been temporarily supported by the upper layer. A further increase in footing pressure which exceeds the combined bearing capacity of the two-layered system will result in the footing punching through the upper soil layer and penetrating into the underlying layer until sufficient bearing resistance is encountered at some lower level, or the applied load is distributed elsewhere. The resulting damage to the rig can range from minor structural damage to the leg and its jacking mechanism to the complete loss of the rig.

If it is required to install a rig at a site with a potential for punch-through type failure it is important that a reliable estimate of the ultimate bearing capacity of the stratified soil profile be made. Two basic methods have been proposed for predicting the capacity of foundations supported by such a soil profile, these being the Meyerhof and Hanna method and the Projected Area method.

Meyerhof and Hanna Analysis

The analysis of soil failure in model tests on circular and strip footings by Meyerhof (1974) and Meyerhof and Hanna (1978, 1980) has led to the development of a semi-empirical method to determine the bearing capacity of a footing punching through a relatively thin, dense layer into a softer underlying deposit.

The method as originally proposed by Meyerhof (1974) considers that at the ultimate load a mass of the upper soil layer, of roughly truncated conical shape, is pushed into the underlying weak deposit (Fig. 1). The forces resisting the applied loading are then taken to be equivalent to the total cohesion force and an integrated passive earth pressure inclined at an angle δ to the horizontal acting on an assumed vertical failure plane in the upper layer which passes through the edge of the footing, plus the ultimate bearing capacity for the foundation supported on a thick bed of the underlying soil at the level of the soil interface.

For the case of a circular footing supported by a strong sand layer overlying a weak clay layer, the ultimate bearing capacity is given by the equation

$$q_u = q_b + 2\gamma' H (H + 2D) K_s S_s \tan\phi / B + + \gamma' D \leq q_t$$

where q_b = ultimate bearing capacity of the underlying clay
$$= 6.0C[1 + 0.2(D + H)/B]$$
q_t = ultimate bearing capacity of the overlying sand layer
$$= 0.3\gamma' B N_y + \gamma' D N_q$$

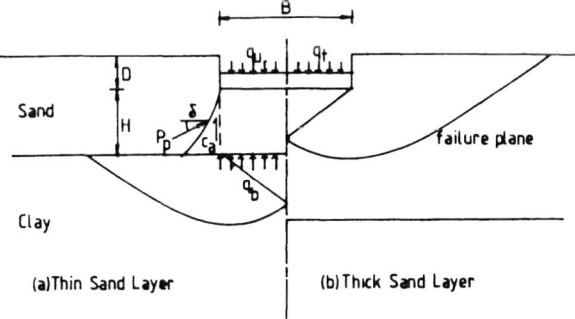

(a)Thin Sand Layer   (b)Thick Sand Layer

ig. 1 Failure of soil below footing on dense and overlying soft clay (after Meyerhof, 1974)

N_y and N_q = dimensionless bearing capacity factors
C = average undrained shear strength of the underlying clay
γ' = submerged unit weight of soil
H = thickness of sand beneath the maximum area of footing
D = depth from sea floor to the maximum area of footing
η = friction angle of sand
B = effective diameter of footing
S_s = shape factor
K_s = coefficient of punching shear resistance
= function of friction angle ϕ and the bearing capacity ratio q_2/q_1
q_1 and q_2 = ultimate bearing capacities of strip footings under vertical load on the surface of homogeneous beds of the upper and lower soil respectively.

The Projected Area Analysis

This method also considers that at the ultimate load a soil mass of roughly truncated conical shape is pushed into the underlying deposit. However, in this method the force resisting the applied foundation load is calculated on the assumption that the applied load is transmitted through a failure zone in the upper layer to a fictitious foundation at the top of the underlying weak layer. The area of this fictitious foundation is related directly to the assumed slope of the failure plane in the upper soil layer (Fig. 2). The ultimate bearing capacity is then given by

$$q_u = q_b \left(1.0 + 2\beta \frac{H}{B}\right)^2 + \gamma' D \leq q_t$$

where q_b = ultimate bearing capacity of lower layer as before
q_t = ultimate bearing capacity of upper layer as before
β = slope of failure plane in upper layer

All other symbols as before.

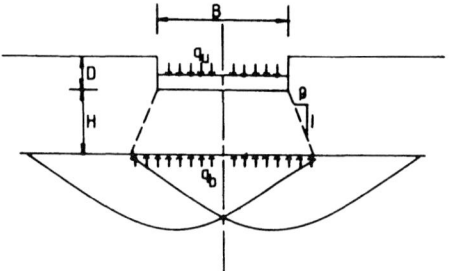

Fig. 2   Projected area method

Over the years a number of different values for the slope of the failure plane in the upper layer have been suggested. One commonly used value is that proposed by Young and Focht (1981) based on the arbitrary assumption that the failure zone in the upper layer has sides with slopes of three (vertically) to one (horizontally). In a second method, as proposed by Jacobsen *et al.* (1977), the slope of the failure plane is given by an empirical equation and is a function solely of the ratio of the bearing capacity of the footing supported by the overlying sand at the measured penetration to that of the footing supported by the underlying clay layer at the level of the soil interface. The empirical equation was developed from a series of model tests with a thin sand layer overlying a bed of soft clay, the slope being given by

$$2\beta = \alpha = 0.1125 \times 0.0344 \frac{q_t}{q_b}$$

Volume of Backfilling

The over-burden term in the above equations ($\gamma'D$) assumes that the soil above the footing is totally displaced by the footing and the hole left above the footing is not subject to infilling by either soil displaced from beneath or the collapse of the hole above. However, for a footing penetrating a given soil profile there is likely to be some backfilling. It is usual to make the conservative assumption that the hole is completely backfilled with soil at its *in situ* density. The effective surcharge will then be limited and

the over-burden term is replaced by $\gamma'V/A$ where V is the volume of the soil occupied by the footing and A is the area of the footing bearing on the soil.

CENTRIFUGE MODEL TESTS

Of all the possibilities for modelling different aspects of jack-up behaviour aboard a centrifuge, only the static surface bearing of spud-can foundations in sand and the punch-through problem of sand overlying clay have so far attracted specific attention.

Herdy and Townsend (1983) have performed four model tests at accelerations of 20–50 g using a 30 mm diameter, near surface footing bearing on a sand layer 45 mm deep overlying clay. In effect they were simulating field prototypes 0.6–1.5 m in diameter. These are large prototypes relative to most of the conventional 1 g model tests on which techniques of foundation design have been based, but they are still an order of magnitude down on realistic spud-can dimensions. It is none the less encouraging to see that, under vertical static loading only, punch-through failures were observed in the three models where the underlying clay was soft and that these were in close agreement with analysis based on Meyerhof (1974).

James and Tanaka (1984) have modelled plane circular footings on sand under horizontal and vertical load combinations with variations in load eccentricity. Using footings 50, 75, 100 mm diameter at accelerations varying between 10 and 60 g they have modelled prototypes up to 3.0 m in diameter. By including data from a test on the largest footing at 1 g they demonstrated the effects of prototype size on the static bearing pressure. Using the same sand in each case they showed that in moving from a 0.1 m footing to a 3.0 m footing under vertical loading only the bearing capacity factor N_γ could drop by a factor of 3 (equivalent to a change in ϕ' of 6°). Combinations of vertical and horizontal loading led to the expected transition from bearing to sliding failure for the surface footings and their results were in broad agreement with those of other researchers working at 1 g. A single test using a conical footing with an apex angle of 120° had approximately one quarter of the capacity of the equivalent flat plate.

Higham (1984) has conducted a series of tests on footings 50 and 75 mm in diameter at 25 g corresponding to prototypes 1.25 and 1.88 m in diameter, aimed at evaluating the theories for punch-through for a limited range of site conditions. The results are reported below.

Experimental Programme

A total of eleven tests were completed in a preliminary study. Except for the first test, in which a uniform sand profile was used, all the tests were carried out using a soil profile with a bed of compacted air-dry Mersey River sand of between 175 and 230 mm overlying a block of remoulded saturated Cowden clay which had been consolidated into a container with internal dimensions 560 mm square × 460 mm deep. A total of three beds of clay were used in the test programme with values of undrained shear strength of 36, 60 and 240–260 kPa. The first two were specially prepared for the study and, being the full size of the container, were large enough to allow for four tests in each. A smaller bed of stiff clay was made available and only two tests were performed in this.

For each clay bed an indication of its undrained shear strength was obtained prior to the preparation of the sand bed by use of a small penetrometer. A more accurate assessment was later obtained by a series of static unconsolidated undrained triaxial tests on samples taken after completion of the test programme. The friction angle of the sand layer was determined by fitting a straight line to the initial part of the bearing capacity profiles for each test and then backfiguring a value from the bearing capacity factors given by Vesic (1975). The calculated values for the angle of friction of 37–38° were within the range expected for he type of sand used, the method of bed

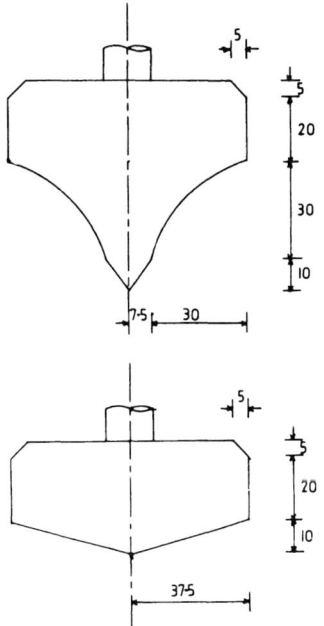

Fig. 3 Model footing configurations (75 mm diameter) (All dimensions in millimetres)

preparation and the measured stress levels; and these were the values used in analysing the test data for punching shear failure. The tests were carried out using four model footings manufactured from mild steel, two having a diameter of 50 mm and the other two a diameter of 75 mm. Two different vertical profiles were machined for each footing diameter (Fig. 3) in order to study the effects of the profile on the measured bearing capacity, with the two 50 mm diameter profiles being 2/3 scale models of the 75 mm footings.

For each test the soil bed was rotated in the Simon Engineering Laboratories centrifuge at 90 rpm, corresponding to a radial acceleration of 25.2 g at the radius of the clay/sand interface. With the centrifuge rotating at the required speed, the model footing was driven into the soil model by the use of a servo-controlled hydraulic jack (Fig. 4) to a depth of 300 to 350 mm measured relative to the point of maximum area of the footing. During the test the load applied to the footing was measured by a load cell on the hydraulic jack and the movement of the

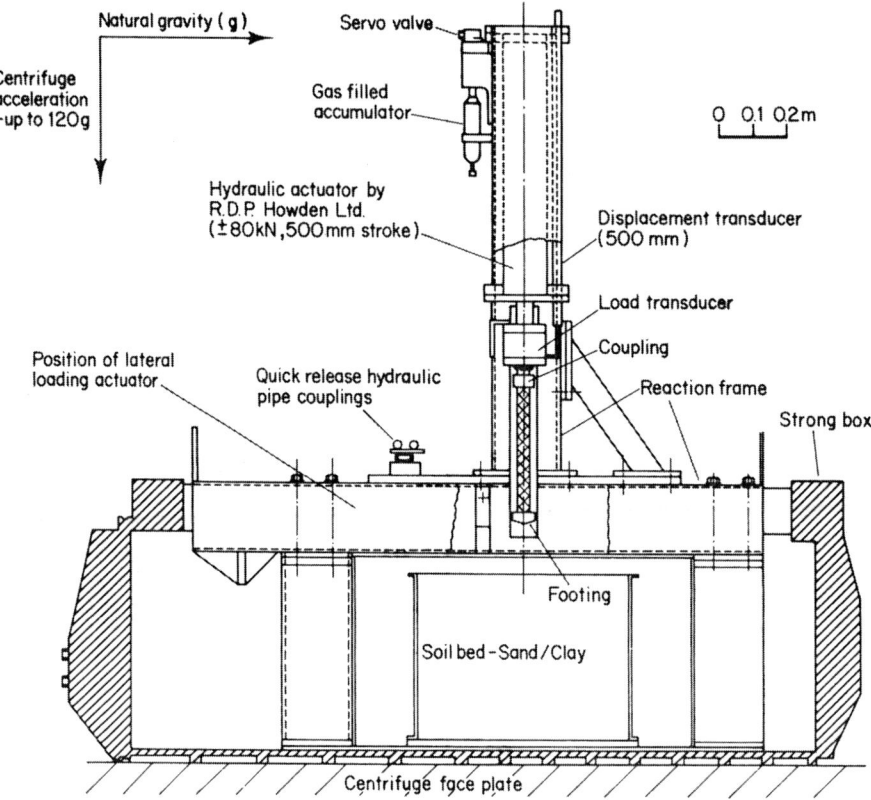

Fig. 4

footing was measured by use of a LVDT with a stroke length of 500 mm. The load and displacement measurements were recorded automatically by use of an on-line computer system. A general description of the centrifuge and a discussion of the various design considerations are given by Craig and Rowe (1981). Full details of the particular experiments are given by Higham (1984).

Analysis of Results

Table 1 summarizes details of each test performed. For all the tests in which the footing underwent punching shear failure, the test data were analysed to determine the value of the bearing capacity factor, $K_s S_s$, and α with regard to Meyerhof and Hanna's and Jacobsen's methods respectively, such that the theoretical bearing capacity profile gave the best possible fit to the plotted test data at

the point of the maximum bearing capacity. The value of the shape factor, S_s, was calculated using the conservative values for the punching shear coefficient, K_s, given by Meyerhof and Hanna (1978). Figure 5 shows a typical bearing capacity versus depth relationship for one test using the clay bed of intermediate strength and a theoretical 'best fit' curve based on Meyerhof and Hanna analysis. Figure 6 shows similar fitting for the same test using Jacobsen analysis and also includes data from a parallel test with a deeply profiled footing. The values of $K_s S_s$ and S_s calculated from a series of such constructions are plotted against the bearing capacity ratio q_2/q_1 in Figs 7 and 8 respectively, with the values of α plotted against the bearing capacity ratios q_t/q_b and q_1/q_2 in Figs 9 and 10 respectively.

The data from the three tests performed by Herdy and Townsend (1983), in which punch-through failures were observed, were

TABLE 1
Summary of centrifuge tests performed

Test no.	Footing diameter (mm)	Footing shape	Sand depth (mm)	Clay strength (kPa)	Maximum bearing stress (kPa) (Continuous increase)
0	75	flat	460	—	—
5	50	flat	230	36	3750
10	50	profiled	230	36	3600
15	75	flat	230	36	2900
20	75	profiled	230	36	2350
25	50	flat	210	60	3700
30	50	profiled	210	60	3200
35	75	flat	210	60	2200
40	75	profiled	210	72	2400
45	50	flat	175	240	4400
50	75	flat	175	260	3600*

*There is some uncertainty in this figure

analysed and plotted in the same manner, and these calculated values are incorporated into the following discussion.

In the analysis of the test data with regard to Meyerhof and Hanna's method (Figs 7 and 8), the calculated values of the bearing capacity factor $K_s S_s$ showed no discernible trend with regard to the bearing capacity ratio q_2/q_1, whereas the calculated values of the shape factor, S_s, decreased for

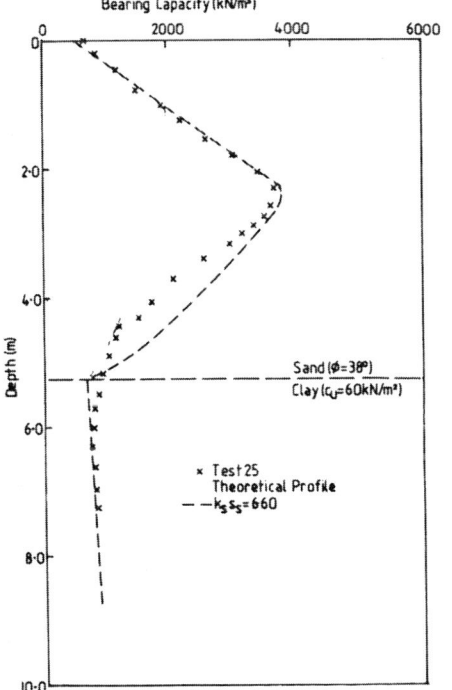

Fig. 5 Bearing capacity *vs.* depth profile for calculated values of punching shear factors

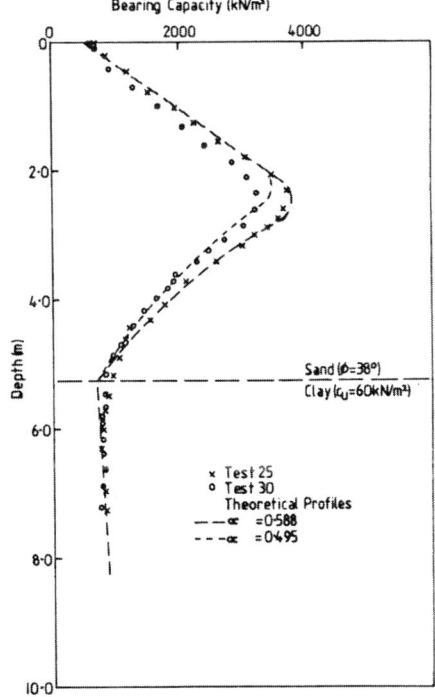

Fig. 6 Bearing capacity *vs.* depth profile for calculated values of projected area coefficients

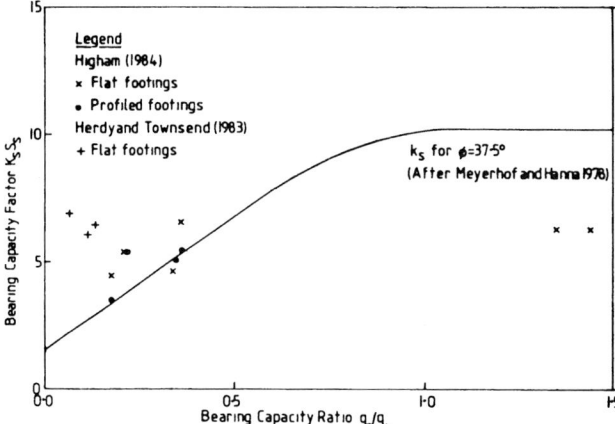

Fig. 7   Variation of bearing capacity factor $K_s S_s$ with the bearing capacity ratio q_2/q_1

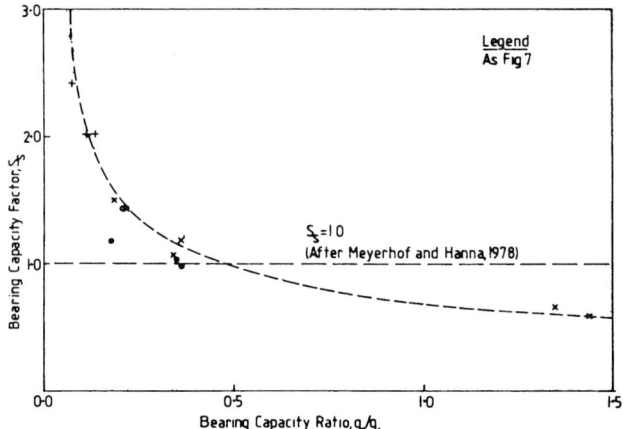

Fig. 8   Variation of bearing capacity factor S_s with the bearing capacity ratio q_2/q_1

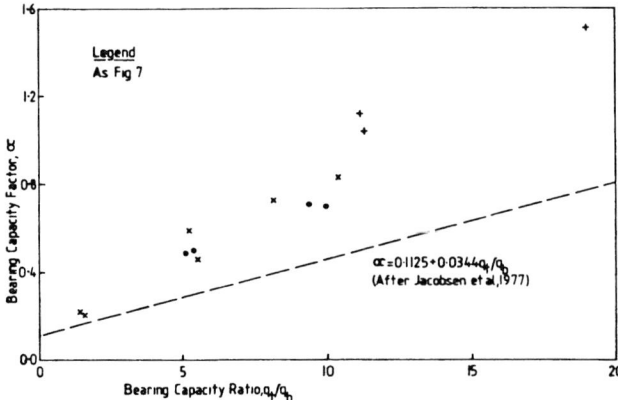

Fig. 9   Variation of bearing capacity factor α with bearing capacity ratio q_T/q_D

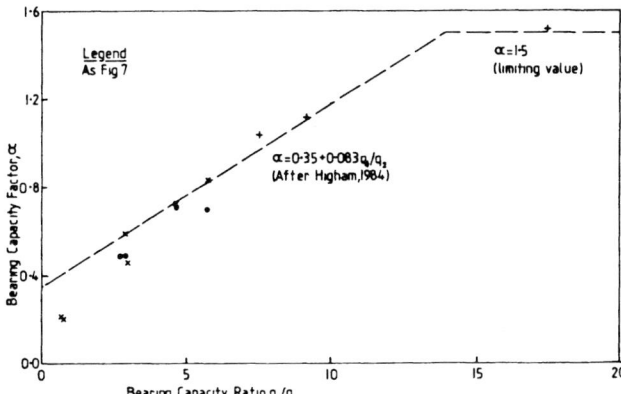

Fig. 10 Variation of bearing capacity factor α with bearing capacity ratio q_1/q_2

an increase in the value of the bearing capacity ratio q_2/q_1, with the value calculated for tests carried out using essentially flat footings being less than unity for values of bearing capacity ratio greater than 0.5. For values of bearing capacity ratio of less than 0.5, the use of a shape factor value of unity, as suggested by Meyerhof and Hanna (1978), should prove to be conservative.

In the analysis of test data with regard to Jacobsen's method, Figs 9 and 10 show that an equation of the form $\alpha = 0.35 + 0.083q_1/q_2$ would give a better prediction of bearing capacity for the majority of tests than that proposed by Jacobsen *et al.* (1977). Considering the form of this equation, it would seem necessary to limit the value of α for high values of the bearing capacity ratio q_1/q_2. From the range of values calculated in analysing the test data a limiting value of $\alpha = 1.5$ should be conservative.

For the final two tests in the test programme the calculated values of the bearing capacity ratio q_2/q_1 were greater than unity. For such cases it has been suggested (Meyerhof, 1974) that the failure of the soil beneath the footing is a combination of punching shear failure and lateral squeezings of the overlying stiff layer. Because the failure mode cannot be considered to be solely punching failure, for these two tests, the calculated values of K_sS_s and α cannot strictly be compared with those values calculated for the preceding tests. However, these two tests indicate that, in those cases

where failure is by a combined punching and squeezing mechanism, the ultimate value of bearing capacity is less than the value given by considering the failure of the soil to be solely due to punching shear.

Theoretical Considerations

For the Meyerhof and Hanna method, the ultimate bearing capacity of a foundation under vertical load is given by the equation

$$q_u = q_b + 2'\gamma H(H + 2D) K_sS_s \tan \phi/B + \gamma'V/A$$

By use of the substitution $D = D' - H$, where D' is the depth to the soil interface, the above equation can be rearranged to give a quadratic of the form

$$q_u = -a_mH^2 + b_mH + c_m$$

where

$$a_m = 2\gamma'K_sS_s \tan \phi/B$$
$$b_m = 4D'\gamma'K_sS_s \tan \phi/B$$

and

$$c_m = q_b + \gamma'V/A$$

Similarly, for the projected area method, the equation for the value of ultimate bearing capacity of a foundation under vertical load

$$q_u = q_b(1.0 + 2\beta H/B)^2 + \gamma'V/A$$

can be rearranged to give a quadratic of the

form

$$q_u = a_p H^2 + b_p H + c_p$$

where

$$a_p = q_b (2\beta/B)^2$$

$$b_p = 4q_b \beta/B$$

$$c_p = q_b + \gamma'V/A$$

The shape of the bearing capacity profile as determined by each of the two basic methods for a stratified soil profile with a strong layer overlying a weak soil is shown in Fig. 11. For all of the tests carried out in this preliminary study, a plot of the test data in the form of a bearing capacity profile had a shape similar to that for a theoretical profile given by the projected area method. Thus, while it was possible to determine a value for the bearing capacity factor $\alpha (= 2\beta)$, and hence a theoretical profile that was a good fit to the plotted test data for the projected area method, it was only possible to determine values for the shape factor S_s that gave a reasonable fit to the peak values

of bearing capacity and not to the whole curve for the Meyerhof and Hanna method.

For the Meyerhof and Hanna method, to give a good fit to the plotted test data would require that the coefficient a_m be negative. This in turn would require that the vertical component of the total passive earth pressure acts on the assumed failure plane in the same direction as the applied foundation load. However, as this force is in effect a friction force, it can only act so as to oppose the movement of the failing soil mass, i.e. it must act in the opposite direction to the applied foundation load.

Therefore the behaviour of a foundation during punching shear failure is governed more by the transmission of the applied load through a failure zone in the upper layer to a fictitious foundation of increased area of the level of the soil interface than by the forces acting on the failure plane in the upper soil layer.

CONCLUSIONS

The main points concluded from the present series of centrifuge tests in this preliminary study are:

(a) For a 'flat' footing supported by a stratified soil profile with a strong layer overlying a weak soil layer:
   (i) At best the Meyerhof and Hanna method can only be used to predict the peak values of bearing capacity. When taking a shape factor value of unity, the values of bearing capacity calculated for bearing capacity ratio (q_2/q_2) values greater than 0.5 should be treated with caution.
   (ii) For the projected area method the use of α proposed by Jacobsen et al. (1977) under predicted the measured value of bearing capacity for the particular test conditions used. A modified equation has been proposed which gives a better prediciton for the value of bearing capacity. In considering the range of values cal-

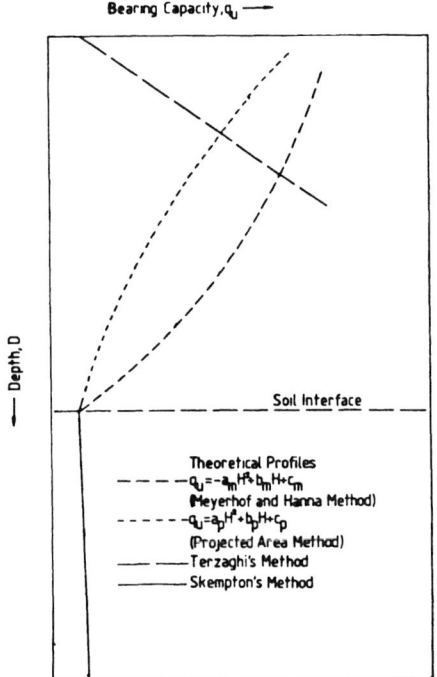

Fig. 11 Comparison of theoretical bearing capacity vs. depth profiles

culated in this study, an upper limit for the value of α of 1.5 is suggested.

(b) For a footing penetrating a cohesionless soil layer, the particular vertical profile used resulted in a reduction of the measured bearing capacity of only some 15% relative to a 'flat' footing — this is in contrast to the results of James and Tanaka (1984).

(c) For a footing penetrating a cohesive soil layer, the shape of the vertical profile of the footing has no appreciable effect on the measured bearing capacity.

ASSESSMENT OF CENTRIFUGE TESTS TO DATE AND FUTURE POTENTIAL

Table 2 summarizes the three sets of centrifuge experiments referred to above. It is clear that to date none of the experimental programme has been carried out with complete similarity to the typical jack-up loads and dimensions considered earlier. The reasons for this are largely practical and stem from three independent studies using existing or readily assembled equipment — each set with its own inherent limitations. Each set of experiments has been acknowledged as preliminary in nature.

In the case of Herdy and Townsend (1983), the centrifuge available was limited

in its capacity and the on-board loading equipment was similarly limited. Thus prototype sizes and loads were low relative to the offshore situation, but stresses were relatively high.

James and Tanaka (1984) used a bigger machine capable of modelling footings much larger than those actually simulated. They could possibly cope with a 100 m² field prototype at 1:100 scale in the future but are currently limited by the capacity of loading equipment. The stresses they have used are again higher than those typical offshore.

In our study the centrifuge was capable of modelling the full field dimensions using 100 mm diameter footings at accelerations up to 110 g. The loading equipment available had a very high capacity (80 kN) and was excessively powerful for the job in hand. In view of the unstable nature of the punch-through mechanism, which causes the massive servo-hydraulic loading actuator to plunge into the model under its own increased self-weight in the centrifuge, it seemed sensible to limit the accelerations used and to utilize relatively high strength soils. This led once more to model stress levels being greater than those in the field. Beds of softer soils are, in a number of ways, harder to manufacture and handle in the laboratory but, with the experience now gained in the way in which all the equipment behaved, there should be no difficulty in simulating

TABLE 2
Comparison of model studies with typical field data from Young et al. *(1984)*

	Herdy and Townsend (1983)	James and Tanaka (1984)	Higham (1984)	Field
Model footing diameter (mm)	30	50–100	50–75	—
Acceleration factor (g)	20–50	1 and 10–60	25	—
Maximum bearing stress (kPa)	450–1150	1100	2200–4400	—
Prototype diameter (m)	0.6–1.5	0.1–3.0	1.3–1.9	12
Bearing stress (kPa)	450–1150	1100	2200–4400	300

realistically both field dimensions and stresses. Moreover, with the knowledge that all these studies — despite these deficiencies — have confirmed recognized analytical procedures for the static vertical bearing case, there is every confidence that useful extensions to such modelling can be carried out into areas less adequately understood at present.

In looking at the potential areas of future activity it is pertinent to consider again the recognized field problems. For spud-can foundations the static punch-through can be adequately modelled and a range of studies can be contemplated, assessing, for example, the effects of relative layer strengths and layer thicknesses and of various footing profiles. Addition of lateral loading capability, as already initiated by James and Tanaka, causes little difficulty for surface or near-surface footings, or for footings installed in model beds at depth before centrifuging. Where gross changes of geometry are contemplated, there are mechanical problems with lateral loading for models involving deep penetration or punch-through. In cohesive soils the need for performing the large penetration at high g may not be pressing, but when structural performance depends upon granular soils the evidence from tests on other deep foundations indicates that 'in flight' installation is paramount (Craig, 1984).

The effects of cyclic vertical and horizontal loads are readily incorporated into the model tests, and herein lies a particular attraction since reliable field data on the effects of cyclic loading are virtually absent from the published record. Although proof-testing by preload may be sufficient for quasi-static vertical loading, provided adequate capacity is available, preload cannot proof-test foundations against lateral loading or the degrading effects of shear-stress reversals. At shallow depths in cohesionless soils such loading may lead to loss of support due to pumping action underwater, but the removal of material by current-activated scour is not realistic — the modelling of sediment transport is difficult and not compat-

ible with the use of prototype particle sizes in model studies which may be desirable from a number of other points of view.

The performance of spud-can foundations in the face of massive sea-floor movement may not be a realistic modelling ambition except in very limited site-specific contexts. Another site-specific possibility is the assessment of the likely performance of foundations placed adjacent to previous spud-can locations or adjacent to other structures, e.g. piles or gravity platform bases where complex interactions can be expected.

The prediction of jack-down and spud-can extraction can be readily assessed for the idealized conditions of a model. To a limited extent the models can also incorporate some of the techniques used by rig operators to free jammed footings, e.g. jetting.

The mat-supported jack-up does not appear to have yet been subjected to any specific modelling. Of particular interest is the performance of structures on soft cohesive soils where strength increases linearly with depth and where possible thin, near-surface sand or silt horizons could lead to localized liquefaction and loss of sliding resistance under cyclic loading.

Limited studies have been started on small surface footings on beds of homogeneous clay with strength increasing with depth. Leung et al. (1984) and Nakase et al. (1984) used strip footings 80 and 50 mm wide respectively under centrifuge accelerations of 40–80 g and 50 g with applied static, vertical bearing stresses rising to 100 and 25 kPa. Similar conclusions can be drawn from these studies as from the jack-up tests reported above: modelled dimensions and stresses may not be in good similarity with the field cases, but the potential for development is clearly demonstrated.

A substantial amount of work has previously been reported at the other end of the foundation spectrum — that of the very large surface and near-surface founded gravity structures, e.g. Rowe and Craig (1980). Models have been subjected to combinations of static and cyclic, vertical and horizontal environmental loadings, and there is no problem in devising suitable equipment modifications to cope with the differences in prototype dimensions and stresses between typical jack-ups and earlier models.

A recent paper on needs in offshore foundation design (Semple *et al.*, 1982) drew attention to the problems of jack-ups and gave some priority to their solution. It stressed the need for research on the application of bearing capacity theory in addition to the need for the capture of more reliable field data. The centrifuge has a role to play, and if the potential of this and indeed of other research tools has not as yet been applied to the jack-up this must be seen as an extension of the land-based approach of adopting a rather cavalier attitude to temporary works while in contrast adopting a conservative attitude to permanent structures.

REFERENCES

1. Craig, W. H. 1984. Installation studies for modal piles. In *Proc. Symp. on Application of Centrifuge Modelling to Geotechnical Design*, University of Manchester, pp. 440–455.
2. Craig, W. H. and Rowe, P. W. 1981. Operation of a geotechnical centrifuge from 1970–1979. *ASTM Geotech. Test. J.* 4(1), 19–15.
3. Hanna, A. M. and Meyerhof, G. G. 1980. Design charts for ultimate bearing capacity of foundations on sand overlying soft clay. *Can. Geotech. J.* 17(2), 300–303.
4. Herdy, A. C. and Townsend, F. C. 1983. Preliminary investigation of the bearing capacity of layered soils by centrifuge modelling. Report, University of Florida, Gainesville.
5. Higham, M. D. 1984. Models of jack-up rig foundations. MSc thesis, University of Manchester.
6. Jacobsen, M., Christensen, K. V. and Sorensen, C. S. 1977. Gennemlokning af tynde sandlag. Vag-och Vattenbyggaren, Stockholm, Svenska Vag-och Vattenbyggaren Riksforbund, pp. 23–25.
7. James, R. G. and Tanaka, H. 1984. An investigation of the bearing capacity of footings under eccentric and inclined loading on sand in a geotechnical centrifuge. In *Proc. Symp. Recent Advances in Geotechnical Centrifuge Modelling*, University of California, Davis, pp. 88–115.
8. Leung, P. K., Schiffman, R. L., Ko, H. Y. and Pane, V. 1984. Centrifuge modelling of shallow foundations on soft soil. *Proc. 16th Offshore Technology Conference*, Houston. Paper PTC4808, Vol. 3, pp. 275–282.
9. McClelland, B., Young, A. G. and Remmes, B. D. 1983. Avoiding jack-up rig failures. In *Proc. Symp. Geotechnical Aspects of Coastal and Offshore Structures*. Balkema, Rotterdam, pp. 139–160.
10. Meyerhof, G. G. 1974. Ultimate bearing capacity of footings on sand layer overlying clay. *Can. Geotech. J.* 11, 223–229.
11. Meyerhof, G. G. and Hanna, A. M. 1978. Ultimate bearing capacity of foundations on layered soils under inclined load. *Can. Geotech. J.* 15, 565–572.
12. Nakase, A., Kusakabe, O. and Wong, S. F. 1984. Centrifuge model tests on bearing capacity of clay. *J. Geotech. Engng* 110(12), 1749–1765.
13. Rowe, P. W. and Craig, W. H. 1980. Applications of models to predictions of offshore gravity platform foundation performance. In *Proc. Int. Conf. on Offshore Site Investigation*. Graham & Trotman, London, pp. 169–281.
14. Semple, R. M., St John, H. D. and Toolan, F. E. 1982. Statement of research needs in offshore foundation design. Report prepared by Society for Underwater Technology, London.
15. Skempton, A. W. 1951. Bearing capacity of clays. Building Research Congress, London, pp. 180–189.
16. Vesic, A. C. 1975. Bearing capacity of shallow foundations. In *Handbook of Foundation Engineering* (Eds Winterhorn and Fang). Van Nostrand, New York, pp. 121–147.
17. Young, A.G. and Focht, J. P. 1981. Subsurface hazards affect mobile jack-up rig operations. Soundings. McClelland Engineers Inc., Houston.
18. Young, A. G., Remmes, B. D. and Meyer, B. J. 1984. Foundation performance of offshore jack-up drilling rigs. *J. Geotech. Engng* 110(7), 841–859.

Summary

C. P. Wroth, Oxford University, UK

I am grateful to the Organising Committee for the invitation to attempt to sum up the proceedings. Summing up is not really the appropriate phrase — it is a difficult job to do justice in a summary to the amount of material that has been presented over the two days of the conference. Clearly, each paper merits further individual attention in order to reflect on its content. What I am going to say must necessarily be an unbalanced critique, because we are considering a whole range of knowledge and experience in a wide diversity of topics, and my comments are bound to be biased by my own interests.

My first general comment is that I found the second day of the conference livelier and of more interest than the first — I think there are two reasons for this. First of all, the second day's theme was concentrated on geotechnical aspects, so it was one part of the tripartite structure of geophysics, geology and geotechnics. Secondly, the papers were focused much more on specific details and specific jobs.

Let me remind the audience of the topics that were covered on the first day. First, we had Dr Riemersma talking about positioning requirements, and it seemed to me to be an unhappy reflection on human frailty that he was concentrating so much on the errors in the system and on the human factors that led to trouble, emphasizing that the techniques are vastly superior to the ability of the human beings who used them. Then, Dr Palmer talked about a fascinating case history of the Ocean Thermal Power Project; this was of particular interest because most of the other stories we heard were not so specific and not about such a novel project.

Dr Schüttenhelm and Mr Stoker, in their presentations, reported significant additions to the general store of knowledge about the geological properties of the North Sea sector around the north of the UK and the Abyssal Plains. This must be very useful background information for site investigations in these areas, and I think it is laudable to attempt to add geotechnical information to the geological data that have been obtained. However, I wonder whether this is a little too dangerous and whether the maps should be issued with a warning to the effect that they are an aid, and not a substitute, for proper site investigation.

We then had three papers that can be grouped together, presented separately by Loevik, Hutchins and Games, which all dealt with important improvements in various aspects of seabed surveys. Mr Green then proselytised about his integrated approach to site investigation, which launched us into a debate about the philosophy of education and some lively discussions. Finally we had Mr Sarginson, essentially making a plea that the contractors providing services should be treated in a more honourable manner.

The second day's programme started with Dr Richards giving a very good review of *in situ* testing and sampling in deep water, which he defined as deeper than 300 m. I found this a fascinating glimpse of developments that are taking place, and the limits of time undoubtedly prevented him from doing justice to a lot of valuable information. This was followed by Christophersen telling us about the site investigation of the Troll Field and the very detailed information obtained about this one site.

Subsequently, our attention was drawn to various *in situ* testing devices. Two papers were about pressuremeter tests — M. Faÿ talking about the French version of the self-boring pressuremeter developed for offshore use, and Mr Powell with a valuable comparison of data obtained from three different pressuremeters at one site in this country. These were followed by Professor Silva discussing the development and deployment of an *in situ* vane, and I must congratulate him on some very commendable real engineering — to have got that instrument to work successfully first time in such a depth of water was a great achievement.

We then diverted to some comments about laboratory testing, and Mr Marsland provided timely reminders of the importance of soil fabric as well as the fact that

soil is not the homogeneous, continuous material that is assumed in so many theories. Messrs Hight and Jardine focused our attention on the problem of attempting high quality laboratory testing, and indirectly underlined the value of *in situ* testing. At the same time, I respond to Mr Lunne's plea that *in situ* testing is not the panacea of all site investigation, and that its interpretation must be backed up with high-quality laboratory testing which must include appropriate reconsolidation of specimens. Mr Lunne gave us a most useful review of *in situ* testing in sand and recent developments in that area. These developments have been backed up by Professor Jamiolkowski, whose very careful work with his Italian colleagues, testing these various instruments in controlled circumstances in laboratory test chambers, provides invaluable verification for their acceptance in practice.

Finally, Mr Craig told us about a very real problem, that of jack-up rig foundations, which he has studied in centrifugal model tests at Manchester. I warmed to his plea that industry should be much more sophisticated in its approach to this important problem and should take account of current work and present understanding.

If I could attempt to draw some more general conclusions about the conference, I should like to repeat what we heard from Mr Zuidberg about the very steady and marked development that has taken place since the last conference in 1980. He commented on the increasing reliability of all the equipment, the maturity of the industry and the growing confidence that people now have in the results that are obtained from any of the surveys or site investigations. I am sure that this will continue as exploration moves steadily into deeper waters. Economists' jargon refers to whether or not developments are due to 'demand pull' or 'technology push'. It seems to me that in the offshore industry at the moment it is a case of 'demand pull': it is the move into deeper waters which is causing the relevant technology to be developed, and not newly invented technology looking for an application.

I wonder whether some techniques are now developed to the stage where they are outstripping advances in interpretation and understanding, and are therefore not subject to adequate scrutiny and validation. I can illustrate this by riding one of my own hobby horses and talking about the interpretation of cone penetration tests, which have been much commended by Mr Lunne. I want to put before you some ideas about the interpretation of sleeve friction readings obtained in tests in clay.

In the standard cone penetration test, independent measurements are made of the sleeve friction, f_s, and of the end bearing, q_c. If these are related to current methods of pile design in terms of total stresses, the sleeve friction would be related to the undrained shear strength, denoted by s_u by means of the 'α-factor' i.e. $f_s = \alpha s_u$, whereas for the end bearing, allowance would be made for the magnitude of total overburden stress, σ_{vo}, so that $(q_c - \sigma_{vo}) = N_k s_u$. The factors α and N_k are dimensionless functions of the overconsolidation ratio of the clay, OCR, and to a lesser extent the friction angle, ϕ'. I suggest that the definition of the friction ratio f_r as

$$f_r = \frac{f_s}{q_c}$$

is an unfortunate one; it has developed naturally from a recognition that the two measurements f_s and q_c are related in a way that indicates the type of soil being tested. It seems to me that it would be better to compare the sleeve friction with the *net* end bearing, having allowed for the overburden pressure. This would be a simple computation, leading to a revised friction ratio f_r^*, defined as

$$f_r^* = \frac{f_s}{q_c - \sigma_{vo}} = \frac{\alpha s_u}{N_k s_u} = \frac{\alpha}{N_k}$$

so that it would be the ratio of α to N_k. As such, for a given clay (with a given value of ϕ') it would be expected to be a unique function of overconsolidation ratio. The revised friction ratio could then be used as an indictor of OCR, which cannot be the case for the original definition f_r, because it includes the irrelevant effect of the overburden pressure.

For example, for a normally consolidated clay, typical values would be: $\alpha \simeq 1$, $N_k \simeq 15$, so that $f_r^* \simeq 6.67\%$. For a heavily overconsolidated clay, the likely values would be $\alpha \simeq 0.5$, $N_k \simeq 20$, giving $f_r^* = 2.5\%$.

If the cone also contains a pure-pressure transducer (i.e. it is a piezocone), then a separate and independent assessment can be made of the OCR of the clay. I have suggested in my Rankine lecture that the pure-pressure parameter denoted as B_q by Professor Janbu, and defined as

$$B_q = \frac{\Delta u}{q_c - \sigma_{vo}}$$

is also a unique function of OCR for a given clay. It should, therefore, be possible in the case of a piezocone to iterate between estimated values of OCR (which affect the appropriate choice for N_k) obtained from (i) the deduced ratio s_u/σ'_{vo}, (ii) the revised friction ratio f_r^*, and (iii) B_q, and so obtain the single most consistent value for OCR of the deposit under test. By this means, the interpretation of cone penetration tests could be considerably enhanced.

The same concept of a revised friction ratio could be applied to tests in sand, with the exception that the end bearing must be corrected for the *effective* vertical stress, instead of the total vertical stress. May I also make a plea that, in any of the empirical relationships used to aid interpretation of cone penetration tests (and other *in situ* tests), it is important to use dimensionless parameters. For example, in the results of other people's work, shown by Mr Lunne, such as end bearing q_c related to friction angle or relative density, the correlations are not satisfactory, and not valid. It is the *corrected* end bearing $(q_c - \sigma'_{vo})$ that should be correlated with other soil properties.

I shall close by making one or two more general comments. About 120 people have registered for OSSI '85, a good number for a conference. We have a good mixture of disciplines within civil engineering, geotechnics, geology and geophysics, and a cross-section of representatives from industry, academia and government research establishments. It is a disappointment that there is no one here from the Headquarters of the Departments of Energy and Environment, who have so much influence on and involvement in the offshore site investigation industry. Dr Richards made the point about the different attitudes to development from industry on the one hand, and from research establishments and universities on the other. It is valuable to exchange ideas, experience and attitudes in a conference such as this.

My final duty — and a most pleasant one — is to express thanks on behalf of all the delegates to a number of people. First, to the Organising Committee, chaired by Dennis Ardus, supported by Sid Green, Eugene Toolan and Tim Freeman. I imagine that it was their idea to hold the conference, their initiative and drive that made it possible. Of course, the conference would not have happened without the organisational ability of the Society for Underwater Technology, of its Secretary, David Wardle, and of the Conference Organiser, Jean Pritchard, and her team. We also must thank the session chairmen, who guided us ably through the discussions; the authors, who prepared the papers and presented them so well; and those who entered the discussion from the floor. Finally, I should like to heartily congratulate all those who have contributed to a most successful conference.

REFERENCE

Wroth, C. T. 1984. The interpretation of *in situ* soil tests. 24th Rankine Lecture *Geotechnique* **34** (4), 449–489.